Project Management ToolBox

Project Management ToolBox

Tools and Techniques for the Practicing Project Manager

Dragan Z. Milosevic

WILEY

John Wiley & Sons, Inc.

To the two ladies in my life: my wife, Dragana, and my daughter, Jovana.

This book is printed on acid-free paper. ∞

Copyright © 2003 by John Wiley & Sons, Inc. All rights reserved.
Published by John Wiley & Sons, Inc., Hoboken, New Jersey
Published simultaneously in Canada

For general information on our other products and services or for technical support, please contact our Customer Care Department within the United States at (800) 762-2974, outside the United States at (317) 572-3993 or fax (317) 572-4002.

Wiley also publishes its books in a variety of electronic formats. Some content that appears in print may not be available in electronic books. For more information about Wiley products, visit our web site at www.wiley.com.

Library of Congress Cataloging-in-Publication Data:
Milosevic, Dragan.
    Project management toolbox: tools and techniques for the practicing
        project manager / Dragan Milosevic.
            p. cm.
    Includes bibliographical references and index.
    ISBN 0-471-20822-1 (cloth)
    1. Project management. I. Title.

    HD69.P75 M54 2003
    658.4'04--dc21

Printed in the United States of America

10 9 8 7 6 5 4 3

Contents

Packing the Theory in the Project Management Toolbox

The power of project management (PM) is at its historic peak, primarily because it has become a business strategy of choice. The evidence is abundant. Large corporations, the locomotive of the American economy, have launched corporate-wide PM initiatives and centers of PM excellence, poised to create a successful environment for PM. To improve their competitiveness, Fortune 500 started a PM Benchmarking Forum, tasked to identify best PM practices. Small businesses are racing to follow suit. Most interestingly, this is an across-the-industry phenomenon. Joining traditional proponents of PM such as construction and aerospace are the powerhouses of new economy, high-technology, and telecommunications companies.

Riding on this wave of PM popularity, the Project Management Institute, the world's largest association of project managers, has been enjoying exponential growth in terms of both membership and certified Project Management Professionals. Management gurus have thrown their weight behind PM as well. Tom Peters calls project manager the job number one in the twenty-first century. Eliahy Goldratt, the pioneer of the theory of constraints, pictures PM as the next frontier of continuous business improvement. As they wish to give credence to the guru's assertions, companies have poured billions of dollars into PM training.

A central place in the PM rise belongs to project process management, an outgrowth of the total quality movement of the 1980s. PM process is about seamless performance of orderly arranged project activities and phases, resulting in project deliverables. The point is simple—we need a controlled, no-nonsense, predictable execution to successfully put out fast, repeatable, high-quality project products. For this to happen, we need mechanisms built into the process that would enforce it on a daily basis. This mechanism is the PM toolbox. It provides a practical and tangible, yet systematic way of planning and controlling our projects. Quite obviously, to strengthen the PM process is to strengthen the PM toolbox, which leads us to the basic purpose of the book: to display a repository of PM tools (methods and techniques) and offer strong criteria to select, customize, and incorporate the most powerful of them in the PM toolbox, which would then be built in the process.

As leading companies continue to beef up their project process, other organizations strive to build their first project process, destined to bring more orderliness to their strategic and tactical projects. The key building blocks for achieving such objectives through PM toolbox are described here. This book is about PM toolbox and its constituent tools, not about managing a project or project process. It significantly differs, therefore, from traditional PM literature whose major emphasis is on the theoretical description of the process of managing a single project, with an occasional glimpse at some of the classic PM tools. In contrast, this book addresses needs of project managers who believe that they have had enough of theoretical talk and feel that what they really need is "packing" the theory in the form of the toolbox that they know how and when to use.

The Book Approach

The construction of such PM toolbox is the focus of this book. As an aid in envisioning the construction, we present its process in Figure P.1.

In short, once Chapter 1 builds the case for PM toolbox that supports the competitive strategy of the company, we shift our attention to Chapters 2 to 15. These chapters act as a repository of 50+ tools, explaining how-tos for each of them, another topic often neglected in the traditional literature. Tools necessary for initiating projects, committing an organization to bringing projects to life, are covered in Part I. Part II features tools for planning projects, necessary to chart a realistic roadmap to prepare the project for the future and accomplish its business goals. Executing, controlling, and closing projects is what the tools in Part III target. Using all of these tools as basic building blocks, Part IV shows how to build a PM toolbox, as well as real-world examples of the toolboxes. More detailed description of these steps follows.

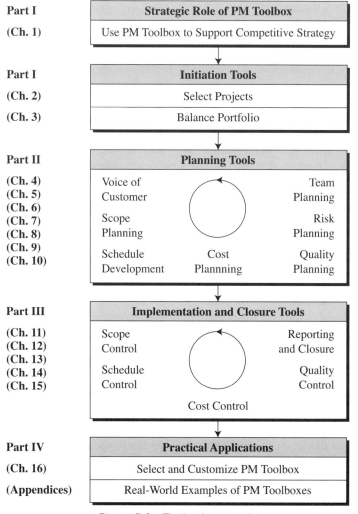

Figure P.1 The book approach.

The initial interest of Part I is the powerful role of PM tools in delivering a project (Chapter 1). Here we seek to shed light on the role, revealing an overarching approach for using the tools to build a PM toolbox to enforce a PM process day in and day out and support competitive strategy of the company. Determined to employ the approach throughout the book, Part I, Chapters 2 and 3, begin our march through the tools, right there at the beginning of a project's life. The attention is centered on tools for selection of the most viable projects out of a pool of project proposals, which can mean the difference between leading or lagging behind the competitive pack. The intention is to maximize the value of the portfolio of chosen projects against an organization's goals and strategy. Also, the aim is to balance the portfolio in order to secure proper risk allocation.

Those whose job is to plan for a project should look carefully at Part II, Chapters 4 through 10. Our times demand that we treat the customers as royalty. That is why Part I begins with the customer intimacy tools, telling us how to learn, internalize, and build the voice of the project customer in the design of project process and product. First zeroing in on the sacred project triad of scope-time-cost, tools in this part show how to scope a project, spelling out its goals, milestones, deliverables, and assumptions. They explain how to identify activities, establish dependencies among them, estimate their durations, and make sure they are resourced, all with the purpose of developing a project schedule. Finally, the tools in this part guide you through determining necessary resources, costing them, and allocating them to specific project activities in specific time periods, thus producing a time-phased project cost budget. The focus, then, shifts to tools for quality planning and facilitating processes—risk planning and team planning.

Part III, Chapters 11 through 15, features several groups of implementation tools, designed to carry out the plan developed with tools from Part II. These chapters offer ways to execute projects, by synchronizing the deployment of people and material resources to make the project plan happen. The controlling tools focus on monitoring and measuring progress of the execution, and taking required actions, to ensure the delivery on project business goals. The role of closure tools is to draw the project to a smooth end and provide for its acceptance. In selecting the implementation tools, we carefully listened to managers who believe that the main role of implementation is to be fact-based proactive, continuously predicting the path to project completion. Consequently, these tools tell how to proactively report and control the scope, schedule, cost, and quality; explain how to exercise proactive control of implementation processes related to risk response and team development; and make sure the project venture is properly phased down.

Developing a framework for selecting and customizing PM toolbox lies at the heart of Part IV. Actually, once we know what tools are available to a project process, as featured in the first three parts, Part IV starts off with Chapter 16, specifying how to select tools for inclusion into the PM toolbox. This is done by means of a unique situational (contingency) framework, almost nonexistent in the traditional literature. Its purpose is to help you select tools that best fit your specific situation. To make the application of the tools in the real world more transparent, several appendices in Part IV display examples of PM toolboxes in various project situations. These situations are from projects of varying size and complexity, as well as from different industries. Also, clear correspondence between the tools covered in the book and the Project Management Institute's popular *A Guide to Project Management Body of Knowledge* is established.

Chapter Outline

In explaining individual PM tools, each of Chapters 2 to 15 follows a similar structure and flow that we proceed to describe in the following (see Figure P.2).

Introduction. This section, typically a one-pager, provides a complete list of the tools covered in the chapter. Also, it defines the purpose of the tools and illustrates their role in the PM process, and lays out goals for the chapter.

Tools. Each chapter groups and reviews several tools, which belong to a certain PM process; for example, risk planning. Like the chapters, these tools also have a similar structure and flow (discussed in the continuation).

Concluding Remarks. The purpose of this section is to provide guidelines for selecting tools within each chapter. It starts with a summary of tools the chapter covers, emphasizing whether the tools complement each other, or if able to perform the same function, compete with each other for the attention of the user. In that context, project situations favoring the use of individual tools are specified.

Figure P.2   Chapter outline.

Tool Outline

Each of the tools in Chapters 2 through 15 is structured in a consistent way described in the following (see Figure P.3).

What Is the Tool? This is usually a paragraph or two describing the purpose of the tool and its main features. The description refers to an illustrated example that is provided to help understand what the tool looks like.

Constructing the Tool. Steps involved in constructing, building, or developing the tool are detailed in this section. They are made up of a series of substeps that describe specific activities. Essentially, this is the how-to of the tool.

Utilizing the Tool. Multiple elements are involved in this section. *When to Use* the tool explains situations in which the tool can be applied. How much time the project manager or project team needs to consume when utilizing the tool is described in *Time to Use*. *Benefits* specifies what value the tool creates for the user. By contrast, *Advantages and Disadvantages* generally concentrates on the simplicity/complexity and ease-of-use issues of the tool utilization. The *Variations* element refers the reader to different names and versions of the tool used. Most of the time, the tool is presented in its generic form, used across industries. Helping the reader adapt the tool to fit his or her project needs is the *Customize the Tool* element.

Summary and Tool Check. At the end of each tool section, a summary reminds the reader of the purpose, use, and benefits of the tool, offering a checklist for appropriately using or structuring the tool.

Common Elements. In addition to the structure of the chapters and tools, other common elements appear throughout the book. These are text boxes that may provide any of the following:

- ▲ Tips
- ▲ Checklists
- ▲ Detailed examples
- ▲ Case studies
- ▲ Background information

Their purpose is to offer readers more depth or breadth of information than the basic text.

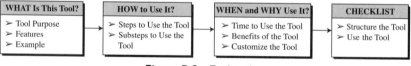

Figure P.3 Tool outline.

How to Read and Use This Book

Today's project managers are busy businesspeople. Few will be able to squeeze in their tight schedules time to read this book cover-to-cover. If they can, the book is shaped to naturally unfold through a project life cycle. If they cannot, the book is designed to accommodate different readers with different needs:

▲ The reader with an interest in a specific PM tool is advised to review the tool in detail.

▲ The reader who is interested in a certain group of tools—the whole group of project selection tools or schedule development tools, for example—will be just fine to study carefully only those tools.

▲ The reader with a keen interest in the industry applications of tools should look at these chapters in detail.

Further, the major purpose of the book is to help the following industry professionals:

▲ Project Team Members. Project professionals of this type often face a challenge of solving a specific project problem for which a certain PM tool offers an easy path to a solution. Without adequate support in an organization where invariably everyone is stretched thin, the book is a reliable guide to draw upon.

▲ Project Managers. Many project managers work in organizations that do not have a formal PM process and toolbox. As a result, they are tired of managing ad hoc, beginning from scratch every time they begin a new project. Here, managers can use the book to identify a blueprint for a PM toolbox from industry applications similar to their environment and adapt it (see the appendices on industry applications). Other project managers may need this book's chapters to find new, better tools for their already existing PM process and toolbox.

▲ Multiproject Managers. For those who manage multiple projects and their organizations haven't developed a shared PM toolbox, the book is a way to create a repeatable, less frustrating PM toolbox to support the PM process. Use the book to identify a blueprint for a PM toolbox in the industry applications appendix that is similar to yours and adapt it. If you already have a PM toolbox that needs enhancement, look through the book chapters for some better tools.

▲ Process Owners/Process Teams. Individual managers or teams tasked with designing a sophisticated PM process or improving the existing ones can use the book to find the tools for the whole new PM toolbox to support PM process or the ones missing in the current toolbox. They can also use one of PM toolboxes in the industry applications part as a blueprint for planning their own toolbox to support a rapid and repeatable PM process.

▲ Academics and Consultants. To academics the book can be a research resource, recommended reading for their PM class, or a resource to teach their students how to use and apply the tools. Consultants on assignments to help companies develop PM processes and toolboxes, train their project professionals, or deploy tools can use the book as a framework for the assignments. In addition, they may find it useful as a text to teach PM tools to students.

Your organizations are in the heat of competitive battles. For them to win, you need to deliver your projects in a fast, repeatable, and concurrent fashion. The way to accomplish this is to equip your PM process with powerful and customized PM toolbox. This book shows you how.

Project
Initiation Tools

Strategic Role of the PM Toolbox

Man is a tool-using animal. Without tools he is nothing, with tools he is all.

Thomas Carlyle

Conventional wisdom holds that individual project management (PM) tools are enabling devices to reach an objective or, more specifically, a project deliverable. While this traditional role of PM tools is more than meaningful, we believe that there is more to the tools than that. In particular, PM tools can be used as basic building blocks to construct a PM toolbox. In this new role, the PM toolbox supports the standardized project management (SPM) process. The goal of this chapter is to

- ▲ Clarify the new role.
- ▲ Explain its strategic importance.

To accomplish these objectives, we first present an overview of the new role. Then, special aspects of the role are explained.

The New Role of the PM Toolbox

As shown in Figure 1.1, a typical SPM process includes process phases, milestones, technical deliverables, and managerial deliverables. Supporting it in its new role is the PM toolbox (at this time, we only focus on managerial deliverables and the PMtoolbox). Two principles are important in the support. First, each managerial deliverable is specifically supported by a specific tool or tools. That means that each of these tools is selected because of its systematic procedure to help produce the deliverable. Second, the PM toolbox is constructed to include all tools necessary to complete the whole set of SPM processes' managerial deliverables, which we have emphasized in Figure 1.1 by shading the deliverables and toolbox. This implies that the PM toolbox is designed for the specific SPM process. If this principle were not met, we would not have the toolbox. Rather, we would have a set of individual tools, performing the conventional role of PM tools. Technical deliverables and their tools are project type-specific and are beyond the scope of this book.

The value of the new role of the PM toolbox is obvious. Designed as a set of predefined PM tools, the PM toolbox supports SPM process by providing a practical and tangible, yet systematic, way of producing a set of the process's managerial deliverables. That this support has a strategic meaning can be seen in Figure 1.2. As shown by upward arrows, the PM toolbox supports SPM process that helps implement PM strategy, which supports the competitive strategy of the company in its quest for survival and growth. For this upward chain of support to work, competitive strategy has to drive PM strategy, which drives SPM process, which drives PM toolbox design (downward arrows). Evidently, the new role of the PM toolbox is played in the wider, strategic context of PM. That calls for dissecting the context, its elements, and their relationships, which we will do next.

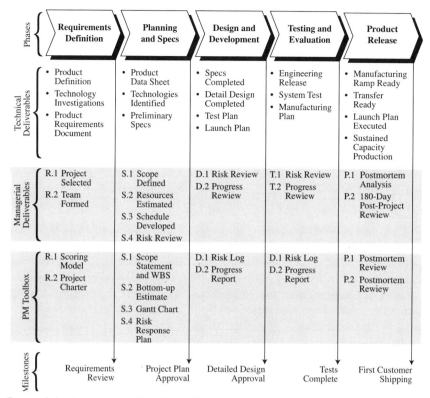

	Requirements Definition	Planning and Specs	Design and Development	Testing and Evaluation	Product Release
Phases					
Technical Deliverables	• Product Definition • Technology Investigations • Product Requirements Document	• Product Data Sheet • Technologies Identified • Preliminary Specs	• Specs Completed • Detail Design Completed • Test Plan • Launch Plan	• Engineering Release • System Test • Manufacturing Plan	• Manufacturing Ramp Ready • Transfer Ready • Launch Plan Executed • Sustained Capacity Production
Managerial Deliverables	R.1 Project Selected R.2 Team Formed	S.1 Scope Defined S.2 Resources Estimated S.3 Schedule Developed S.4 Risk Review	D.1 Risk Review D.2 Progress Rewiew	T.1 Risk Review T.2 Progress Rewiew	P.1 Postmortem Analysis P.2 180-Day Post-Project Rewiew
PM Toolbox	R.1 Scoring Model R.2 Project Charter	S.1 Scope Statement and WBS S.2 Bottom-up Estimate S.3 Gantt Chart S.4 Risk Response Plan	D.1 Risk Log D.2 Progress Report	D.1 Risk Log D.2 Progress Report	P.1 Postmortem Review P.2 Postmortem Rewiew
Milestones	Requirements Review	Project Plan Approval	Detailed Design Approval	Tests Complete	First Customer Shipping

Figure 1.1 An example of the standardized project management process with the corresponding project management toolbox.

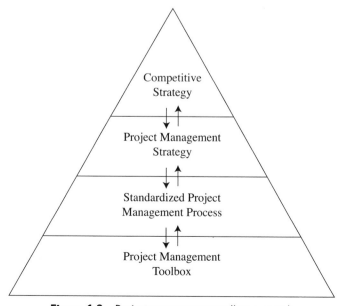

Figure 1.2 Project management toolbox pyramid.

PM Strategy Supports Competitive Strategy

Looking into the strategic context of PM will help us understand the special aspects of the new role of the PM toolbox, which is how the PM toolbox's support of the SPM process has to be aligned with the competitive strategy. Since the alignment is driven from the top of the pyramid (see Figure 1.2), we start from there—the competitive strategy.

The essence of such strategy lies in creating competitive advantages that give a company an edge over its rivals [1]. To equip themselves with the advantages, companies rely on their organizational resources [2]. Visualize, for example, PM as an organizational resource. Useful for this visualization can be the framework of generic competitive strategies (further competitive strategies), shown in Figure 1.3 [3].

The core of differentiation strategies (high differentiation/high cost quadrant in Figure 1.3) is an ability to offer customers something different from competitors. This may include fast time –to market (which we used as an example in Figure 1.3), high quality, innovative technology, special features, superior service, and so on. When striving for product superiority, companies pursuing these strategies build in whatever features customers are willing to pay for. That enables them to charge the premium price to cover the extra costs for differentiating features [4].

		DIFFERENTIATION	
		Low	**High**
COST	**High**		**Competitive Strategy:** **Differentiation** PM Strategy: Schedule Focus SPM Process: Schedule Focus PM Toolbox: Schedule Focus
	Low	**Competitive Strategy:** **Low Cost** PM Strategy: Cost Focus SPM Process: Cost Focus PM Toolbox: Cost Focus	**Competitive Strategy:** **Best-Cost** PM Strategy: Cost-Quality Focus SPM Process: Cost-Quality Focus PM Toolbox: Cost-Quality Focus

Key: SPM–Standardized Project Management
PM–Project Management

Figure 1.3 Examples of project management strategies, processes, and toolboxes supporting competitive strategies.

Companies focusing on low-cost strategies aim at establishing a sustainable cost advantage over rivals (low cost/low differentiation quadrant in Figure 1.3). The intent is to use the low-cost advantage as a source of underpricing rivals and taking market share away from them. Another option is to earn a higher profit rate by selling at the going market price [5]. This is pursued with a good basic product that has few frills, along with a continuous quest for cost reduction without giving up quality and crucial features.

Best-cost companies combine upscale features with low cost (low cost/high differentiation quadrant in Figure 1.3). This should lead to superior value by meeting or exceeding customer expectations on product features and surpassing their expectations on price. At the same time, the aim is to become the low-cost provider of a product that has good-to-excellent features and use that cost advantage to underprice rivals with comparable features. Because such a company has the lowest (best) cost compared with similarly positioned rivals, the strategy is called *best-cost strategy*. The blank quadrant in Figure 1.3 of high cost/low differentiation is not a viable option in competitive battles for survival and prosperity.

Now let's use the framework of competitive strategies to see how PM helps create the competitive advantages. Examples of three companies— Intel, Armstrong World Industries (AWI), and Oregon Anesthesiology Group (OAG)—will help us illustrate the point. Intel's competitive strategy is one of differentiation (see Figure 1.3). The strategy targets innovation and time-to-market speed as competitive advantages. In it, a significant role belongs to product development projects, whose job is to roll out new computer chips faster and faster. This is where PM comes into play, focusing on shrinking the chip development projects cycle times and hitting the market before the competition. We have observed the same passion for schedule acceleration in projects implemented in other non-product development activities of Intel as well, from large fab construction to small quality improvement projects. This certainly is not a random act. Rather, it is a management choice to deploy PM in order to help build the advantages through the accomplishment of ever-shorter project cycle times throughout the company.

Other companies with the differentiation strategy also work hard to create competitive advantage through the reduction of project cycle times. Firms such as General Electric, NEC, Northern Telecom, and AT&T managed to reduce their average cycle times by 20 percent to 50 percent [5]. The power of fast cycle time lies in its consequences. For example, in product development, the company that hits the market before there is competition will often enjoy premium pricing, extended product sales life, higher profit margins, and increased market share [6–8].

AWI's competitive strategy is quite different from Intel's. Instead of differentiation and time-to-market emphasis that Intel relentlessly pursues, AWI has set out to become the cost leader in the industry (low cost/low differentiation quadrant in Figure 1.3). As a plant manager put it: "We have been in the business of manufacturing building materials for over 70 years. Technological change is not a major factor in our industry; rather, it is the ability to compete on low cost. To develop the ability and become the leader in the industry, we have had to streamline every possible manufacturing-related process and continuously lower the bar for our manufacturing cost goals. Part of that effort has been the process for managing our cost cutting and manufacturing process development projects." An equal zeal for cost cutting in other non-manufacturing projects of AWI is also noticeable. Manufacturing and non-manufacturing areas are apparently where PM focused on lowering project cost goals, and they support AWI's low-cost advantages.

This is no secret to other firms using PM to support the same low-cost strategy. The reason is in the increased magnitude of project cost and the cost pressures many leading companies are facing. When an enterprise resource-planning project may cost $300M [9] or fab may cost $4B [10], companies have to reduce the cost and pressures with the goal of creating low-cost competitive advantages [11]. Such PM support helps them build larger market share and higher profits [2].

Intel and AWI's competitive advantages exploit their schedule- and cost-focused PM, respectively. In contrast, OAG (Oregon Anesthesiology Group) relies on a best-cost PM (low cost/high differentiation quadrant in Figure 1.3). In this corporation of some 190 medical doctors, the goal is to have the best cost relative to competitors whose health care services are of comparable quality [4]. Accordingly, their PM aims at accomplishing such cost and quality goals. Says a vice president of OAG, "In a cutthroat market as ours, health management organizations have been putting a relentless pressure on all care providers to cut costs. To stay afloat, we standardized our PM protocols in all of our information systems and continuous quality improvement projects. That helped us bring our projects within cost and quality goals. If not so, our customers will take their business somewhere else." Using its cost/quality advantage, OAG has been able to hold a commanding share of its market.

Other experts confirm that a cost/quality combination is pursued as a project goal in some firms [12]. The bottom line is that the companies focused on the best-cost strategy and related competitive advantages need PM that supports such strategy.

In summary, these examples provide a context that we need to construct a common base of understanding:

▲ Companies choose their competitive strategies.
▲ Companies align their PM strategy with the competitive strategies.

First, companies select competitive strategies as a means of competing with their rivals in the marketplace. Although each type of the competitive strategy has the same goal—create competitive advantages—ways to accomplish the goal differ. One builds the advantages on the basis of differentiation, another on low cost, and still another on best-cost approach. Second, companies strategize their PM. The objective is to align PM strategy with the selected competitive strategy. Consequently, in the case of Intel, AWI, and OAG, each company's perspective of how PM strategy should be focused is different: schedule focus (Intel), cost focus (AWI), and cost/quality focus (OAG). No wonder then that some prominent researchers view PM as one of the crucial threats, as well as an opportunity that managers encounter in their competitive battles [13].

While PM strategy has a significant role to play, it is by no means the only driver in the competitive strategy pursuit. Rather, other business strategies—often referred to as *functional strategies* [2]—are required to make the competitive strategy work. In particular, also contributing are research and development, marketing, manufacturing, human resource strategies, and so on.

So far we have been shedding light on the context of the PM toolbox's support of the SPM process. We need to continue with the context, providing more details on the SPM process, before we return to the details of the role.

Standardized PM Process Supports PM Strategy

As shown in Figure 1.2, the SPM process supports the PM strategy. Or, to put it differently, the SPM process serves as a mechanism for delivering PM strategy. To clarify this, we will offer the following:

- ▲ Some empirical examples that the SPM acts as a support mechanism for PM strategy
- ▲ The SPM process elements through which the support occurs
- ▲ The meaning of *standardized*
- ▲ The way to align SPM process with PM strategy

We begin with the examples of the empirical evidence. In a recent report the Fortune 500 PM Benchmarking Forum asserts that 85 percent of their members use standardized approaches and procedures when managing projects [14]. Similarly, many software organizations utilize the Capability Maturity Model. They aim at delivering their projects through the standardized PM process. Driven by the same goal of standardizing their PM processes, some organizations seek ISO 9000 certification of their PM. Finally, other organizations implement PM maturity models in an attempt to gradually improve management of projects through an SPM process [15].

To summarize, there is a significant interest in SPM process as a delivery mechanism for PM strategy [16]. In delivering PM strategy, SPM process implements several elements (Figure 1.1):

▲ Project life cycle phases

▲ Managerial and technical activities

▲ Deliverables

▲ Milestones

Project life cycle is viewed as a collection of project phases determined by the control needs of the organizations involved in the project. Consequently, a variety of project life cycle models are in use in corporations today. Some of them are traditional models, including phases of concept, definition, execution, and finish. While they may be losing their appeal for their generic nature, the new ones that are more industry-specific have emerged and gained in popularity. One example is the concurrent engineering process, which for ease of understanding is illustrated in Figure 1.1 as being sequential. Another example includes the evolutionary-delivery model in software development projects [17].

Project life cycle phases are composed of logically related project activities. The activities can be divided into two groups: managerial and technical activities. The former are activities by which we manage a project; *develop the project scope* and *construct the project schedule* are typical examples. They are similar across project types—construction, software, marketing, or financial type of project.

Where these project types really vary is with regard to technical activities. For example, technical activities characteristic of software projects would include *requirements definition* or *beta testing*. Such activities do not exist in a construction project, where examples of typical technical activities are *pre-job meeting* and *punch list development*. In short, technical activities take care of managing the project product. Naturally, then, they are project-type-specific and reflect the nature of the project product.

Both PM and technical activities usually culminate in the completion of deliverables—that is, tangible products in the SPM process. Managerial activities produce managerial deliverables (also called PM deliverables) such as *scope defined* or *risk review* (Figure 1.1). Technical activities lead to technical deliverables—for example, *product data sheet*, *system test*, and *manufacturing ramp ready*. As we refer to these deliverables in Figure 1.1, note that we opted to show deliverables only along with milestones that signify the end of a phase. To simplify the figure, managerial and technical activities are left out.

When not explained, the word *standardized* in SPM process may create confusion. What does it really mean? If we define SPM process as a standardized sequence of project activities (that culminate in project deliverables), then *standardized* means the degree of absence of variation in implementing such activities [18]. Let's use Figure 1.4 to explain this. On one extreme, there may be a complete variation in a PM process. Literally, every time the PM process is performed, it is performed in a different way. Obviously, 100 percent variation means that standardization is equal to 0. This is often referred to as an ad hoc approach. On the other extreme, a certain process may be 100 percent standardized, meaning that every time it is performed in the same way. In this case, variation is 0 percent. Between the two extremes lies a continuum of many PM processes with different ratios of the standardization and variation.

Take, for example, process L on the x-axis, one of the many possible PM processes. As Figure 1.4 shows, the degree of standardization and the degree of variation add up to 100 percent. If we go down the diagonal line to other PM processes, while the degree of standardization will increase, the degree of variation will decrease, but their sum will remain constant at 100 percent. Moving up the diagonal line will lead to a higher variation and lower standardization, still with the sum of 100 percent. Using plain language, the lower the variation, the higher standardization, and vice versa, and the more varied the implementation of project activities, the less standardized they are.

Speaking in practical terms, this means that organizations have a host of options when developing an SPM process—they can be more standardized or less standardized. Therefore, they need to decide on the level to which they want to standardize the PM process. The rationale of standardization is to create a predictable process that prevents PM activities from differing completely from project to project, from project manager to project manager. Put simply, SPM process saves project people the trouble of reinventing a new process for each individual project [19]. As a result, the process is repeatable despite changes in customer expectations or management turnover. The higher the standardization, the higher the repeatability.

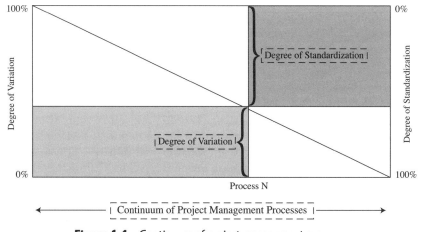

Figure 1.4 Continuum of project management processes.

The decision about how much to standardize the SPM process is the decision about the ratio of standardization and variation (popularly called *flexibility*). It is driven by PM strategy, or more precisely, by the types of projects the PM strategy deals with. Generally, PM strategy for projects of higher certainty will strive for higher levels of standardization and lower levels of flexibility. According to experts, the majority of projects in organizations belong to this group [20]. PM strategy for projects that face high uncertainty dictates lower standardization and higher flexibility. More details about this are provided in Chapter 16. When we use the term *SPM process* in this book, we mean it is standardized over 50 percent.

We started this section arguing that the SPM process supports PM. For this to be possible, the SPM process has to be aligned with PM strategy. In particular, when PM strategy is schedule focused, cost focused, or cost/quality focused, the SPM process will follow suit. That means the SPM process's disciplined and interrelated set of phases, deliverables, and milestones will be schedule focused, cost focused, or cost/quality focused (see Figure 1.3). Closely intertwined with the process are other PM components—project organization, information technology, culture, and leadership. In other words, the SPM process does not act alone but in synergy with these PM components in delivering PM strategy.

PM Toolbox Supports Standardized PM Process

In providing details of how the PM toolbox functions in its new role of supporting the SPM process, we do the following:

- ▲ Define PM tools and toolbox.
- ▲ Describe two options for the PM toolbox.
- ▲ Explain how SPM process drives the alignment of the PM toolbox.
- ▲ Compare benefits of one-tool-at-a-time versus the PM toolbox approach.
- ▲ Clarify the standardization of the PM toolbox.

Defining the PM Tools and Toolbox. PM tools include procedures and techniques by which a PM deliverable is produced. Similarly, *A Guide to the Project Management Body of Knowledge* [21, 22] and other sources use the term "tools and techniques" in place of what we define as PM tools. Two examples of such tools are the Team Charter and Monte Carlo analysis. They differ in the type of information they process. The Team Charter provides a systematic procedure to process qualitative information about authorizing a team to implement a project. On the other hand, Monte Carlo analysis is a risk-planning tool that also specifies the systematic procedure.

This time it does that by means of an algorithm in order to quantify risks. In other words, it is a quantitative tool. The heart of both the qualitative and quantitative groups of tools—and all PM tools belong to one of these groups—is in their systematic procedure. Note that we don't talk about PM software tools here. True, many PM tools that we discuss in this book exist in a software format. However, our focus is not on such formats. Rather, we concentrate on the substance of PM tools: their systematic procedure.

Two Options for the PM Toolbox. We define the PM toolbox as a set of predefined PM tools that a project manager can use in managing an SPM process. Two options are available in using the PM toolbox. In the first option, each tool in the toolbox is chosen to support specific managerial deliverables in the SPM process. For example, two tools in Figure 1.1 numbered S.1 are *WBS (Work Breakdown Structure)* and *Scope Statement*. They support the accomplishment of the managerial deliverable S.1 *Scope Defined* (in this figure other PM tools and managerial deliverables they support are also correspondingly numbered). Also, the PM toolbox is designed to comprise all tools one needs to implement the whole SPM process and accomplish its set of deliverables.

In the second option, the idea is to replace managerial deliverables with the PM toolbox. As for Figure 1.1, that means taking the managerial deliverables out and putting the PM toolbox in their place. The idea here is to conceive individual tools in the PM toolbox as proxies for managerial deliverables. Take, for example, the managerial deliverable *Schedule Developed*. It can be replaced with a specific tool—*Gantt Chart* or *Milestone Chart*. Or, think of the managerial deliverable *Projects Selected*. Instead of this deliverable, we can use a PM tool called *Scoring Model*, which rates, ranks, and selects new projects. Obviously, the second option demands us to think of PM tools in radically new ways. Rather than focusing on the process of the tool, as has been traditionally done, we emphasize its use to produce an output—which is actually the deliverable. Therefore, each tool in the PM toolbox can be visualized as a deliverable. Similarly, the PM toolbox can be envisaged as a set of managerial deliverables in the SPM process. The first benefit here is that we simplify the SPM process by taking one layer out of it—managerial deliverables. Still, the project manager and his or her team have a road map (systematic procedure) to create the whole set of managerial deliverables.

How SPM Process Drives the Alignment of the PM Toolbox. Real-world companies use both options for the PM toolbox, the first one more frequently [23]. Whichever option is chosen, the PM toolbox has to be aligned with the SPM process. As illustrated in Figure 1.3, the focus of PM strategy is aligned with the competitive strategy. Since SPM process is part of this PM focus, it is logical that the PM toolbox will have the same focus. In particular, the SPM process of schedule focus, cost focus, or cost/quality focus will drive the PM toolbox to have schedule focus, cost focus, or cost/quality focus, respectively.

Comparing Benefits of One-Tool-at-a-Time versus the PM Toolbox Approach. Regardless of their competitive strategy, today's companies that are suppliers of projects face a competitive reality: their customers have taken charge. Customers tell the companies what they want, when they want it (as fast as possible—high speed), how they want it (as good as possible—better quality and customer satisfaction), and how much they are willing to pay (as low as possible—low cost) [20]. And the companies listen; satisfied customers are critical for a company's economic returns. In 1997 companies that reached higher customer satisfaction created over 100 percent more shareholder wealth than their competitors, with lower customer satisfaction scores [24]. As a result, there is no such a thing as *the* customer; there is only *this* customer, the one using his or her power and ability to demand. To respond to these demands, leading companies make sure that they build an SPM process capable of delivering projects with the following:

- ▲ Speed
- ▲ Repeatability
- ▲ Concurrency

A crucial role in this effort belongs to the PM toolbox. Let's explore each one in turn.

Speed. This is the ability of an organization to deliver a project in a fast manner. While what is meant by *fast* may vary, it means that fast is competitive. In one case, fast meant that the cycle time had to be reduced from 18 months to 9 months for an organization to keep up with the competition. For this to be possible, many components in the SPM process must be present. For example, there must be within- and across-phase overlapping of project activities, non-speed-adding activities have to be eliminated as well as any other redundancies, and so on. What this really means is having an SPM process that is streamlined to be fast to deliver per-customer demands.

Repeatability. Delivering a project with speed is not enough unless it is repeatable. That means there is the organization's ability to deliver a stream of consecutive projects consistently per customer requirements, one after another, every time. We call this *longitudinal repeatability*. If that requirement is speed, for example, then each delivered project is consistently fast. Repeatable projects minimize variation in how they are executed, improving speed and quality. Improvements in quality lead to lower cost because they result in less rework, fewer mistakes, fewer delays and snags, and better use of time. With higher speed, better quality, and lower cost, the organization can better respond to customer demands and make them satisfied.

Concurrency. In addition to speed and repeatable consecutive projects, responding to the customer demands also requires an ability to deliver a host of simultaneous projects, which are typically interdependent. We call this *lateral repeatability*, a different challenge from the longitudinal one. Here, some projects are large, others small. Since they are interdependent and share the same pool of resources, the challenge is to execute all of them in parallel, as a concerted group. No variation is allowed in any one; each needs to maintain speed and quality. If they don't, they will make others slip, increase cost, and make the customer unhappy. Similar to the longitudinal repeatability, minimizing the variation in each project will trigger speed and quality improvements that lead to lower cost, again contributing to meeting customer demands and satisfaction.

To reach this level of speed, repeatability, and concurrency, the heroic approach of "get the best people, pour all resources they need, and they will produce a great project" cannot help. Rather, a strong SPM process supported by PM tools is necessary. Empirical evidence that PM tools impact project success already exists [25][26]. The problem is when a project manager is given an SPM process but is left with a task of determining which tool and how to use it to support the process.

Selecting PM tools one at a time demands a substantial amount of resources and expertise. It is not reasonable to presume that each project manager—especially if he or she is less-than-experienced, as is the case with many—would have the resources and expertise to quickly, smoothly, and consistently select his or her own set of tools. Rather, such managers end up struggling to find the right PM tools and how to use them, introducing variability in the SPM process. This may lower the speed and make projects less repeatable and concurrency less likely (see Table 1.1). In contrast, project managers given the SPM process and the PM toolbox know exactly which tool and how to use it in order to support the process. In other words, they have a standardized PM toolbox capable of supporting an SPM process with minimum variation. As a result, their projects may have higher speed, more repeatability, and concurrency (Table 1.1)

Table 1.1: One-Tool-At-A-Time vs. the PM Toolbox Approach

Requirement	Impact on SPM Process One-Tool-At-A-Time	PM Toolbox
Speed	Lower	Higher
Repeatability	Less repeatable	More repeatable
Concurrency	Less likely	More likely

Standardization of the PM Toolbox. Very often, project managers assume that the PM toolbox is of a one-size-fits-all nature. This, of course, is incorrect. The PM toolbox can come in many sizes, shapes, and flavors, as we will show in Chapter 16. Logically, this is an issue related to the SPM process and types of projects the process serves. Since the PM toolbox is aligned with the SPM process, it is apparent that the level of standardization of the SPM process impacts the standardization level of the PM toolbox. For example, an SPM process that is highly standardized will probably be supported by a highly standardized PM toolbox, and vice versa.

Concluding Remarks

PM tools serve two roles in supporting the SPM process. First, in their conventional role, the tools are enabling devices for reaching a project deliverable in the SPM process. Second, in their new role, they serve as basic building blocks to construct the PM toolbox, which supports the SPM process.

Many organizations rely on the SPM process when delivering PM strategy. The process is a standardized, disciplined, and interrelated set of phases, deliverables, and milestones that each project goes through. Closely intertwined with the process are other PM components. These include project organization, information technology, culture, and leadership. Together, they help deliver PM strategy.

PM strategy is crucial in supporting the execution of the competitive strategy of an organization. Truly, project management strategy is carefully concocted and aligned with the specific type of competitive strategy. The goal, of course, is to deliver a desired punch—schedule-focused, cost-focused, or cost/quality-focused PM strategy. Combined with other business strategies, project management thus has become a business strategy of choice in today's corporations.

In the world of cutthroat competition, it takes competitive advantage to win. These advantages come in all sorts of shapes. Some are called fast time-to-market advantages [27]. Others are of the low-cost nature. Still others are in having the lowest cost for a certain level of quality. No matter what they are, the advantages do not come about spontaneously. Rather, they are an outcome of organizations executing their competitive strategies. In summary, the role of the PM toolbox is to support SPM processes that help deliver PM strategy. By supporting the competitive strategy, the PM strategy helps create competitive advantages for the company

The ability to design a PM toolbox is closely related to the reader's knowledge of individual PM tools. To help the knowledge get to the appropriate level, chapters that follow will review the tools. Then, in chapter 16 we will offer the methodology to design the toolbox.

References

1. Hamel, G. and C. K. Prahalad. 1989. "Strategic Intent." Harvard Business Review 67(3): 92–101.

2. Harrison, J. S. and C. H. S. John. 1998. *Strategic Management of Organizations and Stakeholders*. St. Paul, Minn.: South-Western College Publishing.

3. Porter, M. E. 1985. *Competitive Advantage*. New York: The Free Press.

4. Thompson, A. A. and A. J. I. Strickland. 1995. *Crafting and Implementing Strategy*. Chicago: Irwin.

5. Adler, P. S., et al. 1996. "Getting the Most out of Your Product Development Process." *Harvard Business Review* 74(2): 134–152.

6. Calantone, R. J. and C. A. D. Benedetto. 2000. "Performance and Time to Market: Accelerating Cycle Time with Overlapping Stages." *IEEE Transactions on Engineering Management* 47(2): 232–244.

7. Nevens, T. M., G. L. Summe, and B. Uttal. 1990. "Commercializing Technology: What the Best Companies Do." *Harvard Business Review* 68(3): 154–163.

8. Smith, P. and D. Reinertsen. 1990. *Developing Products in Half the Time*. New York: Van Nostrand Reinhold.

9. Mclemore, I. 1999. "High Stake Games." Business Finance (5)30–33.

10. Iansiti, M. and J. West. 1997. "Technology Integration Turning Great Research into Great Products." *Harvard Business Review* 75(3): 69–79.

11. Wheelwright, S. C. and K. B. Clark. 1992. *Revolutionizing Product Development*. New York: The Free Press.

12. Clark, K. B. and T. Fujimoto. 1991. *Product Development Performance*. Boston: Harvard Business School Press.

13. Cusumano, M. A. and K. Nobeoka. 1998. *Thinking Beyond Lean*. New York: The Free Press.

14. Tony, F. and R. Powers. 1997. *Best Practices of Project Management Groups in Large Functional Organizations*. Drexel Hill, Pa.: Project Management Institute.

15. Kerzner, H. 2001. *Strategic Planning for Project Management*. New York: John Wiley & Sons.

16. Kerzner, H. 2000. *Applied Project Management*. New York: John Wiley & Sons.

17. Kemerer, C. F. 1997. *Software Project Management*. Boston: McGraw-Hill.

18. Stevenson, W. J. 1993. *Production and Operations Management*. Boston: Irwin.

19. Sobek, D., J. Liker, and A. Ward. 1998. "Another Look at How Toyota Integrates Product Development." *Harvard Business Review* 76(4): 36–49.

20. Hammer, M. and J. Champy. 1993. *Reengineering the Corporation*. New York: Harper Business.

21. Project Management Institute. 2000. *A Guide to the Project Management Body of Knowledge*. Newton Square , Pa.: Project Management Institute.

22. Thamhain, H. J. 1999. "Emerging Project Management Techniques: A Managerial Assessment." Portland International Conference on Management of Engineering and Technology. Portland, Oregon.

23. Coombs, R., A. McMeekin, and R. Pybus. 1998. "Toward Development of Benchmarking Tools for R&D Project Management." *R&D Management* 28(3): 175–186.

24. University of Michigan Business School and American Society for Quality. 1998. *American Customer Satisfaction Index: 1994–1998*. Ann Arbor: University of Michigan Press.

25. Pinto, J. K. and J. E. Prescott. 1990. "Planning and Tactical Factors in the Project Implementation Process." *Journal of Management Studies* 27(3): 305–327.

26. Might, R. 1989. "How Northern Telecom Competes on Time." *Harvard Business Review* 67(4): 108–114.

27. Eisenhardt, K. M. and S. L. Brown. 1998. "Time Pacing: Competing in Marketing That Won't Stand Still." *Harvard Business Review* 76(2): 59–69.

Project
Selection

This chapter is contributed by Dr. Joseph P. Martino.

In research the horizon recedes as we advance, . . . and research is always incomplete.

Mark Pattison, Isaac Casaubon

True as these words may be, managers still need to make choices among possible projects, and either see them through to completion or terminate them when they "turn sour." This chapter is intended to help managers use the following tools to select projects that have a good chance of succeeding and paying off:

- ▲ Scoring Models
- ▲ Analytic Hierarchy Process

> ▲ Economic Methods (Payback Time, Net Present Value, Internal Rate of Return)
> ▲ Portfolio Selection
> ▲ Real Options Approach

These tools are designed to select projects that maximize the value of project portfolio for the company and are aligned with its business strategy. The assumption is that you have more candidate projects than you can support with available resources. To help you select from your project menu, this chapter provides a collection of tools that account for financial, technical, behavioral, and strategic criteria or factors in evaluating and selecting projects. The project mix will be in a state of constant flux, as management reevaluates its selections on an ongoing basis to respond to changing competitive and other needs. Project selection tools allow management to select projects for initiation and termination as conditions change.

The goal of this chapter is to help practicing and prospective project managers and managers accomplish the following:

> ▲ Learn how to use various project selection tools.
> ▲ Select project selection tools that match their project situation.
> ▲ Customize the tools of their choice.

Mastering these skills has a central role in initiating projects and building the SPM process. See Figure 2.1 for where these tools fit into the project management process.

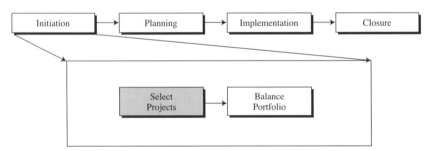

Figure 2.1   The role of project selection tools in the standardized project management process.

Table 2.1: A Comparison of Project Selection Methods

Type of Method	Subtypes Presented Here	Basic Purpose	Major Advantage	Major Disadvantage
Numerical, Ranking Methods	Scoring Models	Rank candidate projects in order of desirability. Managers fund projects in order, until resources are exhausted.	Completely transparent, easy to use, readily understandable	May give impression of false precision. Requires significant input from higher management.
	Analytic Hierarchy Process		Allows criteria to be disaggregated into several levels.	Requires extensive input from functional and higher management.
Numerical, Economic Methods	Payback Period	Evaluate economic payoff.	Simple to use and understand; very robust against uncertainties. Direct comparison with capital budgeting.	Does not account for time value of money. Required data may not be available for some projects such as basic research.
	Net Present Value	Evaluate economic payoff, including time value of money.	Easy to calculate using spreadsheet; direct comparison with capital budgeting.	Required data may not be available for some projects such as basic research.
	Internal Rate of Return			
Numerical, Optimization Methods	Portfolio Selection	Choose portfolio of projects that maximizes some measure of payoff.	Allows use of multiple criteria for selecting an entire portfolio of projects.	Extensive computations required for large project portfolios.
Real Options	Projects as Options	Reduce risk by selecting best combination of alternatives.	Reduces both downside and upside risk associated with projects.	Requires extensive data and analysis.

Scoring Models
What Are Scoring Models?

A Scoring Model includes a list of relevant criteria the decision maker wishes to take into account when selecting projects from the candidate menu. Projects are then rated by decision makers on each criterion, typically on a numerical scale with anchor phrases. Finally, multiplying these scores by weightings and adding them up across all criteria will produce a score that represents the merit of the project. Higher scores designate projects of higher merit. The Scoring Model can be designed specifically for any given selection situation. For better understanding of the position of the Scoring Models in the family of project selection methods, see Table 2.1.

Applying Scoring Models

Collect Information Inputs. Like other methods, Scoring Models follow the basic steps of project selection process (see box, "Basic Project Selection Process"). To be fully functional and meaningful, the models need a candidate project menu to choose from. In reality, what goes with the menu is a set of quality inputs including:

- ▲ Project proposal
- ▲ Strategic and tactical plans
- ▲ Historical information.

Since the purpose of the models is to help maximize the value of the selected portfolio of projects for the company, understanding which of the company's goals a project supports is a key point. While these goals are described in the strategic and tactical plans of the organizations, project proposals offer specifics on projects. To make better decisions, decision makers should also rely on the historical information about both the results of past project selection decisions and past project performance. When these inputs are available, you can move to choose relevant project selection criteria.

Select Relevant Criteria. One of the major reasons for the failure of the Scoring Models is constructing them on inappropriate criteria. Consequently, a key to Scoring Models appears to be in compiling an appropriate list of scoring criteria that well reflect the strategic financial, technical, and behavioral situation of the company. The challenge often seems to be in overcoming the temptation to develop a detailed and, therefore, cumbersome list of criteria that becomes unmanageable. Therefore, narrowing down to the vital few criteria that really matter is rather difficult, especially when the menu of criteria to start from is deceptively long. Consider, for example, criteria from the box coming up in the chapter titled "Criteria to Be Considered in Project Selection." To be used effectively, most of them need to be broken into more specific items (see

Table 2.2 for an example of the breakdown into 19 specific criteria). This further elevates the challenge of sticking with the vital few. That the challenge is real can be seen from experiences of some companies in which models fell in disuse because they included 50+ criteria. A smart approach used by some leading companies is in conceiving the list and refining over time with the intent to reduce the number of criteria.

Construct a Scoring Model. To construct a model, you must understand and resolve several issues:

▲ Form of the model, with certain categories of criteria or factors
▲ Value and importance of criteria
▲ Measurement of criteria

First we will deal with forming of the model. A "generic" scoring model would have the following form:

$$Score = \frac{A\,(bB + cC + dD)\,(1 + eE\,)}{fF\,(1 + gG\,)}$$

In this model, the symbols A, B, C, D, E, F, and G represent the criteria to be included in the score for the project. The *value* of each criterion for a given project is substituted in the formula. The symbols b, c, d, e, f, and g represent the *weights* assigned to each criterion. In the model, the criteria in the numerator are *benefits*, while the criteria in the denominator are costs or other *disbenefits*. The criteria are selected by management, as are the weights. The *values* of the criteria are project-specific and are normally provided by the project team.

Basic Project Selection Process

What types of projects are candidates (e.g., product development, market entry, capital projects, continuous improvement, etc.)? This question is addressed in the beginning step of the project selection process (see Figure 2.2). It creates the candidate project menu and is followed by selecting relevant criteria. The purpose is to include factors that help maximize the value of project portfolio for the company.

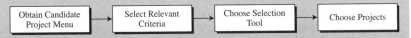

Figure 2.2 Basic project selection process.

Also known at this time should be the constraints such as budget, staffing, and so on. Important in this is to make sure that any policy considerations and strategic plans have been taken into account (e.g., support of existing products or construction of the new fab). The criteria will be plugged into the preferred selection tool to decide which projects from the candidate menu to choose for initiation or continuation.

Criteria to Be Considered in Project Selection

Criteria (or factors) that are relevant in project selection depend on the type of projects and their situation. For example, the following criteria are typically considered in choosing research and development projects. This list is intended to be suggestive rather than complete.

- Cost
- Probability of technical success
- Market Size
- Availability of required staff
- Strategic positioning for project
- Favorability of regulatory environment

- Payoff
- Probability of market success
- Market share
- Degree of organizational commitment
- Degree of competition
- Company policy considerations

Although many of these criteria may be used in selecting different types of projects, the important thing is to include in the analysis whatever criteria are relevant to your project situation.

The model uses three categories of criteria:

▲ Overriding criteria (e.g., A). These are factors of such great importance that if they go to zero, the entire score should be zero. For instance, factors to be included in the model might be measures of performance such as efficiency or total output. A performance measure of zero should disqualify the project completely, regardless of any other "merit."

▲ Tradable criteria (B, C, D, F). These are factors that can be traded against one another; a decrease in one is acceptable if accompanied by a sufficient increase in another. For instance, the designer may be willing to trade between reliability and maintainability, so long as "cost of ownership" remains constant. In this case, the weights would reflect the relative costs of increasing reliability and making maintenance easier. Cost F is shown as a single criterion that is relevant to all projects. Typically this would include monetary costs of the project. This might be disaggregated into cost categories such as wages, materials, facilities, and shipping, if there is the possibility of a trade-off among these cost categories (e.g., using a more expensive material to save labor costs). If no such trade-off exists, the costs should simply be summed and treated as a single factor.

▲ Optional criteria. These are factors that may not be relevant to all projects; if they are present, they should affect the score, but if they are absent, they should not affect the score. Note that either costs or benefits may involve optional factors. For instance, E in the formula represents a benefit that may not be a consideration with all projects. It should be counted in the score only if it is relevant to a project. For example, this might be a rating of ease of consumer use, which would not be relevant to a project aimed at an industrial process. G in the formula represents an "optional" cost that might not be relevant to all projects. Typically this type of cost is one in which the availability of some resource is a more important consideration than its monetary cost. For instance, there may be limits on the availability of a testing device, supercomputer, or a scarce skill such as a programmer. In such a case, the hours or other measure of use should be included separately from monetary cost and should apply only to those projects requiring that resource.

The second issue focuses on value and importance of criteria. Once the form of the model is selected, the designers of the model need to distinguish between the *value* of a criterion and the *weight* or *importance* of that criterion. In the preceding formula, B, C, and D are the values of their respective factors for a specific project, while b, c, and d are the weights assigned to those factors, reflecting the importance assigned to them by the decision maker. In the case of the tradable factors, the ratio b/c represents the trade-off relationship between factors B and C. If B is decreased by one unit, C must be increased by at least the amount b/c for the sum of the tradable factors to remain constant or increase. That is, the decision maker is willing to trade one factor for another according to the ratios of their weights, so long as the total sum remains constant or increases.

A very simple model used to score the project in Table 2.2 (see column *item*) is as follows:

$$\text{Score} = B + C + D$$

Simply, the score is equal to the sum of B, C, and D, the values of their respective criteria such as "Unique Product Functionalities" or "Technical Complexity" (Table 2.2).

Finally, the third issue to resolve is one of measurement of criteria. Some criteria are objectively measurable, such as costs and revenues. Others, such as probability of success or strategic importance, must be obtained judgmentally. Scoring Models can readily include both objective and subjective or judgmental criteria. It is helpful if the judgmental criteria are estimated with a scale with anchor phrases, to obtain consistency in estimating the magnitude of the factor for each project. The estimates should be made on some convenient scale, such as 1 to 10. For an example see the box that follows titled "An Example of the Scale to Measure Magnitude of the Criterion—Availability of the Skills." A similar scale should be devised to aid in making estimates of each of the criteria to be obtained judgmentally, as was done in the example from Table 2.2. Objective factors such as costs can be measured directly, so anchored scales are not necessary.

Table 2.2: Rating *Scanner* Project with a Scoring Model

Criteria/Factors (scored 0 – 10)

Criteria/ Factor	Item	Scores Out of 10 (Points)	Average Criterion/ Factor Score (Points)
Strategic Positioning	Degree of project's alignment with business unit strategy	8	8.0
	Strategic significance	8	
Product/ Competitive Advantage	Unique product functionalities	8	
	Provides better customer benefits	9	
	Meets customer value drivers better	7	8.0
Market Appeal	Market size	8	
	Market share	8	
	Market growth	6	
	Degree of competition	6	7.0
Alignment with Core Competencies	Marketing alignment	8	
	Technological alignment	7	
	Manufacturing alignment	6	7.0
Technical Merit	Technical gap	9	
	Technical complexity	6	
	Technical probability of success	9	8.0
Financial Merit	Expected Net Present Value (NPV)	9	
	Expected Internal Rate of Return (IRR)	9	
	Payback Time (PT)	7	
	Probability of NPV, IRR, PT	7	8.0
Total Project Score			**46 out of a possible 60 points, which is 77%**

Most Scoring Models are more complex than a simple sum of criteria. Suppose the factors we wish to include in the score are Probability of Success, Payoff, and Cost. Suppose further that we are willing to trade payoff and probability of success (e.g., we are willing to accept a project with lower Probability of Success if Payoff is high enough), and we think Payoff is twice as important as either Probability of Success or Cost. Probability of Success and Payoff are benefits, while Cost is a cost or disbenefit. Then the Scoring Model will be as follows:

$$Score = \frac{\left(P_{Success} + 2 * Payoff\right)}{Cost}$$

This more complex model will be used later to score projects shown in Table 2.6. The designer of the Scoring Model is free to include whatever factors are considered important, and to assign weights or coefficients to reflect relative importance.

Rank Projects. When the criteria for the model have been selected, the form of the Scoring Model chosen, weights established, and measurement scales defined, you are ready to rank the candidate project menu. Note that while the decision maker must obtain the criteria and their weights from management, this is a one-time activity. The project-specific data for individual projects will in most cases come from those proposing the project. They will provide either objective data (e.g., costs, staff hours, machine use) or ratings based on the scales the decision maker has established. In some cases, project-specific data may be obtained from sources other than the project originators. For instance, data on probability of market success or payoff might be obtained from marketing rather than from R&D. While this data must be obtained for each project being ranked, the criteria and their weights will remain fixed until management decides they must be revised.

An Example of the Scale to Measure Magnitude of the Criterion— Availability of the Skills

10   All skills in ample supply

9   All skills available with no excess

8   All technical skills available

7   Most professional skills available

6   Some technical skills retraining needed

5   Some professional retraining needed

4   Extensive technical retraining needed

3   Extensive professional retraining needed

2   All technical skills must be hired

1   All technical skills and some professional skills must be hired

0   All professional and technical skills must be hired

In most cases, the project data will be in units that vary in magnitude: probabilities to the right of the decimal, monetary costs to the left of the decimal, scale rankings in integers, and so forth. It is necessary to convert all the factors to a common range of values. Consider Table 2.6, with a portfolio of proposed projects. The magnitudes of the factors range from 3 digits to the left of the decimal to 2 digits to the right of the decimal. It is customary to *standardize* the project data by subtracting from each value the average of that value over all projects, and dividing that by the standard deviation of the value over each project. In terms of Table 2.6, this means subtracting from every value in a column the mean value of that column, then dividing the result by the standard deviation of the same column. Once the project-specific data have been loaded into a spreadsheet, this standardization process is trivial. It is as easy for a thousand projects as for a dozen.

Assuming the project-specific values are approximately normally distributed, the result should be standardized values ranging from about –3 to about +3. These must now be restored to positive values. If any of the original values in a column was zero, the standardized value for that factor should also be zero. Add to every value the absolute value of the most negative number in the column resulting from the subtraction and division process. This will result in standardized values ranging from zero to approximately 6. If none of the original values was zero, add to every number 1 plus the absolute value of the most negative number. This will give values ranging from 1 to approximately 7. These standardized values should then be substituted in the model. Each project then receives a score based on weights supplied by management, and project data supplied by the project originators (and possibly by other departments).

Table 2.3 shows the results of the model applied to standardized scores from Table 2.6. The rows have been re-ordered in decreasing magnitude of the score. If the standardized values are available in a spreadsheet, the process of computing project scores is trivial. Likewise, sorting the projects in order of scores is readily accomplished using a spreadsheet. In this example, Project 4 is the highest-ranked project. The other projects fall in order of their score. The next step would be to approve projects starting from the top of the list and working down, until the budget is exhausted. Note that the difference between projects 8 and 1 comes only in the third significant figure. Since most of the original data was good only to one or two significant figures, this difference should not be taken seriously.

Utilizing Scoring Models
When to Use. While these methods can be used for any type of project, they are especially useful in the earlier phases of a project life cycle when major project selection decisions are made. Take, for example, new product development projects. In the earlier project phases, market payoff is distant or even inappropriate as a measure of merit. In such projects, considerations such as technical merit—a frequent criterion in the Scoring Models—may be of greater significance than economic payoff. Selection of other types of projects, large and small, widely relies on Scoring Models as well. The final score is typically used for two purposes:

Table 2.3: Results of Applying Scoring Model to Example Data

Project	Cost $K	P(Sxs)	Payoff $M	Score
4	1.89	2.67	3.35	4.95
6	2.12	3.38	2.78	4.22
3	1.00	2.13	1.00	4.13
5	2.17	3.33	2.78	4.10
8	2.51	3.88	2.37	3.43
1	1.45	2.13	1.42	3.42
12	3.58	3.56	4.22	3.35
11	3.70	3.61	3.34	2.78
2	1.62	1.00	1.58	2.57
7	2.49	3.67	1.34	2.56
14	4.44	3.78	3.54	2.45
16	6.39	3.88	5.74	2.40
9	3.11	3.56	1.91	2.38
10	4.13	3.65	2.92	2.30
13	4.11	3.98	2.53	2.20
15	5.43	3.86	2.88	1.77

Go/kill decisions. These are located at certain points within project management process, often at the ends of project phases. Their purpose is to decide which new projects to initiate and which of the existing ones to continue or terminate.

Project prioritization. This is where resources are allocated to the new projects with a "go" decision, and the total list of new and existing projects, which already have resources assigned, are prioritized.

For a comparison of Scoring Models and the Analytic Hierarchy Process (AHP), two project ranking tools that we cover, see the AHP section later in this chapter.

Time to Use. Although the principles behind the Scoring Models are relatively simple, developing an effective Scoring Model can be a very lengthy and time-consuming endeavor. In an example, it took a company several years to carefully select, word, operationally define, and test each criterion for validity and reliability. This detailed Scoring Model included 19 criteria and was designed to rate projects of strategic stature. Applying a model like this to sort through, rate, and rank a larger group of projects can easily take more than day or two of each decision maker's time. In contrast, some decision makers use off-the-shelf Scoring Models for the evaluation of their smaller, tactical projects, typically spending several hours on the scoring effort (see the "Tips for Scoring Models" box that follows).

Benefits. The value of the Scoring Model is that it can be tailored to fit the decision situation, taking into account multiple goals and criteria, both objective and judgmental, which are deemed important for the decision [2]. This prevents putting a heavy emphasis on financial criteria that tend not to be reliable early in the project life. With such an approach, decision makers are forced to scrutinize each project on a same set of criteria, focusing rigorously on critical issues but recognizing that some criteria are more important than others (by means of weights). Finally, the Scoring Model can be subjected to sensitivity analysis, by determining how much change in a weight would be required to change the priority order significantly.

Advantages and Disadvantages. Views about advantages and disadvantages of Scoring Models are not in short supply. Consider the following advantages:

- ▲ *They are conceptually simple.* They trim down the complex selection decision to a handy number of specific questions and yield a single score, a helpful input into a project selection effort. This is perhaps a major reason for models' wide popularity.

- ▲ *They are completely transparent.* This is in the sense that anyone can examine the model and see why the priority ranking came out as it did.

- ▲ *They seem to work.* Several studies showed that they yield good decisions. Procter & Gamble claims that their computer-based Scoring Model provides 85 percent predicative ability [1].

- ▲ *They are easy to use.* Managers involved in project selection rated scoring models best in terms of ease of use and deemed them highly suitable for project selection.

Tips for Scoring Models

Empirical evidence suggests that leading organizations favor approaches like this in using Scoring Models [1]:

- ● *Have senior managers do scoring.* There seems to be major bias in the selection process when project teams do the scoring, perhaps for favoring their projects.

- ● *Make it discussion forum.* The use of the model in a meeting format provides a benefit of executive buy-in, because senior managers examine each project together, discuss it on each criterion, rate it, and build a group consensus.

- ● *Use a visible scorecard.* When each decision maker has an electronic keypad, he or she can vote anonymously, have his or her votes instantly fed into a computer, and group results displayed on a large screen.

When it comes to disadvantages, there are beliefs that Scoring Models suffer from the following [1]:

- ▲ *Imaginary precision.* Using the models implies that the scores are highly precise, which simply is not the case—scoring is often arbitrary. This is why they should not be overused nor their results taken for granted.

- ▲ *Potential lack of efficiency of allocation of scare resources.* The issue here is that the models do not constraint the necessary resources when maximizing the total score for the project, as some Economic Methods do.

- ▲ *Time-consuming.* This disadvantage should be viewed in the light of the importance of the decisions. Were the decisions really crucial, extra time should be spent on them.

Variations. Major variations in using the models are closely linked to their time-consuming nature. In attempts to reduce the necessary time for project selection, companies took different paths. Unlike the approach of executive scoring of projects in the meeting, some companies first ask decision makers in the meeting to sort through projects and divide them into several groups of importance, then go to individual voting within the groups. Others rely on individual evaluators scoring before they come to the meeting. The purpose of the meeting is no more than seeing the results and discussing them. While this is a very time-efficient approach, it deprives the decision makers of the benefit of walking through each factor as a group, debating it, and building a consensus. Rarely, some organizations have project teams do prescoring of the project and present it to the decision makers, who accept or disprove the scores.

Customize Scoring Models. Although very beneficial tools, Scoring Models that we covered are of generic nature and, therefore, need to be adapted to account for a company's specifics and projects. Following we offer several ideas how to go about the adaptation.

Customization Action	*Examples of Customization Actions*
Define limits of use.	Use Scoring Models for both large and small projects.
	Use Scoring Models in new project selection and go/kill reviews of the existing projects.
Adapt a feature.	Present briefly each project proposal to the decision makers before they engage in group discussion and rating of projects (helps busy decision makers).
Leave out a feature.	Do not use weights for criteria in very small projects in multiproject environment.

Scoring Models Check
- Check to make sure that a Scoring Model is appropriately structured and applied. It should:
- Be based on project proposals, strategic and tactical plans, and historical information.
- Include relevant criteria with weights and rating scale.
- Score each project on each criterion, multiply scores by weights and sum them across all criteria.
- Yield a single score for each project.
- Provide ranking of projects.

Summary

Presented in this section were Scoring Models, a project selection tool. These tools help assign scores to projects on a list of criteria. The score represents the merit of the project. Higher scores designate projects of higher merit. While these methods can be used for any type of project, they are especially useful in two situations: first, in the earlier phases of a project life cycle when major project selection decisions are made and projects prioritized, and second, in go/kill decisions. The key points in structuring and applying the Scoring Models are provided in the box that follows.

Analytical Hierarchy Process
What Is the Analytic Hierarchy Process (AHP)?

The Analytic Hierarchy Process (AHP), like the Scoring Model, provides a means for ranking projects. However, the Scoring Model is a "one-level" process. If one or more of the criteria are made up of subcriteria that are combined to get the value for that factor, this combining must be done outside the model. By contrast, the AHP specifically includes means for incorporating the combination of subcriteria. The procedure thus uses a "hierarchy" of criteria, with each criterion being disaggregated into subcriteria corresponding to one's understanding of the project ranking situation (see Figure 2.3). Such breakdown offers an opportunity to seek cause-effect explanations between goal (e.g., select the best project), criteria (e.g., technical merit), subcriteria (e.g., probability of technical success), and candidate projects. Once this hierarchical structure is established, AHP continues with weighting of criteria and subcriteria and scoring that computes a composite score for each candidate project at each level as well as an overall score. The overall score signifies the merit of the project with higher scores representing projects of higher merit.

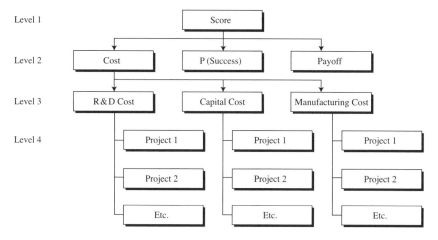

Figure 2.3.   An example of the AHP decision hierarchy.

Applying AHP

The description of the AHP is, unfortunately, highly abstract. To make it more concrete, it will be illustrated using the same project menu and project selection criteria as were used for the scoring model.

Define the Problem and Goal. AHP's first step is to define the problem and determine the goal [3]. In our case the goal is known—we want to rank the new and existing projects so that we can select the best new ones to initiate or continue/terminate them if they are already in existence.

Collect Information Inputs. Like Scoring Models, AHP starts with the need to have a list of candidate projects to select from and quality information inputs to perform ranking. The inputs are as follows:

- ▲ Project proposal
- ▲ Strategic and tactical plans
- ▲ Historical information.

As explained in the "Scoring Models" section, these inputs ensure understanding of the company's strategy and goals, project goals and scope, and the results of past project selection decisions and past project performance.

Construct the Hierarchy. As with any other project selection method, it is necessary to decide on the criteria that will be used to rate projects. The unique feature of AHP is that the criteria are disaggregated into as many sublevels as needed for each criterion. To illustrate this, we will use the same criteria as we did for the Scoring Model. We wish to rank projects according to their merit, taking into account Cost, Probability of Success, and Payoff, with Payoff having twice the weight of Probability of Success and Cost. The hierarchical model is given in Figure 2.3. To illustrate that subcriteria can be included, the Cost criterion has been subdivided. However, since that data is not available in Table 2.3, the actual computation will not involve the subcriteria. Instead, the projects will be evaluated against the *each* of the lowest-level criteria, in our case Cost, Probability of Success, and Payoff. That is, projects would be evaluated against all three elements of cost (if these were available), as well as against Probability of Success and Payoff.

Construct Comparison Matrices. At each level of the hierarchy, a matrix must be developed whose elements are the relative preference (weight, significance, value, etc.) for each criterion at that level as compared with all the other criteria at that same level. The entries in the matrix show the degree to which the *Row* factor is preferred to the *Column* criterion.

To continue the example using data from Table 2.3, we develop the first-level comparison matrix. At the first level of the matrix below the total score, the three factors are Cost, Probability of Success, and Payoff. The matrix is as follows:

		1	1	2
		Cost	P(Sxs)	Payoff
1	Cost	1	1	0.5
1	P(Sxs)	1	1	0.5
2	Payoff	2	2	1
	ColSums	4	4	2

The relative weights of the three factors are shown in the top row and the extreme left column. The degree of preference is shown in the cells inside the double line. In the row for Cost, it is shown as having half the preference as Payoff, in the Payoff column. The diagonal elements, of course, are all 1, and the elements below the diagonal are the reciprocals of their corresponding elements above the diagonal. Thus, in the row for Payoff, the cell entry in the column for Cost is 2, showing that the preference for Payoff is twice that for Cost.

In this example we had definite numbers for the relative preferences. However, in many cases, the elements of the matrix must be developed judgmentally. The process of developing each matrix can be very time-consuming. It is important to note, however, that not all the matrices in the hierarchy need to be developed by the same people. For instance, in this model, the relative likelihood of project success matrix might be filled out by technical management, the payoff matrix by marketing, and the cost matrix by executives from personnel, purchasing, and other relevant departments.

Next, a new matrix is formed in which the elements are the elements of the above matrix, all divided by their column sums:

	Cost	P(Sxs)	Payoff	RowAvg
Cost	0.25	0.25	0.25	0.25
P(Sxs)	0.25	0.25	0.25	0.25
Payoff	0.5	0.5	0.5	0.5

The row averages are the weights that will be used to multiply the respective results of a similar operation on the values of Cost, Probability of Success, and Payoff for each of the projects.

A similar matrix of preferences is developed for each factor, over all projects. A portion of the Cost matrix follows. Note that since a higher cost is less preferred than a lower cost, the column cost is divided by the row cost to obtain the cell entry of preference of row for column for each project. To save space, only the upper left corner of the entire array is shown. Similar arrays for Probability of Success and Cost follow. Since for these factors a larger value is preferred to a smaller value, row weight is divided by column weight to obtain preference of row factor for column factor.

Cost		43	44	16	30
	Project	1	2	3	4
43	1	1.00	1.02	0.37	0.70
44	2	0.98	1.00	0.36	0.68
16	3	2.69	2.75	1.00	1.88
30	4	1.43	1.47	0.53	1.00
49	5	0.88	0.90	0.33	0.61

Here again we have definite numbers to start with, and the computation of the matrix is trivial if a spreadsheet is used. In many cases, however, the matrix must be filled in judgmentally. Since this is a 16 x 16 matrix, $n(n-1)/2$ or $(16 \times 15)/2 = 120$ judgments would be required. (The entries below the diagonal should be the reciprocals of their mirror-image elements above the diagonal, and diagonal elements are always 1.) This can be a major effort on the part of the people responsible for rating the projects on this particular criterion

A portion of the matrix for relative preferences of Probability of Success is shown in the following. Note that since a higher probability is preferred to a lower probability, *row* probability is divided by *column* probability to obtain relative preference for a row.

P(Sxs)		0.7	0.64	0.51	0.73
	Project	1	2	3	4
0.7	1	1.00	1.09	1.37	0.96
0.64	2	0.91	1.00	1.25	0.88
0.51	3	0.73	0.80	1.00	0.70
0.73	4	1.04	1.14	1.43	1.00
0.9	5	1.29	1.41	1.76	1.23

The Payoff matrix, which follows, is developed in the same manner. Like the Probability of Success matrix, higher payoff is preferred to lower, so *row* payoff is divided by *column* payoff to compute a cell entry.

Payoff		255	113	244	870
	Project	1	2	3	4
255	1	1.00	2.26	1.05	0.29
113	2	0.44	1.00	0.46	0.13
244	3	0.96	2.16	1.00	0.28
870	4	3.41	7.70	3.57	1.00
885	5	3.47	7.83	3.63	1.02

Compute the Rating. As was the case with the array of preferences for the three criteria, now each column of the three project preference arrays is divided by its column sum, and the row averages are computed. Table 2.4 shows the row averages for each factor (inside the double lines), the weights for each factor computed at the first step (across the top), and the final scores for each project in the rightmost column. The scores are computed by multiplying the top row by the row for the project and summing. The projects have been sorted in decreasing order of score. As with the Scoring Model, once the criteria matrices have been generated by management and the project-specific data generated by the project teams or other appropriate groups, the final computation of the scores is trivial if a spreadsheet is used.

Utilizing AHP
When to Use. AHP was not designed specifically to assist in project ranking. Rather, it was developed to facilitate sound decision making, whatever the decision problem is. Consequently, its realm in projects goes far beyond project ranking to include ranking alternatives and making decisions in all knowledge areas of project management. Consider, for example, selecting the best among several candidates for the project manager position. Choosing the best among multiple vendor bids is another example. Still

another would be selecting the best among four possible ways to accelerate the project that is significantly late. Apparently, the opportunities to apply AHP in project management are numerous. In all of them multiple alternatives are available and AHP's job is to help rank them and choose the best. In this sense, AHP's purpose is to reduce the risk by selecting the best alternatives. This is why AHP is considered a decision and risk analysis tool, included here because its major current application is in project selection.

Aside from this, AHP is generally used in larger, more important projects for new project selection and go/kill reviews of the existing projects [4]. Helpful in this are specialized software tools such as Expert Choice or Automan, or Excel. While in smaller decision problems using AHP may be an individual exercise, the real value of AHP lies in a group decision-making environment (see the box titled "Tips for AHP" coming up in the chapter). This is particularly important when the data for different subcriteria will be generated by different groups. Instead of a group of experts in disparate disciplines trying to come to a consensus, AHP allows experts in each discipline to fill out the relevant matrices, using their disciplinary expertise. The results can then be combined with the results from other groups, using the AHP.

Table 2.4: Final Project Rankings Obtained with the Analytic Hierarchy Process

Project	Cost	P(Sxs)	Payoff	Score
	0.25	0.25	0.5	
6	0.150	0.072	0.096	0.104
4	0.085	0.062	0.104	0.089
5	0.052	0.077	0.106	0.085
16	0.037	0.068	0.111	0.081
12	0.032	0.063	0.103	0.076
10	0.027	0.045	0.105	0.070
11	0.038	0.055	0.091	0.069
7	0.095	0.066	0.052	0.066
3	0.160	0.043	0.029	0.065
15	0.040	0.065	0.051	0.052
8	0.053	0.083	0.024	0.046
14	0.038	0.065	0.039	0.045
1	0.059	0.060	0.030	0.045
9	0.041	0.048	0.028	0.036
13	0.035	0.073	0.017	0.035
2	0.058	0.055	0.013	0.035

Comparison of the Scoring Models and AHP

How do the results of Scoring Models and AHP compare? The following table gives the rank order of each project in the example, as determined by the two methods (repeating the scores shown previously):

Rank	1	2	3	4	5	6	7	8	9	10	11	12	13	14	15	16
AHP (project #)	6	4	5	16	12	10	11	7	3	15	8	14	1	9	13	2
Scor Modl (project #)	4	6	3	5	8	1	12	11	2	7	14	16	9	10	13	15

Although in many cases the two methods are in good agreement, that is not the case in this example. The methods are in close agreement on the first and second rank projects. From then on, however, the two methods disagree significantly. The correlation coefficient for the two sets of ranks is only 0.14. Which method is right? There is no way to tell. If we had some objective measure of merit for projects, we wouldn't be using these methods. The important consideration is that each of these methods has strengths and weaknesses. The Scoring Model is easier to use. However, it is a one-level process that ignores finer divisions of factors. The AHP is more complex to use (although with a spreadsheet the calculations are quite simple), but it does take into account the hierarchy of factors. If ease of use is important, the Scoring Model is the better choice. If the decision involves factors that can be disaggregated into lower-level factors, and this desegregation is important, the AHP is the better choice. Recognize, however, that there is no way of knowing which is "right."

Time to Use. The time required depends on whether objective data (monetary costs, staff hours, machine time, etc). can be available, or whether the various matrices must be filled out judgmentally. Where objective data can be made available, the time required is only that for gathering the data. Computing the matrices using a spreadsheet is trivial. If the matrices must be filled out judgmentally, however, the effort can be very time-consuming on the part of management, who are responsible for deciding on the criteria and the relative importance of these criteria. For example, developing and applying a simple hierarchy with three levels and several factors per level may take a group of a few decision makers several hours. Situation changes radically when hierarchy has five levels with tens of criteria involved, driving the necessary time to tens of hours for each involved decision maker. Note, however, that in either case once the matrices are available, the only *new* information required every time the method is used is project-specific data. Management effort is thus a one-time activity and need be repeated only when external conditions demand a change in the matrices. Making use of a well-trained facilitator significantly improves time performance.

Benefits. AHP provides value on multiple levels beginning with the ability to marry simplicity with complexity. By breaking complex problems such as the project ranking into levels, AHP helps focus on smaller parts of the problem, thinking about one or two criteria at a time. As evidence from the proverbial Miller's Law suggests, humans can compare only 7+-2 items at a time. AHP's focus on a series of one-on-one comparisons of criteria, for example, then synthesizing them further improves the quality of decision. Such an approach enables the decision maker not only to arrive at best decision but also provides the rationale that it is the best. AHP's ability to handle complex situations in an easy way is based on the use of multiple criteria, some of which are subjective, while others are objective aspects of a decision. The subjective aspects may include qualitative judgments based on decision makers' feelings and emotions as well as their thoughts. On the other hand, the objective aspects may address quantitative criteria such as profitability numbers. AHP's flexibility of juggling simultaneously both objective and subjective, quantitative and qualitative aspects is unmatched. It truly ensures a systematic, all-angle project ranking, or more generally, risk reduction in decision making.

Advantages and Disadvantages. AHP offers powerful advantages captured in:

- ▲ *Simple procedure.* The procedure lays out a simple and effective sequence of steps to arrive at project ranking, even in group decision making when diverse expertise and preferences are involved.
- ▲ *Visual hierarchies.* Complex decision problem of project ranking is represented with graphical hierarchical structures. Even when the structures are elaborate, they make the problem visual, simplifying potential risk and conflict [5].
- ▲ *User friendliness.* The availability of computerized tools has simplified the mathematical procedure, eliminating the once insurmountable process of computations.

Disadvantages are related to AHP's:

- ▲ *Subjective nature.* This is because different decision makers can assign different levels of importance to a particular criterion.
- ▲ *Complexity.* As the number of criteria increases, the tabulation and calculations become complex and also tedious.
- ▲ *Difficulty.* Some users feel that quantifying the importance of some criteria may turn out difficult at times.

Variations. The approach used in AHP is very similar to another tool called MOGSA (Mission, Objectives, Goals, Strategies, Actions), an example of which is shown in Figure 2.4 [6]. MOGSA also uses the hierarchical breakdown of the decision problem to build a network of relationships among three major levels of the hierarchy: impact level (M, O), target level (G), and operational level (S, A). Where MOGSA differs from AHP is in some computational aspects.

Tips for AHP

The following approaches can help in using AHP:

● *Have senior managers select the criteria.* The criteria reflect the firm's strategy and goals.

● *Make it a discussion forum.* If the relative preferences in the upper-level matrices must be obtained judgmentally, using a meeting to generate them helps obtain executive buy-in, because senior managers examine each criterion and judge relative importance.

● *Use disciplinary teams for judging specific matrices.* Since the AHP allows disaggregation of criteria, using experts on each criterion improves the results of judgmental comparisons.

● *Use a visible display.* When all participants have electronic keypads, they can vote anonymously, have their votes instantly tallied by a computer, and group results displayed on a large screen. Reasons for altering the relative preference can then be discussed.

Customize AHP. Think of AHP described here as a generic one. You can significantly enhance its value by adapting its use to the specifics of your project situation. A few simple ideas given in the following are no more than clues in that direction.

Summary

AHP was the topic of this section. Although most widely used in larger, more important projects for new project selection and go/kill reviews of the existing projects, AHP is used for ranking alternatives and making decisions in all knowledge areas of project management. In all of them multiple alternatives are available, and AHP's job is to help rank them and choose the best. The value of AHP lies in its ability to deal with complex situations in an easy way. In doing so, AHP handles both objective and subjective, quantitative and qualitative aspects. The following box highlights key points in structuring and applying AHP.

Customization Action	Examples of Customization Actions
Define limits of use.	Use AHP in larger, strategic projects for new project selection, and go/kill reviews of the existing projects (use professional software and facilitator).
	Use AHP as a risk-planning tool in evaluating several alternative approaches within the project (schedule, cost, hiring alternatives).
Adapt a feature.	Stick with simple hierarchies (levels 1, 2, and 3) when using AHP as a risk planning tool within the project (rely on Microsoft Excel for this).
	Use anchored scales for making judgmental preferences because it aids participants in making consistent judgments.

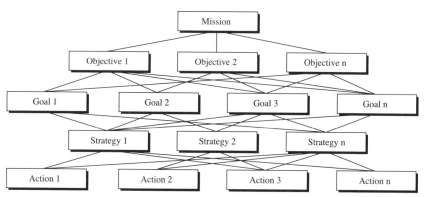

Figure 2.4   An example of the MOGSA (Mission, Objectives, Goals, Strategies, Actions) decision model.

Economic Methods
What Are Economic Methods?

Economic Methods are designed to take into account monetary cost and return from projects. They thus tend to ignore technical merit and similar considerations. Rather, they are intended for use when economic considerations are paramount. However, only financial data are required, which simplifies their use.

Three different economic models will be presented: Payback Time, Net Present Value (NPV), and Internal Rate of Return (IRR). Table 2.5 presents three hypothetical candidate projects, with their associated cost and revenue streams. These projects will be evaluated using each of the three Economic Methods.

AHP Check

Check to make sure that AHP is appropriately structured and applied. It should:

- Be based on project proposals, strategic and tactical plans, and historical information.
- Structure the hierarchy: the top (goal: select best projects)—intermediate levels (selection criteria, subcriteria)—lowest level (candidate projects).
- Construct a set of pairwise comparison matrices for each of the lower levels.
- Do relative scores within each level of the hierarchy.
- Yield a single score for each project on each level and an overall score.
- Rank projects according to final rating.

Payback Time

Payback Time is simply the length of time from when the project is initiated until the cumulative cash flow (revenue minus costs) becomes positive. At that point all the funds invested in the project have been recovered. As can be seen from Table 2.5, the cash flow for Project 1 turns positive in 6 years, for Project 2 in 5 years, and for Project 3 in 8 years. By this criterion, Project 2 is the best choice.

Payback Time is a very conservative criterion and provides more protection against future uncertainties than do either of the other Economic Methods. However, it is insensitive to project size, since a project with massive investment requirements may still have short Payback Time. Moreover, it takes no account of future potential once payback is reached. Both Projects 1 and 3 have greater future income than Project 2, but the initial investment before payback is greater and the time for cash flow to turn positive is greater.

Net Present Value

Net Present Value (NPV) takes into account the time value of money, in that a dollar a year from now is worth less than a dollar today. NPV *discounts* both future costs and revenues by the interest rate, according to the formula:

$$NPV = \sum_{i=1}^{n} \frac{values_i}{(1 + rate)^i}$$

In this formula, *values* represent the sum of costs and revenues in year i, and *rate* is the discount rate or the rate the company pays for borrowed money, expressed as a decimal fraction.

Excel and other spreadsheets compute NPV directly. You only need to enter the discount rate and the values (or a vector of cells) into the NPV function. The result is computed and displayed in the cell holding the function. Using the NPV function, the NPV for each of the three projects, at discount rates of 5 percent, 10 percent and 15 percent, are as shown in the following table:

Project 1	*Project 2*	*Project 3*
NPV at	NPV at	NPV at
5%= 5243	5%= 2320	5%= 6400
10%= 2841	10%= 1254	10%= 3275
15%= 1563	15%= 688	15%= 1679

Table 2.5: Candidate projects for Payback Time Evaluation

Year	Project 1			Project 2			Project 3		
	Cost	Revenue	CumCash	Cost	Revenue	CumCash	Cost	Revenue	CumCash
1	20		-20	30		-30	20		-20
2	50		-70	80		-110	50		-70
3	80		-150	105	147	-68	78		-148
4	120		-270	110	154	-24	80		-228
5	200	300	-170	125	175	26	150		-378
6	450	630	10	150	210	86	375	525	-228
7	475	725	260	160	220	146	525	735	-18
8	500	775	535	170	223	199	600	840	222
9	450	700	785	175	230	254	800	1120	542
10	450	650	985	170	228	312	750	1200	992
11	350	490	1125	150	226	388	625	1000	1367
12	300	420	1245	120	220	488	550	880	1697
13	250	350	1345	110	215	593	500	800	1997
14	150	220	1415	95	180	678	400	640	2237
15	75	105	1445	70	120	728	200	320	2357
16	50	70	1465	40	60	748	80	128	2405

The more the future is discounted (i.e., higher discount rate), the less the NPV of the project. At all discount rates, Project 3 is superior to the other two projects. That is, the higher the NPV, the higher the ranking of the project.

NPV takes into account the magnitude of the project and the discount rate. It is not a particularly conservative measure, however, since it incorporates estimated future revenues that may not actually materialize.

Internal Rate of Return

Internal Rate of Return (IRR) is simply the discount rate at which NPV for the cash flow is zero. There is no closed-form formula for it. IRR must be computed iteratively, "homing in" on the exact discount rate that produces an NPV of zero. Most spreadsheets, however, have an IRR function that allows the user to obtain IRR. You need enter only the list of values (or a vector of cells) and a guess value of IRR. The function then carries out the iterative calculation.

For the set of candidate projects, the IRR values are as follows:

Project	IRR
1	42%
2	40%
3	36%

By the IRR criterion, Project 1 is superior to the other two projects. While IRR discounts future values, it takes no account of the size of a project. Project 3 promises a significantly greater total return than Project 1, but those returns are farther in the future and follow a longer course of investment before cash flow turns positive. Hence, the IRR for Project 3 is lower than that for Project 1.

Choice of Economic Method

As this example shows, the three methods can give differing results on the same set of projects. Which method should be chosen depends on the considerations important to the decision maker. NPV favors projects with large payoffs but gives little protection against future uncertainties. IRR likewise gives little protection against future uncertainties and may tend to give preference to a project with modest total payoff but high return on a modest investment. Payback time is a very conservative method, giving more protection than the other two methods against future uncertainty, but it takes no account of size of project payoff, nor of the discounted value of future costs and revenues. Whichever consideration is most important will govern the choice of methods. Moreover, since these methods treat projects as capital investments, it would be appropriate to use whichever method the company uses for evaluating its other capital investments, thus allowing a direct comparison between projects and other investments.

Tips for Using Economic Methods

● Use the same method as is used by the firm for other capital investment decisions.	This provides a direct comparison between the project and alternative investments.
● Use the firm's standard hurdle rate or discount rate.	This neither favors nor works against the proposed project, by comparison with alternatives.
● Do sensitivity testing.	Check what happens with different discount rates or different scenarios of costs and revenues.

Utilizing Economic Methods

When to Use. Economic Methods require data about revenues and costs expected to result from the project. They are thus appropriate primarily for capital projects and projects intended to improve existing products or develop new products. In such cases they allow a direct comparison of such projects with alternative capital investments. For a comparison of the three methods, see the box on page 44, "Choice of Economic Method."

Time to Use. The time required to use any of these Economic Methods is primarily that required to gather the future cost and revenue data. That can take hours for smaller projects or tens of hours in larger projects. Once the data have been obtained, any of the methods can be calculated readily using a spreadsheet.

Benefits. Economic Methods are readily understandable. They enable the decision maker and project managers to communicate more readily about financial considerations of projects. They also make it easier to compare projects with other opportunities for capital investment. Once the necessary data are obtained, they are easy to compute. They also make sensitivity testing easy. Alternative scenarios about future costs and revenues can be compared to see how robust a selection of projects is against future uncertainty. See the "Tips for Using Economic Methods" box above.

Advantages and Disadvantages. The primary advantage of Economic Methods is that they are their relatively straightforward. Like in all capital investments, their disadvantage is that they require data about future costs and revenues. Not only may these be difficult to obtain, but also they may be subject to considerable uncertainty.

Summary

Presented in this section were three Economic Methods for project selection: Payback Time, Net Present Value, and Internal Rate of Return. They are appropriate primarily for capital projects and projects intended to improve existing products or develop new products. In such cases they allow a direct comparison of such projects with alternative capital investments. They are readily understandable and once the necessary data are obtained, they are easy to compute. We recapture the key points for applying the Economic Methods in the box that follows.

Economic Methods Check

Check to make sure that Economic Methods are appropriately applied. You should:

● Obtain estimates of costs and revenues for the project.	These estimates should be made as far into the future as the project is likely to be generating either costs or revenues.
● Select the appropriate economic method.	Use the method that is used by the firm for other capital investments.
● Carry out the calculations.	For a single project, a spreadsheet is adequate for computing any of the three Economic Methods.
● Check sensitivity to assumptions.	Test the results against more optimistic and more pessimistic streams of costs and revenues, and against higher or lower interest rates on borrowed funds.

Portfolio Selection Methods

What Are Portfolio Selection Methods?

The methods presented in the previous sections treat each project in isolation, without regard to interactions with other projects. In fact, there may be constraints that limit choice of projects. The most obvious constraint is the budget. However, there may be other constraints operating besides the financial budget. Staff limitations, limitations on supporting activities such as model shop time or computer time, and other considerations such as company policy may constrain the choice of projects. To deal with these issues, Portfolio Selection methods must be used [7]. These take into account all the constraints that may be operating to limit the choice of projects.

Applying Portfolio Selection Methods

Collect Inputs. The major inputs include the following:

▲ Data about candidate projects

▲ Company policies

The first step in applying Portfolio Selection methods is to identify the set of projects that are candidates for inclusion in the portfolio. It is then necessary to collect data about each of the candidate projects. The data *must* include those items that represent potential conflicts or interactions among projects. Overlapping resource requirements, company policies about types of projects to be included in the portfolio, and other items that might limit the possible combinations of projects must be collected.

Construct the Portfolio Model. Consider the project menu in Table 2.6. In addition to the selection criteria used in the Scoring Model and AHP examples, we will take interactions among the projects into account to obtain a portfolio with no conflicts. Assume there is a budget limitation of $600 K, there are only 4500 Model Shop hours available, and only 700 hours of computer time available. Within these constraints, you wish to select projects that will maximize Payoff.

Spreadsheets such as Excel can perform this "optimization" for small-sized problems. Table 2.7 shows an Excel spreadsheet set up to carry out the Portfolio Selection that maximizes payoff while not violating any of the constraints. An entry of 1 in the "Selected" column indicates which projects have been selected as part of the portfolio. Cell entries of 0 mean the project has not been selected. The Total Cost cell entry is simply the Excel SUMPRODUCT function of the Selected column and the Cost column. Similarly for the Model Shop Hours and the Computer Hours.

Once the problem is set up using Excel SOLVER (or the equivalent in other spreadsheets), the optimum portfolio can be calculated directly. As can be seen in the table, the Total Cost, Model Shop Hours, and Computer Hours satisfy the constraints imposed. The remaining constraints need some explanation. We do not want the spreadsheet to try to "buy" resources by selecting a "negative project." Hence, we must constrain the entries in the Selected column to be equal to or greater than zero. We also do not want the "optimum" solution to contain multiple copies of the most profitable project. Hence, we must constrain the values in the Selected column not to exceed 1. Finally, we can't buy half a project; hence, the values in the Selected column must be constrained to be integers. The three constraints acting together ensure that the values in the Selected column will be either 1 or 0, or in other words, that whole projects will be selected a maximum of one time each.

It is possible that there might be two versions of the same project in the candidate list, a "crash" version and a "normal" version. We would want to select only one of these; hence, another constraint would have to be added:

$$X + Y <= 1$$

where X and Y are the cells in the Selected column corresponding to the two versions of the same project.

Table 2.6: Candidate Projects for Scoring Models and Portfolio Optimization Methods

Project	Cost $K	P(Sxs)	Payoff $M	ShopHrs	CompHrs	CurrProd	NewProd
1	43	0.7	255	311	70	1	0
2	44	0.64	113	213	70	0	1
3	16	0.51	244	489	43	0	0
4	30	0.73	870	375	47	0	0
5	49	0.9	885	116	49	1	0
6	17	0.85	807	375	55	0	1
7	27	0.78	437	463	54	0	0
8	48	0.98	204	374	59	0	1
9	63	0.56	231	114	50	1	0
10	96	0.53	879	372	64	0	0
11	67	0.64	762	225	50	0	0
12	79	0.74	866	476	42	0	1
13	74	0.86	141	323	40	0	0
14	68	0.76	330	176	38	1	0
15	64	0.76	427	212	49	0	0
16	70	0.8	927	493	43	0	1

Suppose further that there is a company policy that at least one project must support an existing product and at least one project must support a new product. The two rightmost columns in the table have entries of 1 to indicate whether a project supports a new or existing product or whether it is not product-related (it might be a manufacturing improvement project). We could add constraints that the SUMPRODUCT of the Current Product and the SUMPRODUCT of the New Product columns with the Selected column must equal or exceed 1. In the preceding example, this would not alter the outcome, since projects that support current and new products are already in the solution set. However, had that not been the case, and we added those two constraints, some other projects would be forced into the solution set. The result would most likely have been to reduce the total payoff, but to satisfy company policy.

In this example, the budget turns out to be the binding constraint. The projects in the solution set consume almost the entire budget. By contrast, there is a considerable margin left in both Model Shop hours and Computer hours. This opens the possibility of some sensitivity testing. How much more would the total return increase if the projects budget were increased? No matter which is the binding constraint, use of the optimization capability of a spreadsheet will allow the decision maker to determine how much additional return can be obtained from relaxing the binding constraint, or conversely, how much return will be lost if a constraint is made more binding.

Table 2.7: Portfolio Selection that Maximizes Payoff While Not violating Any of the Constraints

Project	Selected	Cost $K	P(Sxs)	Payoff $M	ShopHrs	CompHrs	CurrProd	NewProd
1	0	43	0.7	255	311	70	1	0
2	0	44	0.64	113	213	70	0	1
3	1	16	0.51	244	489	43	0	0
4	1	30	0.73	870	375	47	0	0
5	1	49	0.9	885	116	49	1	0
6	1	17	0.85	807	375	55	0	1
7	1	27	0.78	437	463	54	0	0
8	0	48	0.98	204	374	59	0	1
9	0	63	0.56	231	114	50	1	0
10	1	96	0.53	879	372	64	0	0
11	1	67	0.64	762	225	50	0	0
12	1	79	0.74	866	476	42	0	1
13	0	74	0.86	141	323	40	0	0
14	1	68	0.76	330	176	38	1	0
15	1	64	0.76	427	212	49	0	0

continued

Table 2.7: (Continued)

Project	Selected	Cost $K	P(Sxs)	Payoff $M	ShopHrs	CompHrs	CurrProd	NewProd
16	1	70	0.8	927	493	43	0	1
Sums		855		8378	5107	823		

TotlPayoff 7434

Constraints

		Payoff $M	ShopHrs	CompHrs	CurrProd	NewProd
B2:B17	<=	1				
B2:B17	>=	0				
B2:B17			integer			
TotalCost				583	<=	600
MdlShopHrs				3772	<=	4500
CompHrs				534	<=	700

Utilizing Portfolio Selection

When to Use. Portfolio Selection should be used when there are multiple constraints on the selection of projects (see the "Tips on Portfolio Selection" box that follows). Since some measure of payoff is required for the optimization, it is most easily used when the objective is to maximize revenue. However, it can also be used for types of projects that do not involve direct revenue generation. For instance, a list of candidate projects could have figures of merit developed through either a Scoring Model or the AHP method. The Portfolio Selection could then maximize the sum of the figures of merit subject to constraints on financial budget, staffing, support activities, and other resources. That is, the figure to be maximized need not be financial revenue. It can be any measure of goodness of the candidate projects, whether they are larger or smaller projects.

Time to Use. The time required for use of Portfolio Selection is primarily that of collecting the data for the candidate projects. Setting up the problem on a spreadsheet is fairly simple, and the time required for carrying out the optimization is negligible. However, spreadsheets are suitable only for small optimization problems. If the candidate list is large or there are many constraints, a spreadsheet may be inadequate. In this case some more powerful optimization procedure may be needed. If the task is large enough or complex enough, there may be no alternative but to use a specialized optimization program on a mainframe computer. In such cases, the time to set up the problem, run the program, and check validity can become significant.

Benefits. As compared with other project selection methods, Portfolio Selection takes into account the interactions among projects and allows the decision maker to specify the constraints the portfolio must satisfy [8]. These may include limits on the utilization of particular resources or conformance to company policy about the types of projects that must be included in the portfolio. As a result, a portfolio of projects does not have any built-in conflicts over resource use or conformance to policy.

Tips on Portfolio Selection

- *Higher management should set the optimization criteria.* Only higher management is in a position to know what should be optimized (financial return, technical merit, or other).

- *The responsible departments should establish the constraints.* Since the constraints may be other than financial, the departments responsible for the limiting resources should determine the constraints.

- *Higher management should establish any policies that must be satisfied.* If there are policy considerations to be taken into account by the portfolio optimization, these must be established by higher management, which is the only authority capable of establishing them.

Advantages and Disadvantages. For small optimization problems, Portfolio Selection is a relatively easy application. By contrast, optimization problems with a large list of projects and many constraints have the disadvantage that more data may be required for them than for other project selection methods. This makes the data collection process more time-consuming and costly.

Customize Portfolio Selection. Like with our tools, Portfolio Selection will provide more value if the generic version we have covered is customized to your specific project situation. A few simple ideas given in the following may help provide direction for such an effort.

Customization Action	Examples of Customization Actions
Define limits of use.	Use Portfolio Selection when there are multiple constraints on the selection of projects (e.g., budget and resource conflicts).
	Include both revenue-generating and non-revenue-generating projects (convenient for both small and large companies).
Adapt a feature.	Keep the optimization problem small (rely on Excel spreadsheet for this).
	Keep the number of constraints to a minimum (this is convenient for small companies with limited resources).

Portfolio Selection Check

Check to make sure that Portfolio selection is appropriately applied. You should:

- *Collect cost and other relevant data on each project.* For each candidate project, collect data on each of the criteria by which the projects might be evaluated.
- *Identify potential conflicts among projects.* Identify uses of limited resources over and above the financial budget that might constrain the possible combinations of projects.
- *Identify company policies that might constrain combinations of projects.* Determine whether company policies require certain types of projects to be included in the portfolio, or forbid certain combinations of projects.
- *Set up the optimization problem.* If the problem is sufficiently small, use a spreadsheet. Otherwise, use a more powerful program on a mainframe.
- *Identify the optimum portfolio.* Run the optimization, and determine which combination of projects satisfies all the constraints.
- *Test sensitivity.* Determine how much change in any constraint is required to alter the project portfolio.

Summary
This section reviewed Portfolio Selection, a project selection tool used when there are multiple constraints on the selection of projects. It is most easily used when the objective is to maximize revenue. However, it can also be used for types of projects that do not involve direct revenue generation. Thus, it can be any measure of goodness of the candidate projects, whether they are large or small. Compared with other project selection methods, Portfolio Selection has an advantage of taking into account the interactions among projects and allowing the decision maker to specify the constraints the portfolio must satisfy. Recapturing the core of the application are checkpoints in the box that precedes.

Real Options Approach
What Is the Real Options Approach?

Basic research projects are inexpensive enough that they can be carried as overhead. Product development projects are expensive enough that they must be treated as a capital investment and looked at as being in competition with other capital investments projects the company can make. There is, however, an intermediate type of project that is too large to be treated as overhead but not yet ready to be treated as a capital investment. The appropriate approach to projects of this type is to treat them as *options*.

Analogy with Financial Options. The Real Options Approach is a relatively new method of project selection that uses analogy with financial options. To make it clear, we will first explain the analogy, before moving to lay out its use. An *option* is a financial instrument used to "lay off" risk to some other party willing to bear that risk *for a price*. The two option types are the *put* and the *call*.

Suppose you hold some shares of a specific stock. You want to continue holding them for the long term, but if the investment turns sour, you would like to limit your loss. One way to do this is buy a *put option*. This allows you to sell the stock at a specified price during a specified period of time. If the stock drops below the "strike price" before the option expires, you can sell it at the specified price, regardless of how much lower the market price has fallen. Clearly, the person selling you the put is betting that the price of the stock will not drop below the price specified in the put. For assuming the risk that the price will drop below that value, he or she is charging for the put. The price for a put depends on both the buyer's and seller's estimates of the risk that the option will be exercised. From the buyer's standpoint, the put is a way of reducing downside risk.

Suppose conversely that you believe there is some probability, but not certainty, that the price of a specific stock will increase. Or alternatively, you already own some of that stock on the basis of your belief that its price will increase, but you are hesitant to buy more. Still, in either case you would like to benefit from a price increase if it does occur. You can buy a *call option*. This gives you the right to purchase a specified number of shares of that stock at a specified price, during a specified period. If the price of the stock rises above the strike price during the option period, you can buy the stock at the specified price and resell it at a profit, or you can hold it for further profits. Clearly, the person selling you the call is betting that the price of the stock will not rise above the price specified in the call. For assuming the risk that the price will rise above that value, he or she is charging for the call. The price for the call depends on both the buyer's and the seller's estimates of the risk that the option will be exercised. From the buyer's standpoint, investing in the call buys an opportunity to make a further investment if the investment turns out to be profitable.

In the case of financial options, the key to how much a call or a put is worth is the *volatility* of the stock or other financial instrument in question. The more volatile the price of the instrument, the greater the chance that it will rise above, or fall below, the strike price, in which case the holder will exercise the option. An instrument with a low volatility—that is, one that is not likely to rise or fall very much in price—would involve only a low price for either a put or a call. In the case of financial instruments, the measure of volatility is the standard deviation of the price over some suitable period of time.

Applying the Real Options Approach

An intermediate project of the type described in the preceding text, too big to be carried as overhead but not ready to be treated as a capital investment, is analogous to a call in a financial market. Investing in the project amounts to buying an opportunity to make a further investment *if that further investment turns out to be profitable.* As with a financial investment, there are both upside and downside risks with such a project. The downside risk is that the project will not pan out. Everything spent on it will be lost. The upside risk is that it will turn out to be even more favorable than you anticipated, and you will miss out on some payoff that you could have obtained had you invested more.

Identify the Risks Associated with the Project. There are three kinds of risk the company may face in treating a project as an option. Each of these types of risk must be offset to the extent possible by one or another of the options described previously.

- ▲ Firm-specific risk. The risk that the firm will not have the funding, the technical skills, or other resources needed to carry out the project.
- ▲ Competition risk. The risk arising from actions under the control of competitors, such as preempting with a similar project.

▲ Market risk. The risk due to uncertain factors such as customer demand, regulatory changes, or emergence of a cheaper or substitute technology.

Offset Risks: Analogies to Puts and Calls. The put is a way of limiting losses from downside risk. The call is a way of gaining benefit from upside risk. If we are to treat a project as a call, we need to look for analogies to the financial put and call for reducing downside and upside risks.

Some measures that are analogous to puts are as follows:

▲ Defer.   Postpone the project while gathering additional information about risks and payoffs.

▲ Multistage.   A project that can be done in stages can be stopped if things look bad, and resumed if new information justifies resumption.

▲ Outsource.   A contract with a third party to carry out the project. The contract can be terminated early (perhaps with a penalty fee) if the project begins to look bad. (Note that this creates the risk that you will be creating a competitor.)

▲ Explore.   Start with a pilot or prototype project and expand it if it looks favorable.

▲ Lease. lease facilities from a third party, terminating the lease early (possibly with penalty fee) if the project begins to look bad.

▲ Abandon. Choose facilities or other equipment that have a high salvage value, so that something can be recouped if the project must be terminated.

▲ Flexible scale: design the project so that it can be contracted if conditions turn out to be less favorable than anticipated, but not so bad the project should be terminated.

Each of these measures serves to limit the downside risk of the project. Each costs something, but each either reduces risk or transfers that risk to someone else.

While the project itself is analogous to a call, there are other measures, also analogous to calls, that can enhance payoff if the project turns out to be more favorable than anticipated.

▲ Hedging.   Make low-cost investments in additional facilities so they will be available if things turn out even better than expected.

▲ Flexible scale.   Design the project so that it can be expanded if conditions turn out to be more favorable than anticipated.

There is yet another kind of option that the decision maker should consider:

▲ Strategic growth. This is a project that is a link in a chain of projects or is a prerequisite to other projects. Such a project can be considered a call, since it makes possible greater payoffs from favorable developments. However, it does not in itself produce a payoff.

Structure the Project to Manage Risk. The steps the decision maker should take in managing the risk are as follows:

▲ Define the project and identify the risks to which it is exposed. These must be specific risks, not generic ones (e.g., a specific competitor may come out with a comparable product).
▲ Recognize shadow options (i.e., options that the company might take) to reduce each of the specific risks identified in the previous step.
▲ Devise alternative ways to structure the project, each way involving different combinations of the shadow options identified in the previous step.
▲ Identify the combination of options that results in the most valuable project structure.
▲ Convert one specific set of shadow options into real options.

Defining the project means determining its goals and objectives, and identifying the specific risks that these goals and objectives will either not be met or that they will turn out to be inadequate if results are even more favorable than expected. Each specific risk must be controlled by choosing an appropriate option—a project put or call.

For each risk, the decision maker must identify at least one shadow option that the company might take in order to minimize or control that risk. The potential cost and benefit (degree of risk reduction) from each shadow option must be determined.

Once the shadow options have been determined, the decision maker must devise various project structures, each with a different combination of shadow options. The extent to which this development of alternative structures goes will depend on project size. The bigger the project and the greater the risks, the more effort is justified in defining alternative structures.

Once the alternatives have been devised, the decision maker must examine all the possible project structures and determine which is most favorable. That is, the decision maker chooses the project structure that provides the greatest opportunity to make a profitable investment at a later time.

The final step involves converting the shadow options of the chosen combination into real options. For instance, the shadow option to lease facilities becomes real when the company draws up bidding specifications and submits them to potential bidders. Each of the shadow options in the chosen project structure must be converted into real options. This may be costly, but the set of options, and the project structure, have been chosen to provide maximum protection against risk, as well as maximum benefit from the project itself. Failure to convert one or more shadow options to real options means exposing the project to the risks identified at the outset.

How much is it worth to the company to offset the risks accompanying the "call" of starting a project? In the case of financial instruments, the volatility of the price affects the price of the call. In the case of a project, the equivalent of volatility is the maximum potential for the project. The higher that maximum potential, the more the project is worth as a "call." As the project progresses, the company must continually update its estimates of maximum potential. As the estimate increases or decreases on the basis of new information (e.g., a new government regulation decreases the size of the potential market and thus reduces the estimated potential, or a partnership with a foreign firm opens an additional market and increases the estimate), the project may require a new decision to terminate or continue.

This concept can be illustrated in Figure 2.5 [9]. Corresponding to the volatility of a financial instrument, the *volatility* of a project is the *range of possible outcome values*, that is, the maximum possible value minus the minimum possible value (which may, of course, be zero). The greater this range of possible outcomes, the greater the uncertainty associated with the project, but likewise the greater the possible payoff. A project with a large range of possible payoffs is thus a candidate with attractive possibilities for *future* further investments.

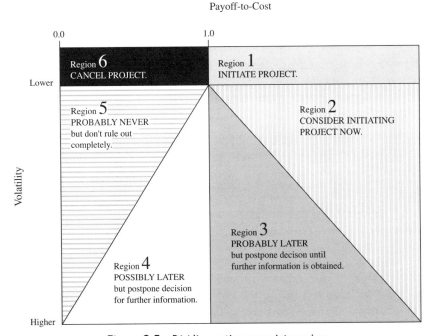

Figure 2.5 Dividing option space into regions.

Another measure of the value of a project is the ratio of NPV of returns from the project *if it achieves maximum payoff* divided by the NPV of the costs to pursue the project. Call this ratio Payoff-to-Cost. The greater this ratio, the more attractive the project is.

Figure 2.5 illustrates the various combinations of Volatility and Payoff-to-Cost, and the decisions to be made in each case. Volatility increases from top to bottom; Payoff-to-Cost increases from left to right. The figure is divided into six regions, based on the combination of Volatility and Payoff-to-Cost. Each region calls for a different decision.

Regions 1 and 6 are the "easy" regions. In these regions, volatility is nearly zero (minimum payoff is close to or equal to maximum payoff), which implies that the payoff value is known with good precision. In Region 1, NPV of payoff is greater than NPV of cost, hence the project is a clear winner. In Region 6, NPV of payoff is less than NPV of cost; hence, the project is a clear loser.

In Region 2, Payoff-to-Cost ratio exceeds 1, and the volatility is low compared to Payoff-to-Cost. However, there is still some uncertainty. The project can be postponed, to await further information (perhaps a reduction in volatility); but if there is some benefit to starting it now as compared to waiting, then consider starting it now.

In Region 3, Payoff-to-Cost ratio exceeds 1, but the volatility is very high by comparison to Payoff-to-Cost. The project should be postponed to see if further information reduces volatility.

In Region 4, Payoff-to-Cost ratio is less than 1. This weighs against the project. However, volatility is high, and if maximum payoff turns out to be high enough, the project might be moved into Region 3 or Region 2.

In Region 5, Payoff-to-Cost ratio is less than 1. This weighs against the project. Moreover, volatility is low, implying that there is little chance the project will ever become attractive. It should be kept as a potential project until further information resolves the issue of volatility.

Having viewed the implications of considering projects as options, we now look at an illustrative example of how to apply the concept of options to a project.

An Illustrative Example

Steps that we have described in applying the Real Options Approach will now be demonstrated in a real-world example. The Research Laboratory of Company X has shown the technical feasibility of a new consumer product. It is within the state of the art. However, considerable engineering work will be required before the idea is ready for product development. The decision has been made to investigate the possibility of treating the engineering development project as an option. If the apparent value of the project is positive after all risk-minimizing options have been exercised, the project will be initiated.

Project Value. Extent of consumer demand is not known. Another product appealing to the same market segment has annual sales of $90M. This is the most plausible estimate of the total market for the new product. If Company X is first to market with the new product, it should achieve at least a 2/3 market share and will enjoy a 100 percent market share until a competitor arrives.

Current engineering staff can complete the engineering development work in three years. Assuming product development takes two years after engineering development is completed, total sales will be $90M per year (no competitors), profit margin is 10 percent, the discount rate is 15 percent, and sales continue for an additional six years after introduction, the potential value of the project is $17M (present NPV).

Identify Risks. The following risks have been identified with the proposed project:

Firm-specific risk:

- ▲ Engineers may have to be diverted to a high-priority project that is having problems, delaying engineering development by two years. Assuming no competitors, project NPV is $13M. Assuming first to market, project NPV is $10M.

Competition risk:

- ▲ Company Y is known to have been doing research in the same technical area, and their researchers have presented relevant papers at scientific conferences. If they get to market first, project NPV is $6M.
- ▲ Foreign Company Z is known to be working on an alternative approach that would require an extension of the state of the art to achieve, but it would allow lower production costs. This cuts profit margin to 5 percent to meet their competition. Assuming first to market, project NPV is $8M.

Market risk:

- ▲ The market segment may not respond to the new product. Maximum sales will be $40M. Assuming no competitors, project NPV is $8M (downside risk).
- ▲ If additional market segments are attracted and foreign sales realized, project NPV may be $25M (upside risk).

Expected costs:

- ▲ Engineering development: Three engineers @ $100K/year each including overhead for three years, NPV is $685K.
- ▲ Product development: Four engineers @ $100K/year each including overhead in years 4 and 5, NPV is $428K.

- ▲ New production facilities: $4M in year 5, NPV is $2M.
- ▲ NPV of total costs: $3113K.

Under all apparent scenarios, the project has a positive NPV and should be pursued. However, by appropriate risk management, the project risks may be reduced. Moreover, getting to market first is critical.

Identify Possible Options. These include the following:

- ▲ A: Hire three additional engineers for engineering development. This reduces development time to 18 months. NPV of additional salaries, overhead, and hiring costs is $575K. NPV of salaries of existing engineers is reduced to $487K, for total engineering development NPV of $1062K. Additional 18 months of sales brings NPV of profit to $24M if no competitors, $20M if first to market.

- ▲ B: Outsource engineering development of noncritical elements, keeping only critical elements in-house. Reduces engineering development time to two years. Outsourcing cost NPV is $476K. In-house engineering development NPV is reduced to $487K. Earlier sales increase NPV of profit to $21M if no competitor, $16M if first to market.

- ▲ C: Concurrent engineering development and product development. Increases product development engineering costs but shortens time to market. NPV of product development engineering increases to $794K. Profit NPV is $21M if no competitor, $17M if first to market.

- ▲ D: Prepare specifications for leased facility instead of building production facility. NPV of specification preparation is $25K. Leasing brings profit margin down to 8 percent. Profit NPV is $14M if no competitor, $10M if first to market. Lease penalty fee for early termination is $1M.

- ▲ E: Initiate research project in same area as Foreign Company Z, to attempt to match their product if they are successful, or bring out own product in any case. Assume four years for research, three years for engineering development, two years for product development, and sales lasting six more years. Profit margin is 15 percent if Company Z is not successful, 10 percent if we bring new technology to market first, additional six years of sales after introduction. If the research project fails, NPV of research project is $1.05M. If research project succeeds, NPV of R&D costs is $1.2M. NPV of profits from both proposed product and higher technology product is $27M if no competitor, $19M if Company Z also succeeds on the same schedule.

Select the Combination of Options. For selecting the best combination of options, it will be assumed that all risks are realized (everything that can go wrong does go wrong), except that the company does get to market first. Combinations of options will then be compared.

- (A+C+D+E fails): Cost NPV = $3.07M. Profit NPV = $4.05M. Net = $0.98M.
- (B+C+D+E fails): Cost NPV = $3.2M. Profit NPV = $4.05M. Net = $0.85M.
- (A+C+D): Cost NPV = $2.02M. Profit NPV = $4.05M. Net = $2.03M.
- (B+C+D): Cost NPV = $2.17M. Profit NPV = $4.05M. Net = $1.75M.
- No options: Cost NPV = $927K. Profit NPV = $2.8M. Net = $1.873M.
- (A+C+D+E succeeds): Cost NPV = $3.226M. Profit NPV = $2.5M. Net = −.73M.

In the worst possible case, the combination of options A+C+D provides the best result. If any of the risks fail to occur, net profit will be much higher. Clearly, Option E is not worth pursuing on the assumption that the research project will fail. Even if Option E succeeds, but everything else goes wrong, Option E is still not worth pursuing. If other things go right, Option E is a potential winner.

Not all options need to be exercised at the outset of the project. As between hiring extra engineers and outsourcing, hiring is a better option. If engineering resources do not need to be diverted to another project, the extra staff will accelerate progress. If some engineering resources must be diverted, the extra engineers will keep the project on the original schedule. The option to do concurrent engineering with production does not need to be exercised until the second year of the project on the accelerated schedule. This gives an extra year to determine whether competition from Company Y and Company Z is likely to materialize. The option to prepare specifications for leasing does not need to be exercised until year 3 on an accelerated schedule and year 5 on a normal schedule. This gives time to get a better estimate of the size of the market. If the size of the market is much larger than initially expected, the leasing option supplements in-house production, and the leased facilities absorb any fluctuations in the market. If the market turns out to be much smaller than expected, leasing avoids in-house capital costs.

One further possibility is to exercise Option E at the outset, but be prepared to cancel it if either success appears unlikely, or Company Z does not appear to be successful. Option E can itself be considered a *call*, in the sense that it amounts to buying an opportunity to make a profitable investment *if it succeeds*.

Utilizing the Real Options Approach
When to Use. As already mentioned projects should be treated as options when they are too big to be carried as overhead, but not yet ready to be treated as capital investments. Theoretically, this is possible with many types of projects. Most successful applications of the approach have been

seen in the new product development, oil exploration, real estate development, and so on. It is fair to say that the approach has not been used enough. The reason perhaps is the lack of awareness on the part of project managers or their thinking that Economic Methods, NPV for example, are more acceptable to senior management because of their clarity and straightforwardness. While NPV is certainly more appropriate in case of routine projects, when it comes to highly uncertain projects of longer duration Real Options Approach may have an advantage (see the box that follows, titled "Tips on the Real Options Approach").

Time to Use. The variety of possible applications of the Real Options Approach makes it difficult to pick a typical application and assess its time requirements. Our assessments, therefore, are of general nature. The time required to structure a project as an option can be very extensive. A thorough search must be made for shadow options that would reduce upside or downside risk. Considerable ingenuity may be required to identify some powerful but low-cost options. Structuring the project to maximize its value will also take considerable time and management effort.

Benefits. The benefits of structuring a project as a set of options can be very extensive. It may make an expensive but potentially very profitable project affordable when otherwise the risks would be too great. It may also confirm that what appears to be a very attractive project is really too risky to undertake. The benefit here is in reducing the risk associated with a project, and maximizing the value of a risky project. Perhaps the greatest benefits is that Real Options Approach forces decision makers and project managers to confront risks and develop a full range of possible decisions that might be considered over time.

Advantages and Disadvantages. One of the major advantages of the approach is in translating real project phenomena into visualizable effects. For example, how will your option space change if you add a concurrent engineering approach to your product development? The disadvantages are the time, managerial effort and expense necessary to develop a satisfactory structure of puts and calls to maximize the project's value.

Tips on the Real Options Approach

Several important considerations must be taken into account in developing project options:

- Upper management must be involved in defining the risks and determining what kinds of puts and calls are acceptable for reducing the risk.

- Shadow options must be realistic in the sense that they could be realized if desired.

- Estimates of potential payoff must be updated as conditions change, to determine whether to terminate the project.

- At the appropriate time, the selected shadow options must be exercised to protect the project against risk.

Real Options Approach Check

Make sure you apply the Real Options Approach appropriately. You should

- *Identify the risks associated with the project.* Firm-specific risk, competition risk, and market risk.

- *Identify shadow options to offset each specific risk.* For each risk, there must be at least one shadow option defined to offset that risk.

- *Structure the project.* For each possible combination of project and shadow options, estimate the value of the project. Choose the structure that maximizes project value.

- *At appropriate times, convert shadow options to real options by exercising them.* Take the steps necessary to convert the shadow options into real ones by putting resources into them.

- *Update estimates as needed.* As time passes, continue to update estimates of project value.

- *Terminate the project if appropriate.* Terminate the project if at any time the potential payoff minus the cost of continuing, including cost of options to minimize risk, becomes negative. Otherwise, continue the project.

Customize Real Options Approach. To derive more value from the approach, start off with the generic model covered here and develop one that fits your project situation. Two simple ideas mentioned in the preceding table may indicate the direction of such customization.

Customization Action	Examples of Customization Actions
Define limits of use	Use the Real Options Approach in highly uncertain projects with which NPV method may be risky (e.g., in product development).
Add a feature	Define option space by two metrics: Payoff-to-Cost and Volatility. This helps visualize it in a simple matrix.

Summary

This section featured the Real Options Approach, a relatively new tool for project selection that uses analogy with financial options. It treats projects as options when they are too big to be carried as overhead but not yet ready to be treated as capital investments. The benefit here is in forcing decision makers and project managers to confront project risks and develop a full range of possible decisions that might be considered over time. Where the Real Options Approach has an advantage over other selection tools is with regard to highly uncertain projects of longer duration. To summarize this section, we present the key points for applying this tool in the box that follows.

Concluding Remarks

Five tools are presented in this chapter: Scoring Models, Analytic Hierarchy Process, Economic Methods, Portfolio Selection, and the Real Options Approach. All five are designed to help select projects, each one with a different approach that fits certain application situations. We have listed below a number of such situations, indicating which one of them favors which tools. Go through the list and identify which of our situations matches your specific situation. If ours are not sufficient to characterize your situation, add more of them and mark how they favor the tools use. Overall, the tool that has the highest number of marks is probably the most appropriate for you to deploy.

A Summary Comparison of Cost Planning Tools

Situation	Favoring Scoring Models	Favoring AHP	Favoring Economic Methods	Favoring Portfolio Selection	Favoring Real Options Approach
Rank candidate projects in order of desirability	√	√			
Fund projects in order, until resources exhausted	√				
Evaluate econ.payoff			√ (Payback Time)		
Evaluate econ. payoff, time value of money			√ (NPV, IRR)		
Need direct comparison with capital budgeting			√		
Organizations with small projects	√			√	
Organizations with large projects	√	√		√	√
Highly uncertain projects, longer duration					√

Situation	Favoring Scoring Models	Favoring AHP	Favoring Economic Methods	Favoring Portfolio Selection	Favoring Real Options Approach
Choose portfolio of projects				√	
Group decision making		√			
Reduce both downside/upside risk					√
Emphasize financial criteria			√		
Emphasize nonfinancial criteria	√	√		√	
Use criteria disaggregated into levels		√			
Simple to use and understand	√		√ (Payback Time)		
Easy to calculate using spreadsheet			√ (NPV, IRR)		

References

1. Cooper, R. G., S. J. Edgett, and E. J. Kleinschmidt. 1998. "Best Practices for Managing R&D Portfolios: Lessons from the Leaders-II." *Research Technology Management* 41(4): 20–33.

2. Henrickson, A. D. and A. J. Traynor. 1999. "A Practical Project-Selection Scoring Tool." *IEEE Transactions on Engineering Management* 46(2): 158–170.

3. Saaty, T. L. 1983. "Priority Setting in Complex Problems." *IEEE Transactions on Engineering Management* 30(2): 140–155.

4. Liberatore, M. 1987. "An Extension of the Analytic Hierarchy Procedure for Industrial R&D Project Selection and Resource Allocation." *IEEE Transactions on Engineering Management* 34(1): 12–21.

5. Saaty, T. L. 1980, *The Analytic Hierarchy Process*. New York: McGraw-Hill.

6. Kocaoglu, D. F. 1983. "A Participative Approach to Program Evaluation." *IEEE Transactions on Engineering Management* 30(3): 112–118.

7. Gear, A. E. 1974. "Review of Some Recent Developments in Portfolio Modeling in Applied Research and Development." *IEEE Transactions on Engineering Management* 21(4): 119–125.

8. Dickenson, M. W., A. C. Thornton, and S. Graves. 2001. "Technology Portfolio Management Optimizing Interdependent Projects over Multiple Time Periods." *IEEE Transactions on Engineering Management* 48(4): 518–527.

9. Luehrman, T. 1998. "Strategy as a Portfolio of Real Options." *Harvard Business Review*. 76(5).

Project
Portfolio
Mapping

Balance *is beautiful.*

Miyoko Ohno

Major topics in this chapter are tools for mapping project portfolio:

- ▲ Traditional Charts
- ▲ Bubble Diagrams

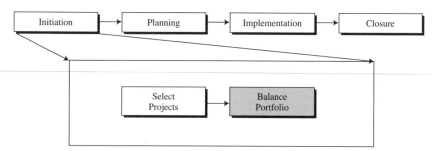

Figure 3.1   The role of portfolio mapping tools in the standardized project management process.

They have a simple purpose: to help companies obtain a balanced portfolio of projects, including existing and newly selected ones (see Figure 3.1). The logic here is that the investment firms are seeking to balance their growth versus income stocks and domestic versus international investments, and spread them out across industries, all in an attempt to acquire a well-diversified portfolio of investments. The analogical use of this concept in project-driven organizations strives for a portfolio of projects that are optimally diversified in terms of a certain number of key dimensions. By supporting such diversification, portfolio planning tools also facilitate the alignment of projects with the organization's strategy while balancing risks involved in projects.

This chapter's goal is to help practicing and prospective project managers

▲ Learn how to use various project portfolio mapping tools.

▲ Select those tools that account for their project situation.

▲ Customize the tools of their choice.

Internalizing these skills has a powerful and positive impact on initiating projects and building a standardized PM process.

Traditional Charts for Portfolio Management

What Are Traditional Charts for Portfolio Management?

These charts are numerous forms of traditional tools, such as histograms, bar graphs, and pie charts, shown in Figure 3.2, that can successfully picture project portfolio balance. Flexibly designed, they do not set limits on project dimensions/parameters to display [2]. Rather, the charts can present various dimensions from project size to resource allocation to earned value to timing. Presenting relative proportions of these dimensions, the charts indicate their distribution, offering managers an opportunity to balance it according to their goals.

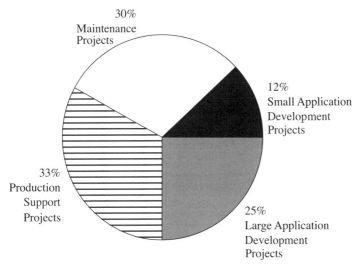

Figure 3.2   Pie chart for spending allocation per project type.

Constructing the Traditional Charts

Prepare Information Inputs. The act of balancing a project portfolio is deeply rooted in quality information about the following:

- ▲ Strategic and tactical plans
- ▲ Active projects

To determine which dimensions matter and will be presented on a chart, you need to look at the strategic and tactical plans of an organization. Then information about projects from the active roster can be fed into building the chart.

Select the Type of Chart. What portfolio dimensions do you need to balance? The answer to this question dictates the type of chart you pick. For example, if there is a need to portray the spending allocation across types of projects, a good choice is the pie chart. In one case of a software applications development group, management wanted to verify that their resource allocation is in tune with their strategic goals. The pie chart, they decided, can do this job.

There are no hard-and-fast rules for selecting the type of chart. For example, instead of the pie chart from Figure 3.2, management could have chosen a histogram. Apparently, their belief was that the pie chart's simplicity and visual impact more clearly projected where project dollars go than the histogram. As in this example, personal preference and past experience are influential in deciding which charts to use. Another concern is the number of charts to prepare. Go for the minimum number that meets information needs of portfolio balancing.

Draw the Chart. A look at Figure 3.2 reveals that we need to draw a pie with its wedges. While the total pie represents total resource allocation, the wedges reflect spending for each project type. Therefore, in drawing the chart, either manually or by computer, you need to enter the project type information and appropriate percentages. Similarly, drawing other chart types is straightforward.

Interpret the Chart. Interpretation of the chart is made possible by comparing the current balance with the desired balance, which is typically part of the strategic goals set by management. As for the current balance, the message from Figure 3.2 seems to be rather clear:

- ▲ The majority of resources (63 percent) are allocated to maintenance and production support projects.
- ▲ The minority of resources (37 percent) is consumed in applications development projects, small and large.

The desired or right balance is that 80 percent of resources be allocated to the applications development projects, the remaining 20 percent to maintenance and production support projects. Apparently, there is a huge gap between the current and the right balance. A balancing action is a necessary element—one that the chart is not designed to provide. Rather, it is management's job to determine and take such action.

Balance. This is the step directed at taking a management action to close the gap between the right and current balance. In the example from Figure 3.2, what are the options available to management? Specifically, they can

- ▲ Increase the number of applications development projects. For this approach to be possible, there must exist a sufficient number of quality project proposals in the pipeline of project selection.
- ▲ Decrease the number of maintenance and production support projects. The assumption here is that another approach can be put in place to avoid the disruption of production support.

In this case, it is quite clear that balancing actions must be closely synchronized with the processes of project selection and production support. What this really hints at is that portfolio balancing is not a simple, stand-alone operation. Instead, it is often part of a wider system including strategic planning, project selection, project implementation, and operations management.

Utilizing Traditional Charts
When to Apply. Traditional charts fit both purposes, presenting small and large projects in a portfolio. In deciding when to use them, an important factor is that their information richness is lower than that of bubble diagrams. Consequently, simpler displays of the project portfolio favor the application of the charts over the diagrams. On a similar note, managers are much more familiar with the charts than the diagrams, giving the

charts an edge in comparable display situations and especially when deciding where to start when balancing project portfolio for the first time (see the box that follows, "Start Where the Money Is Being Spent Now").

Time to Complete. Drawing a Traditional Chart is a quick and effective action. Once the necessary information is prepared, it takes minutes to get the chart done. Balancing activities following the preparation of the chart may take a group of managers hours or even a day or two.

Benefits. The value the charts create for the users is in their capacity to clearly display information important to balancing. In that sense, however, the charts are even more imperfect than Bubble Diagrams. For example, while the diagrams can hint which of their quadrants are more or less desirable for projects, the charts cannot do that. To a significant extent, this is the consequence of their lower information richness compared to the diagrams.

   Advantages and Disadvantages. Traditional Charts' advantage is in their

- ▲ *Simplicity.* Preparing the charts or reading them poses almost no challenge to their users. As a consequence, their simple design and appearance offer no-training-required opportunity to use them.
- ▲ *Familiarity.* Since they have been in use for a long time, managers are accustomed to them, a nice advantage when they can be used for the portfolio balancing.
- ▲ *Flexibility.* The charts function like shells: while their external appearance doesn't change, inside you can insert any dimension that fits. This enables using the charts to display almost any project dimension or parameter.

Start Where the Money Is Being Spent Now

Many managers did not consider what is *the right balance for their portfolio of projects.* If you are one of them, ask yourself, do I know the *current breakdown of either my projects or spending?* We asked this of a manager of an applications development group and she could not tell what percent of funding was going to her project categories/types, new software development projects versus upgrades versus support of the existing products. One way to begin balancing the portfolio of projects is to assess the current breakdown of project spending:

- What is the total project spending now?
- What percentage of the spending goes to each category of projects?

You can also consider the breakdown by project timing (completion dates or in which phase is the project currently). This information can be collected from project teams and charted. That is your *current* portfolio, a good start with which to approach your management to determine the *desired* portfolio. Since the essence of management's job is resource allocation, setting adequate spending goals is a crucial issue.

A disadvantage identified by some experts is their

▲ *Oversimplification.*   Charts such as pie charts contain such basic information that some experts question their ability to display information meaningful to balancing a portfolio of projects.

Variations. Variations of Traditional Charts remind us of life forms in that they are strikingly numerous. Figure 3.3 presents a histogram for the timing of project completions with five parameters: each project in the portfolio, the allocation of resources to each project, resources per quarter, resources per product line, and when the projects will be completed. This chart is very convenient in balancing a project pipeline to produce a continuous stream of projects, without sudden peaks and valleys. Also, it is helpful in balancing resources across the product lines.

Another traditional chart, the bar graph, is shown in Figure 3.4. Again, the focus is on resource allocation, indicating four parameters: projects, budgeted, actually consumed, and earned value resources. The major purpose of the graph is to detect which projects are using resources slower or faster than planned, so the resources may be moved from where they are not needed to where they are needed. For example, project Z is overspending and possibly may need more resources. By contrast, project X may use fewer resources than planned, which may make it possible to reallocate the resources to project X. Where on the bar graph do we see this? We see it in the ratio of earned value bar over the actual bar (also called *cost performance index,* see Chapter 13).

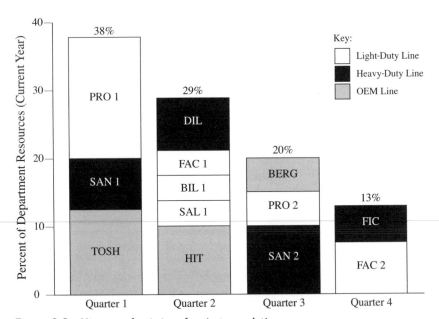

Figure 3.3   Histogram for the timing of project completions.

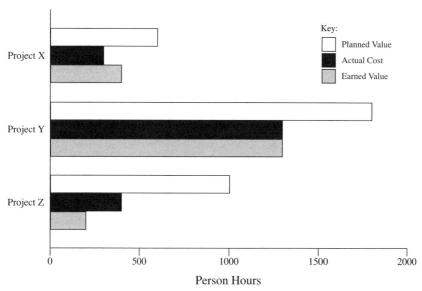

Figure 3.4 Bar graph for resource allocation.

Customize Traditional Charts. Although the charts presented here are beneficial, they may not exactly match your needs. That creates a need to analyze the available formats and features of the charts, and select and adapt them to maximize their value for your projects. Following are a few ideas regarding how to go about the adaptation.

Customization Action	*Examples of Customization Actions*
Define limits of use.	Choose to use a few charts only. Coordinate the selection with the selection of Bubble Diagrams so they are complementary.
	Make sure they best fit your strategic and tactical plans and goals.
	Select dimensions for each of the chosen charts.
	Tie the use of the charts with project selection tools. Purpose: to select and balance projects simultaneously.
	Use the charts in portfolio review meetings and project phase reviews.
Adapt a feature.	Fit any project dimension in the charts that you need, since they are designed to display almost any project parameter.

Traditional Charts Check

Make sure you prepare the Traditional Charts properly. They should show the following:

- Chosen project dimensions or parameters
- The scale for dimensions

Summary

This section dealt with forms of traditional tools, such as histograms, bar graphs, and pie charts, that have the capacity to provide simpler displays of information important to balancing a project portfolio. They are also helpful when organizations are using portfolio displays for the first time. Although these charts are convenient for presenting both small and large projects in a portfolio, understand that they are not capable of suggesting the actions to balance an unbalanced portfolio. To be truly effective, the charts must be customized for your specific situation. The key points about what the charts should show are highlighted in the box above.

Bubble Diagrams
What Is the Bubble Diagram?

Bubble Diagrams are information displays that visually show projects' key parameters necessary to successfully balance a project portfolio (see Figure 3.5). Typically, x and y axes of the diagram represent two of the key dimensions/parameters. Adding the bubble to the diagram indicates the position of the project on the two dimensions, while its size and color may point to additional parameters such as size, percent complete, project type, and so forth. Spread around the diagram, bubbles visualized as projects help managers decide whether they are rightly distributed along key dimensions. If not, managers can take a balancing action.

Constructing the Bubble Diagram

Prepare Information Inputs. Aiming at balancing a project portfolio calls for quality information about the following:

- Strategic and tactical plans
- Project selection criteria
- Project roster with projects' numerical scores

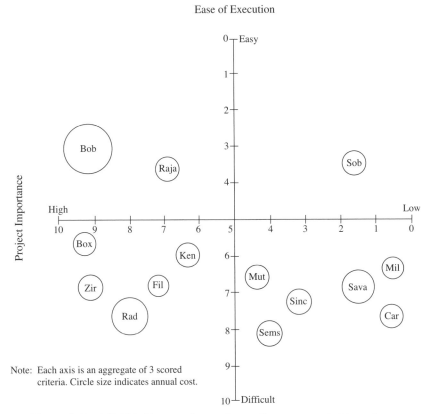

Ease of Execution

Figure 3.5 A Bubble Diagram of ease of execution versus project importance.

These plans and project selection criteria help frame the balancing problem in terms of which dimensions will be on the diagram. The product of the project selection process includes both the list of the existing and newly selected projects; the actual project roster is what are displayed on the Bubble Diagrams and then balanced. We say *accurate roster* because some managers have the tendency to informally add their pet projects to the formal, official roster. If all of these are not included in the balancing act, there is a real risk of mismatch of the required versus available resources and related schedule delays. Project scores are inputs in the diagram that are obtained by means of scoring models for each of the dimensions of the diagram.

Select the Type of Chart. Making a choice of chart begins with clear understanding of dimensions on the axes of the Bubble Diagram (see the "This Is Not the Old BCG Matrix" box that follows). Options are numerous, as the following sample list indicates [3, 4]:

- Return (financial, multiple criteria-based)
- Life cycle phases (concept, planning, execution, closure)
- Completion date (months, years)
- Strategic fit or importance (low, medium, high)
- Quality of resources (low, medium, strong)
- Project costs (dollars, resource hours)
- Probability of success (technical and commercial expressed as percentages)
- Ease of execution (difficult-easy continuum)
- Project categories (R&D, engineering, manufacturing, marketing, maintenance, etc.)
- Project types (e.g., product development—platforms, derivatives, enhancement, joint ventures, etc.)
- Market segments (market A, market B, etc.).

Any pair can be used as x and y axes on the bubble chart to help describe the portfolio and depict the balance. When charts' dimensions are selected, they are likely to belong to one of these groups:

- Scoring models-based charts (e.g., project importance versus ease of execution)
- Risk-return charts (e.g., adjusted NPV versus the probability of technical success)
- Other Bubble Diagrams (e.g., project costs versus life cycle phases)

This Is Not the Old BCG Matrix

When first exposed to Bubble Diagrams many people are quick to notice that these are the same as Boston Consulting Group matrix (or GE or McKinsey matrices) of the 1970s [1]. Actually, they are not. Old matrices were used to plot the *existing strategic business units* on market *attractiveness versus competitive position* matrix, capturing the present state of the units. Project Bubble Diagrams add *newly* selected *projects* to the existing projects, focusing on the *future*. In addition to this difference in unit of analysis and time horizon, dimensions of the diagram are also very different, ranging from *probability of technical and commercial success* to the *ease of project execution*. Many dimensions are used as shown in the preceding list.

Do You Agree with This Portfolio Verdict?

The first thing Peter, Star Tech's new software development (SD) manager, did was to convene a meeting to overview the status of SD projects. There he learned that projects were taking far too long. His feeling was that the pipeline was clogged with too many projects—28. Apparently, there were some really good projects. What perplexed him was the large number of sub par projects. No two of them were in the same product performance/market segment area. This approach was strategically scattershot. There was really no glue to bond all those projects; the total lack of focus was more than evident.

All of the projects held a promise of high reward, obviously a good sign. The problem is that all of them also were highly risky, with a reasonable probability to fail technically or commercially. Almost all projects were long term. The absence of "quick hits" to balance the long-term development projects was apparent. Also, almost all projects were in the programming phase, with no new projects in the requirements definition or closure phase. "Okay," Peter thought to himself. "Now, the groups of issues seem clear. In order to talk to my reports, I need to summarize this into a few words, describing where our portfolio is as of now. Would the following description make sense?"

- Overall, our portfolio stinks!
- Our portfolio is unfocused!
- Too many projects; resources spread too thinly!
- Too many poor projects!
- Our portfolio is poorly balanced!
- Our portfolio does not support our business strategy!

Agree?

Apparently, no single chart can comprehensively characterize a portfolio. That prompts the question of how many charts do you need? And which ones? Multiple charts may be necessary, but you must also avoid proliferation of charts and possible information overload. Therefore, shoot for a maximum of several charts that can accurately capture major strategic requirements of project portfolio. Consider, for example, the "Do You Agree with This Portfolio Verdict?" box above. To catch graphically problems related to the portfolio, the manager would need Bubble Diagrams indicating the following:

- Project types versus market segments
- Completion date versus life cycle phase
- Risk (probability of success) versus return

Also, a traditional chart comparing required resources and available resources would help detect the lack of a resource balance. Once the inputs have been prepared and charts selected, it is time to chart projects on the Bubble Diagram.

Chart Projects. Take a look at Figure 3.5. Drawing the x and y axes, labeling them, and entering the scale is a logical first step. Labels we have chosen are dimensions of project importance and ease of project execution. This type of the diagram is based on scoring models for project selection. In this example, each dimension's scale is a numerical average of three constituent scales. For project importance those include strategic importance to company goals), organizational impact (e.g., profitability or customer impact), and economic benefits (e.g., dollar savings or revenue growth). Ease of execution, on the other hand, relates to cost of execution, project complexity (e.g., difficulty of implementation), and resource availability. The intent here is to assess the project in a comprehensive, multicriteria fashion. When scales are set, use scores for each dimension's three elements to develop averages and chart projects from the roster on the diagram (see the box that follows, "Use Project Scores from Project Selection").

Interpret the Diagram. Look for projects in favorable quadrants (the upper left of the diagram in Figure 3.5), dissecting those in the "low importance and difficult to do" quadrant, striving to balance ease and importance of projects. In particular, in Figure 3.5:

▲ Easy-to-do projects are in short supply, just three out of 14. We need more of these "low-hanging fruit" projects when seeking a balance between easy-to-do and difficult-to-do projects.

▲ There are only two projects in the most sought after upper left quadrant, an intolerably low number. Perhaps, too few proposals for these diamond projects are submitted for project selection.

▲ Another headache looms on the horizon: too many projects, six precisely, are in the lower right quadrant. This is the most unpleasant place to be in. Consider which to eliminate.

▲ Five projects are of high importance but difficult to do. Perhaps there are ways to remove some barriers to ease of implementation.

In this relatively simple example, the Bubble Diagram demonstrates its forté, the ability to display project information well. It also demonstrates what it cannot do: suggest what action to take to balance the portfolio.

Balance. Managers must make a decision to act and balance the portfolio. After an in-depth analysis of the Bubble Diagram and pertaining project information related to Figure 3.5, the following balancing actions could be taken:

▲ Increase the number of easy to do projects. How? Review again projects on hold, searching for easy opportunities to upgrade them to an easier-to-do status.

▲ Increase the number of the most sought after projects in the upper left quadrant by reviewing again all promising project proposals that didn't make the cut in the project selection process. Can their ease of doing or project importance be improved by modifying their scope, resource requirements, and implementation plan? Also, encourage more project proposals to be submitted that fit this quadrant.

- ▲ Kill three projects with the lowest importance.
- ▲ Increase the ease of implementation of all five projects in the high-importance, difficult-to-do quadrant. Review and improve their resource commitments. Analyze and overcome the most difficult barriers in milestone reviews (e.g., the barriers may be in technology integration, production, assembly, or testing).

These are sample actions, not magical solutions. They are here to hint at the essence of balancing, a need for conscious, well-informed decisions to spread out the risk. Often that will require going back to the project selection step and strategic plans, reexamining all choices, and changing some of them to achieve the desired balanced portfolio. Strategic planning, project selection, and portfolio balancing steps are most effective when intertwined.

Utilizing Bubble Diagrams

When to Apply. Bubble Diagrams are versatile tools that can be used in multiple ways. One of them is the hierarchical format, starting on the top of the company (a Bubble Diagram for top strategic projects) and cascading through portfolios of different organizational levels (e.g., a Bubble Diagram for the marketing or R&D department) all the way down the group level (a Bubble Diagram for a multiproject manager). In this application, each lower-level Bubble Diagram supports one above it, enabling the integration of all company projects into an overall portfolio. Although effective, the hierarchical format is also very demanding, which is perhaps why its use is not high. More frequently used is a decentralized format, where different groups in the company construct Bubble Diagrams for their own projects, independently of each other. Good examples are new product, information technology, and manufacturing groups. In this context, the Bubble Diagrams are employed for periodic reviews of the project portfolios, quarterly or semiannually. Those who favor project phase reviews (also called go/kill points) for major projects may use Bubble Diagrams to compare the reviewed project with other projects when deciding to continue it or terminate it. An innovative application of the diagrams is in the setting of multiple projects management. Faced with the pressure of managing six or seven projects at a time, multiproject managers tend to use the Bubble Diagram to control their workload across dimensions of the project size and project phases. (Don't have two large projects simultaneously in the planning stage!)

Use Project Scores from Project Selection

When considering a portfolio approach, many managers go beyond using only financial metrics. Rather, they rely on multiple criteria. A good option for this situation is to build a risk-reward Bubble Diagram, in which axes are based on the same scored criteria that were used in the scoring model to rate, rank, and select projects (see Chapter 2). Furthermore, project scores obtained in the selection process become inputs to your risk-reward Bubble Diagram. That provides a seamless consistency between the project selection and portfolio balancing.

Time to Complete. Provided that information about projects is readily available, constructing a Bubble Diagram is a quick action. Constructing a diagram with 20 projects may take a group of decision makers only a few minutes, especially with the advent of the powerful computer programs. Where time is really consumed is in analyzing the diagram— debating its strengths and weaknesses, opportunities and threats—and making decisions to balance. These steps may take several hours to a full day of the group time.

Benefits. It is difficult to argue against the basic value of the diagrams and its ability to act as a meaningful information display. To many experts this is the reason for the popularity of Bubble Diagrams. The diagram indicates projects that are in more preferred quadrants, for example, high-importance, easy-to-do projects. Similarly, it helps discern projects in less desired quadrants, for example, low importance, difficult-to-do projects. Additionally, the sense for the utmost importance of balancing projects across the desired quadrants comes across pretty clearly. Such a visual demonstration of the project portfolio is unmatched by other tools such as economic methods or scoring models.

Despite benefits of the Bubble Diagrams, some users see some serious imperfections in them [5, 6]. The diagram is designed to act as an information display of a project portfolio, not as a decision model. Therefore, it does not contain a mechanism to help make decisions. In particular, scoring models are designed to be decision models to generate project rankings, whereas bubble diagrams cannot balance the portfolio. Rather, they only offer a starting point for management to make decisions and define actions. Bubble Diagrams cannot tell what is the right balance. Management must define the right balance and compare it against the actual balance. In addition, Bubble Diagrams do not spell out how they should be used; management needs to decide how to use them, for instance, when to change the balance, terminate some projects, or add new ones. However, while some might view these imperfections as design flaws, others may argue that this tool does what is designed for well: display information about portfolio balance.

Advantages and Disadvantages. Bubble Diagrams offer substantial advantages through their

  ▲ *Simplicity.* Their user-friendliness and ease of use are so high that many first-time users need little or no training to construct or read them.
  ▲ *Data-based view.* Bubble Diagrams offer an unbiased view of the data, not tainted by someone's opinion.

The problems with the Bubble Diagrams may result from the following:

  ▲ *Information overload.* Because there are so many different Bubble Diagrams, too many of them may be used, increasing rather than decreasing the complexity of the balancing assignment.

▲ *Inappropriate data.* If the choice of data used to create the diagram is not carefully made, the resultant diagram may be misleading.

Variations. Also called portfolio maps and portfolio matrices, Bubble Diagrams come in all kind of varieties [7, 9]. Figure 3.6 illustrates a risk-return version of the diagram that is very popular in research and development projects. The diagrams indicate the following:

▲ Too much spending for the bread-and-butter projects. These should be small, simple projects with a high probability of success, but low reward. Instead, an inordinate amount of work is being spent on trivial projects.

▲ There is good number of "pearls," projects with high likelihood of success able to produce high returns. These are potential star projects.

▲ Two "oysters" are probably a good number of projects, viewed as long shots that may generate high returns but with low probability to succeed.

▲ The "white elephant" quadrant has only one project, a good number. These are low-return projects with a low likelihood of success.

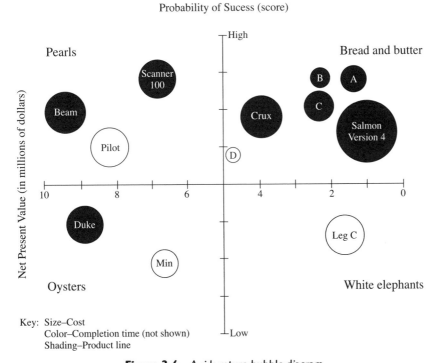

Figure 3.6   A risk-return bubble diagram.

Another popular variation is completion date versus life cycle phase Bubble Diagram, shown in Figure 3.7. The diagram indicates the following:

- ▲ All smaller projects will be completed around the end of year 1. All large projects will be completed around the end of year 2. This is a very poor balance, needing an action to spread projects along the pipeline. A balance of projects of different sizes is also desired.
- ▲ There is a solid balance of projects across life cycle phases, both small and large projects.

Customize Bubble Diagrams. We have only shown a few generic types of Bubble Diagrams. They may be customized a myriad of ways. We suggest you study the generics and adapt them to fit your project situation. Following are a few ideas to spur your customization action.

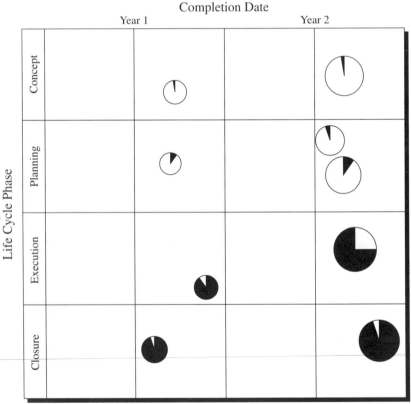

Key: Circle Size–Project budget
     Pie Wedge–Percent complete

Figure 3.7 A Bubble Diagram of completion date versus life cycle phase.

Customization Action	Examples of Customization Actions
Define limits of use.	Select dimensions that best fit your strategic and tactical plans and goals.
	Use Bubble Diagrams to display and balance projects (corporate strategic projects, all departmental projects, and project load of a multiproject manager).
	Use Bubble Diagrams in conjunction with project selection tools. Purpose: to select and balance projects simultaneously.
	Use Bubble Diagrams in portfolio review meetings and project phase reviews.
Adapt a feature.	In uncertain projects that beg a probabilistic view, use ellipselike bubbles to show high/low estimates for both dimensions (e.g., NPV versus probability of success).
Add a feature.	Create three-dimensional bubble diagrams. Procter & Gamble uses NPV, probability of success, and time to launch [5, 6].

Summary

This section was about Bubble Diagrams, information displays that visually show projects' key parameters necessary to successfully balance a project portfolio. They are also helpful in project phase reviews when deciding to give a project the go-ahead or terminate it. Similarly to the Traditional Charts, Bubble Diagrams cannot suggest actions to balance an unbalanced portfolio of projects. However, they can indicate the importance of balancing projects across the desired quadrants. Customizing the diagrams for specific situations is the most effective way for their use. Following is a checklist to help you construct a Bubble Diagram.

Bubble Diagram Check

Be sure to adequately construct the Bubble Diagram. It should show the following elements:

- Chosen dimensions on the x and y axes
- The scale for dimension
- Projects as bubbles in quadrants, balanced as determined by management

Concluding Remarks

We have covered two distinct families of project portfolio mapping tools in this chapter: Traditional Charts and Bubble Diagrams. Each one features a host of forms. When designed to display two or three parameters, Traditional Charts can be used alongside Bubble Diagrams, for instance, using a pie chart similar to Figure 3.2 and a Bubble Diagram like Figure 3.5. Here the lower information richness of the pie chart shows well the big picture of the resource allocation across project types. This information is finely complemented by the higher information richness of the Bubble Diagram, with its details of the balance of individual projects in terms of their importance and ease of implementation.

When there is a requirement for higher information richness of the portfolio display, Bubble Diagrams and Traditional Charts are not compatible and compete against each other. For example, information in the histogram from Figure 3.3 can easily be displayed on the Bubble Chart. In that case, the Bubble Chart would provide a better focus on individual projects, showing them as standalone bubbles with different sizes. On the other hand, the histogram outperforms the diagram in showing cumulative percentages for aggregated projects in each quarter. Overall, Traditional Charts are simpler and better known to managers, suggesting that when a company wants to start balancing its portfolio of projects, the charts are an easier choice. The following table summarizes this information to help you select the proper tools in different project situations.

A Summary Comparison of Project Portfolio Mapping Tools

Situation	Favoring Traditional Charts	Favoring Bubble Diagrams
When Using Bubble Diagrams and Traditional Charts in Combination		
Need lower information richness	√	
Need higher information richness		√
When Bubble Diagrams and Traditional Charts Compete		
Focus on individual projects in the portfolio		√
Focus on expressing percentages for aggregated projects	√	
Start balancing project portfolio for the first time	√	

References

1. Harrison, J. S. and C. H. St. John. 1998. Strategic Management of Organizations and Stakeholders. 2d ed. Cincinnati: South-Western College Publishing.

2. Brenner, M. S. 1994 "Practical R&D Project Prioritization." *Research Technology Management* 37(5): 38–42.

3. Buss, M. D. J. 1983. "How to Rank Computer Projects." *Harvard Business Review* 1(71): 145.

4. Cooper, R. G., S. J. Edgett, and E. J. Kleinschmidt. 1998. "Best Practices for Managing R&D Portfolios: Lessons from the Leaders — Part II." *Research Technology Management* 41(4): 20–33.

5. Cooper, R. G., S. J. Edgett, and E. J. Kleinschmidt. 1997. "Portfolio Management in New Product Development: Lessons from the Leaders — Part I." *Research Technology Management* 40(5): 16–19.

6. Cooper, R. G., S. J. Edgett, and E. J. Kleinschmidt. 1997. "Portfolio Management in New Product Development: Lessons from the Leaders—Part II." *Research Technology Management.* 40(6): 43–52.

7. Archer, N. P. and F. Ghasemzadeh. 1999. "An Integrated Framework for Project Portfolio Selection." *International Journal of Project Management* 17(4): 207–216.

8. Marheson, D., J. E. Matheson, and M. M. Menke., 1995. "Making Excellent R&D Decisions." *Research-Technology Management* 37(6): 21–24.

9. Matheson, J. E. and M. M. Menke. 1994. "Using Decision Quality Principles to Balance Your R&D Portfolio." *Research-Technology Management* 37(3): 38–43.

PART

II

Project Planning Tools

Voice of the Project Customer

This chapter is contributed by Jose Campos.

It is only if the customer says *it is*

(Anonymous)

This chapter focuses on tools for the identification and translation of voice of the project customer, including the following:

- ▲ Customer Roadmap
- ▲ Focus Statement

▲ Sample Selection

▲ Discussion Guide

▲ Quality Function Deployment (QFD)

In particular, these tools—often referred to as customer intimacy tools—orchestrate the activities to define the type of customer information needed by the project and obtain input from the project customers. Also, they ensure that the project team contacts the right type of customer and plans the various contacts with customers. Once the information about customer requirements is acquired, the tools provide a basic framework to translate it into the project scope. Apparently, because their use is part of scope planning, there is a need for close coordination with the tools of schedule and cost planning as the project baseline is defined (see Figure 4.1).

This chapter aims at helping practicing and prospective project managers to

▲ Learn how to use various tools when dealing with voice of the project customer.

▲ Choose those tools that are most appropriate for their project situation.

▲ Customize the tools of their choice.

Clearly, getting a hold of these skills is an enabler of understanding and internalizing customer requirements, and building them in the project products. Consequently, the skills have a premier place in project planning and building a standardized PM process.

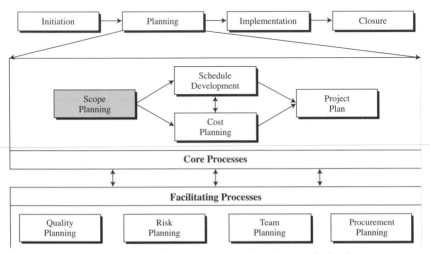

Figure 4.1   The role of voice-of-the-customer tools in the standardized project management process.

Customer Roadmap
What Is the Customer Roadmap?

The Customer Roadmap is a tool for setting up a systematic approach for listening to the voice of the customer in projects (see Figure 4.2). To carry out the approach, the roadmap defines a series of steps and their timing that outline the process to obtain, document, and use customer input. When streamlined and consistently executed, the process has a high potential to help establish a culture of customer involvement in the design and implementation of projects.

The logic of having the Customer Roadmap in place is straightforward and powerful. Any reasonable effort to obtain input from the project customer requires a plan, much like any other project work package or activity does. By carefully developing the roadmap, the project manager can ensure that there is enough time to listen to the customer and effectively use the obtained information. In plain terms, the Customer Roadmap is the plan that prescribes the use of tools in this chapter and their time line.

Constructing a Customer Roadmap

The development and use of Customer Roadmaps is a relatively new phenomenon in project organizations. This lack of experience automatically raises the red flag, pointing to the need to identify risks of missteps. A way to avoid missteps is to methodically design and deploy a Customer Roadmap with well-synchronized and integrated sequence of steps. We describe these steps in the text that follows.

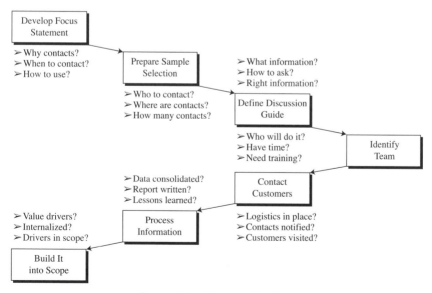

Figure 4.2 Customer Roadmap.

Prepare Information Inputs. Two quality inputs significantly impact the successful development of a Customer Roadmap:

- ▲ Project proposal
- ▲ List of team members

Typically prepared as part of the project selection process, the project proposal provides the information about the proposed goals, scope, and implementation strategies of the project. This information will help identify customers, information that needs to be obtained from them, and how to use it afterwards. Knowing who is on the team lets you get them involved in the roadmap development process. In this way their buy-in is obtained along with their commitment to deploy the map later.

Convince the Project Team to Visit the Project Customers. Often, there is a tendency to claim that the project team has all the information it needs to complete the project. In other words, it is common to encounter resistance from the team members because of overconfidence or tradition (i.e., they did not contact the project customers in previous projects). The project manager must be aware of this resistance and work toward persuading the team members that there is much that is not known. For example, the project manager may demonstrate what is unknown by asking key questions. Who are the customers of this project? Do we know all the needs from the project customers? Do we know the unarticulated needs from the project customers? Answers to these questions can help establish the need to visit customers and better understand their needs. The ultimate argument here is that understanding the needs of the project customer is the first step to customer satisfaction, a powerful strategy to make more money (see the box that follows, "Voice of the Customer Is about Return on Investment").

Develop the Focus Statement. The highlight of this step is the construction of the Focus Statement, which defines the purpose of customer visits. Building on the sufficient interaction in their meeting, team members that may be part of making the actual customer contacts should prepare good answers to the following questions:

- ▲ Why do we need to contact customers of this project?
- ▲ When do we need to start contacts to yield maximum benefit to the project?
- ▲ How are we going to use the information obtained during the contacts?

Voice of the Customer Is about Return on Investment

Customers have taken charge. It is this competitive reality that drives the increasing importance of identifying the voice of the project customer. They tell suppliers what they want, when they want it, how they want it, and how much they are willing to pay. And suppliers listen. As a result, there is no such a thing as *the* customer; there is only *this* customer, the one using his or her power and ability to demand [4]. Leading organizations feed this customer information back into their projects, products, and service experiences to keep customers happy and loyal.

Apple Computer offers an example. Customer service staff uses its phone dialogue with customers to identify the top ten customer issues of the week. Design engineers use the top ten list to improve projects producing new releases and also to identify customers' wishes for possible new projects [5]. Satisfied customers are critical for a company's economic returns; in 1997 companies that reached higher customer satisfaction created over 100 percent more shareholder wealth than their competitors with lower customer satisfaction scores [7, 8]. No wonder, then, why the world's most successful software firms rate customer involvement as the single most important success factor in software project development—customer satisfaction begins with customer involvement [9]. The message here is that ignoring customers' voices is a poor business decision.

Prepare the Sample Selection. Once the purpose of contacting customers has been phrased, it is time to define the type or profile of customers to be contacted (see the "Every Project Has a Customer" box that follows). This is typically done in a meeting involving the team or those that may be selected to make the actual customer contacts. To shape the profile, they need to ask these questions:

- ▲ Who do we need to contact to get the right type of information?
- ▲ Where are these contacts located?
- ▲ How many contacts will be needed to get a valid sample?

This information describes the profile of customers to contact in the Sample Selection tool.

Define Discussion Guide. Now that we know whom to contact, the team needs to create and select questions or topics for discussion with the project customers. In doing so, the team must answer these questions:

- ▲ What information do we really need?
- ▲ How will we phrase questions to obtain the right information?
- ▲ Will the questions yield the right type of information?

Every Project Has a Customer!

While we may not be accustomed to viewing those that will derive "value" from the project as "customers," they are and share the primary characteristic of customers: "One that patronizes project products." Generally, they may be either external or internal customers [3]. *External customers* buy the project product, providing financial support to the organization. Apparently, it is of vital importance to satisfy their needs and keep the cash flow stream going.

When other employees of the organization derive the value from the project, the term *internal customer* is commonly used. Inside the company, the project team passes on its work to other employees. Also, the team receives work passed on from other company employees, the internal suppliers. Similarly, external suppliers are vendors outside the organization who sell their products, services, materials, and so on to the project team. Each project, therefore, is the customer of the preceding people, and each has customers to whom it passes on its work [6].

Internal or external customers show very similar patterns of behavior. For example, their definition of "value" is shaped by their own needs, their value set, and the current set of pressures acting on them. A common error is to assume, because everyone works for the same organization, that the needs are clearly understood. Not so, and this is a source of much aggravation in projects. In short, internal customers exhibit the same traits as external ones. The project manager must learn to listen to voice of the project customers, regardless of where these customers are located. (Note: The terms *project stakeholders* and *project customers* are often used interchangeably.)

Assuming that we may not have more than a one-hour meeting with the project customers, the Discussion Guide should contain three to four questions or topics of discussion. Everyone involved in customer contacts must use the same set of questions and in the same sequence to ensure consistency. Remember that after all interviews have been completed, all the information needs to be consolidated into a single report, consequently note taking is a critical step.

Identify the Team. This step is where we identify the team members that will do the actual interviews or customer contacts. The focus is on three major questions:

- ▲ Who from our team is best qualified to make the contacts?
- ▲ Do these team members have the time to make the contacts and process the information?
- ▲ When will these team members be trained to make effective contacts?

A useful strategy is to create teams of two members. For example, if we selected eight customers to contact, we create two teams of two members, with each team conducting four interviews. Having more than two members meet with one customer may inhibit the conversation. In addition, with a two-person team, one can ask the questions while the other takes notes. The prerequisite to being appointed to the team is to ensure that those selected have the time available to collect and process the information. Also, they need to be trained to the level required. (According to one expert, in a typical company, few project team members are properly trained to do these interviews, possibly a major reason for so many user-unfriendly products.)[10]

Finally, the interviewers need a proper budget to cover travel and other expenses, which is, of course, more critical when traveling to other cities or countries.

Contact Customers. Using the discussion guide, selected teams make the actual contact and conduct an interview with the project customers. Preparing for the interviews includes asking these questions:

- Are all logistics in place to ensure successful contacts?
- Have the customer contacts been notified?
- Has the method of capturing the information been defined and agreed to?

When preparing, we must ensure that all logistics have been taken care of so as not to interfere with the actual interviews; for example, travel time, venue, availability of the customer, and so on. Also, taking notes is an absolute requirement, since it is the only way to keep track of valuable information. Imagine trying to keep track of eight interviews without notes, particularly if several weeks have passed.

Process Information. Once all the interviews have been done, it is time to process the information into a cohesive report that will benefit the project. This can be done in a meeting where all those that interviewed customers collaborate in generating the report. The purpose of this meeting is to make sure that we get positive answers to the following questions:

- Has all the information from all the contacts been consolidated?
- Will a final report be written and circulated to all team members?
- Have lessons learned been captured to improve the next step of customer contacts?

A typical scenario for the meeting may look like this: Begin by asking for highlights, using the Discussion Guide. Each member will articulate findings, surprises, facts, customer needs, and so forth. Proceed to organize all the information into logical categories as agreed by the team members, for example, customer needs or customer frustrations. To rate and prioritize the needs, use the focus statement to establish the criteria for prioritization. Rated and prioritized needs will become so-called customer value drivers. At the end of the meeting we should have anywhere from three to five categories, with all items having a priority. The last step is to assign the responsibility of taking all information and generating a written report.

Build It into the Project Scope. Although this step is the real payoff from listening to the voice of the customer, it is sometimes forgotten. Keep in mind that all gathered information is useless unless it is built into the project. For example, one approach is to call a meeting of the project team, discussing the final report and identifying the implications or actions needed as a result of the discussion. Sharing the final report with management is also a viable option. In short, ensure that the message from the project customers is put to use by building it into the scope and design of project products and process. These questions may help you successfully complete this step:

- ▲ Have customer value drivers been identified?
- ▲ Have customer value drivers been internalized by the project team?
- ▲ Have customer value drivers been built into the scope and design of project products and process?

One structured way to translate the voice of the customer into the project scope and design is with QFD (Quality Function Deployment), discussed at length near the end of this chapter. Starting from customer value drivers ("What does customer want and in which order of priority?"), QFD lays out sequential actions ("How do we meet what customer wants?"), building the drivers in the project scope or sometimes the project specification.

Utilizing the Customer Roadmap

When to Use. The Customer Roadmap should be considered a must in larger projects, typically at the beginning of the project implementation when the scope is being defined. In fact, many product development teams deploy it even in preparing the project proposal for the project selection process. This allows the team to scope the project and incorporate the needed resources into the project plan. In some cases, there may be several sets of customer contacts. Each should have its own roadmap.

When you are developing the roadmap for smaller projects, the process can be very informal and flexible because of the tight budgets involved. An increasingly accepted practice in building maps is to develop a template for the company and quickly adapt it for individual projects. Regardless of the project size and complexity, developing the roadmap or adapting a template is a team activity, meant to ensure clarity and commitment.

Time to Develop. The size of the project dictates the time requirements for the Customer Roadmap. A typical Customer Roadmap in a smaller project can be built in as little as one hour of project team time, whereas more complex programs of customer contacts in large projects may take up to three or four hours to complete. In addition, efforts to obtain customer input require many resources and consume considerable time in large projects. For example, a set of customer visits by the members of the team may consume up to two days per week over six to eight weeks. By contrast, these same efforts in smaller projects may only take an hour or so.

Benefits. The Customer Roadmap aids in clarifying the needs of the project customers. Most team members concur that understanding the needs of the customers is vital, yet they may not be aware that there is a formal process to obtain the right information. The roadmap helps ensure that the right steps are taken in the right sequence. It also ensures that the proper rigor is taken to receive quality input from the project customers.

Another benefit becomes apparent when considering that what customers really want is often not well understood at the outset of a project. In particular, experience shows that in many software projects, 50 percent of the development cycle time is spent on the rework needed to include customer needs [9]. Having an adequate Customer Roadmap in place in a timely manner may help create a more meaningful project scope, reduce the rework, and help reduce time to completion.

Advantages and Disadvantages. There are two major advantages of the Customer Roadmap. It is

- ▲ *Visual.* Its visual depiction of the process to obtain customer input makes the process clear and simple.
- ▲ *Educational.* Because the Customer Roadmap is a relatively new tool for the project management community, it is helpful in educating team members of this new method to obtain quality input from the project customers.

Practice indicates the roadmap's disadvantage. It may be

- ▲ *Ineffective when complex.* This is the case when the Customer Roadmap is applied in planning complex customer contacts—for example, visiting many customers around the world.

Variations. Many variations of the Customer Roadmap are used in the industry, for example, McQuarrie's seven-step procedure [2]. Some are more comprehensive and detailed than others. The steps may be sequential or they may be designed to take place concurrently. The number of tools included may also vary.

Customize the Customer Roadmap. We have described the roadmap as a general-purpose tool, designed to match the needs of a variety of organizations managing projects. We can certainly increase its value if we adapt it to the project needs. Following are a few ideas about potential adaptation of the roadmap.

Customization Action	Examples of Customization Actions
Define limits of use.	Use the roadmap in all large projects, whether for external or internal customers.
	Use the roadmap in small projects flexibly and informally but following steps in the roadmap.
	Use separate roadmaps for new product development projects and other distinct application areas.
Adapt a feature.	Append a Gantt Chart for clarity of roles, responsibilities, and time line.
Add a feature.	Create a glossary for new words associated with project customers to document new terminology.

Summary

This section presented the Customer Roadmap, a tool for systematic approach for listening to the voice of the customer in projects. In larger projects, the roadmap is typically applied at the beginning of the project implementation when the scope is being defined. Smaller projects can also make a use of the roadmap, although in an informal and flexible manner. In essence, the value of the Customer Roadmap is in clarity of understanding the needs of the project customers, helping make an impact on the project outcome. The following box offers this tool's key steps.

The Focus Statement

What Is the Focus Statement?

The Focus Statement is an action guide for customer visits in order to obtain the meaningful voice of the customer. Used in preparation for the visits, the statement's essential purpose is to set the focus during the time that will be spent with the customer (see Figure 4.3). In particular, the statement first defines the reason for the visits and identifies the information to collect that will most benefit the alignment of project scope with the customer needs. It also establishes clear roles and responsibilities in the course of the visits, as well as their time line and the budget.

Customer Roadmap Check

Make sure that Customer Roadmap is properly structured. It should include steps such as:

- Develop Focus Statement.
- Prepare Sample Selection.
- Define Discussion Guide.
- Identify team.
- Contact customers.
- Process information.
- Build processed information into the project scope.

FOCUS STATEMENT

Project Name: Map_____   **Date:** July 1, 01_____

Project Scope:
- Design and deploy a new process to provide accurate, timely repair of laptop computers in the three service centers in the U.S.
- Time: July 1, 2001 – July 1, 2002
- Cost: $100,000
- Process Accuracy: Fewer than 0.5 defects per 100 repairs

Purpose of Visits:
- To discover the stated and unstated needs of customers for the repair
- To identify the unmet needs that customers have for the repair
- To understand the customer definition of "Fast Turnaround"

Objective of Visits:
To write a report outlining the opinions, needs, requirements, and expectations of customers for the repair

---

Scope of Visits:
We feel there is no clear understanding of the actual needs that our customers have for the repair. For many years we have made a number of assumptions about the needs. These customer visits should help us better understand these needs.

Key Dates:
- Start developing visits by July 15, 2001
- Start visits by August 15, 2001
- Complete sample selection and discussion guide by August 10, 2001
- Prepare final report by October 15, 2001

Budget for Visits:
Budget: $17,000

Who Approves/Disapproves the Final Report:
Richard McBee, General Manager of the Service Organization

Team for Visits:
- Mary McCarthy, Team Leader Responsible for Visits Success, Service Manager
- Eileen George, Marketing
- Patty Fouler, Service
- Brad Cox, Service
- Al Frank, Outside Consultant

Figure 4.3   The Focus Statement.

Developing the Focus Statement

Collect Necessary Information. The development of a quality Focus Statement is based on details of the following:

- ▲ Project proposal
- ▲ Project team roster

The primary purpose of the project proposal here is to provide information on the preliminary project scope and objectives that will be used to construct the statement. We say preliminary because most of the proposals are preliminary in the nature of their information. To further detail the statement and create the visits team, you must look at the project team roster.

Restate the Project Scope. The ultimate purpose of customer visits is to enable the project scope of work that is in line with what customers want.

Consequently, the whole effort should start and be led all the way throughout the visits by clear understanding of the following:

- ▲ What are project outcomes?
- ▲ What are project objectives?

To keep reminding the visits team of this, the project scope statement and objectives should be restated in the first box of the Focus Statement, with the particular emphasis on those areas that are most relevant to project customers (see Figure 4.3). In essence, this serves as the anchor for the entire focus statement and links the project scope with the reasons for visiting customers.

Clarify the Reasons. In this step we define the purpose, objectives, and scope of the visits. Together we call them the reasons for the visits. We begin by asking the following questions:

- ▲ What is the business reason for this set of visits?
- ▲ Why are we doing this set of visits?
- ▲ What is the mission of the visits team?

These questions help identify the purpose or the "why" of the proposed set of customer contacts. Continuing along this line of thought, you should ask the following question:

- ▲ What is the verifiable indicator that will show that we have accomplished what we desired from the visits?

Here the focus is on the objective for the visits (for an example, see the text box that follows, titled "From Dissatisfaction to Sheer Delight: Kano Model"). In other words, given the objectives of the project, develop a set of objectives for visiting customers (see Figure 4.3). This forces the visits team to clearly explain why they must visit customers in order to achieve the objectives of the project. A good question to ask at this point is this: What do we wish to learn from the customers?

Think of the scope of visits—the third element—as an insurance policy to validate the reasons for visiting customers. Accordingly, its purpose is to ensure that there is a full set of statements that explain reasons to visit customers. If the purpose and objectives for the visits are clear and sufficiently tangible, this box can be left unfilled (see Figure 4.3). Otherwise, answers to these questions may help fill it in:

- ▲ What specific project areas will be addressed by these visits?
- ▲ What are the specific areas of study for the visits?
- ▲ What are the specific areas that will not be addressed by the visits?

From Dissatisfaction to Sheer Delight: Kano Model

Japanese professor, Noriaki Kano, suggested three classes of customer requirements [1]:

- *Dissatisfiers.* Requirements that are expected in a project. In a car, a radio or steering wheel are examples, which are generally not stated by customers but assumed as given. If these features are not present, the customer is dissatisfied. Similarly, if we propose the project completion date and miss it, project customer will be dissatisfied. Dissatisfiers are "minimums," or as some call it, the "must-haves" in a project.

- *Satisfiers.* Requirements that customers state they want. For example, a customer may want a moon roof. Typically, these requirements are options, but fulfilling them creates satisfaction. If a customer asks us to deliver the project by a certain date, and we do deliver, this is a satisfier.

- *Delighters.* The presence of features that customers do not expect may lead to their excitement and delight. This is what happened to many buyers when they first met with satellite navigational systems in automobiles. Beating the aggressive deadlines set by customers may delight them.

We can define objectives of our visits to include asking the project customers to identify the dissatisfiers, satisfiers, and delighters for our project. For example, if we meet our completion date, the customer may see it as a 7 in a scale of 1 to 10 (the highest). Thus, the obvious question is, what will it take to make it a 10? The answer may well be what we need to do to produce a "delighter."

Set the Key Dates and Budget. This is where we define the classic elements of any action guide—the time line and the budget (Figure 4.3). The time line may typically include several key dates:

- The expected start and completion date of the visits
- Milestones such as the completion of the discussion guide, sample selection, final report presentation, etc.

Obviously, the intent is to harmonize the timing of the customer contacts with the overall time line of the project in a manner that helps complete the project and keep it in tune with its objectives. Establishing the budget for the visits is another act of sound management that is closely coordinated with the Work Breakdown Structure development and cost/resource estimating.

Establish Roles and Responsibilities. After defining the reasons for customer visits, we need to know who does what regarding the visits (see Figure 4.3). Several questions may help produce such information:

- Who is the decision maker who approves or disapproves the final report?
- Who will lead the visits team?

▲ Who are individuals making up the team?

▲ Who will be in front of the customers?

▲ Who will process the collected information?

▲ Who will generate the final report?

This question of who approves the report rightfully assumes that there will be a final report associated with the customer contacts. It is important to recognize who will have the final say on the report. Generally, this should be the person responsible for the project, typically the project manager. Next, the person who leads the visits team is responsible for planning, executing, and delivering the results of customer contacts. It is important to ensure that the person selected for the job has the time and skills to deliver the expected results. It can be one of the team members or even an outside facilitator.

Another issue is the visits team composition. While listing the members in the Focus Statement is a recommended approach, there is no need to address the remaining three questions in the statement. Rather, address them outside of the statement in order to select who is responsible for being in front of the customers and for other tasks such as writing the final report. Make sure that the persons selected have the necessary time and skills.

Document the Focus Statement. Using a team approach to develop the Focus Statement is a very efficient strategy. At the end of team meeting, the project manager ensures that all documentation is properly identified, saved, and distributed to all those involved.

Utilizing the Focus Statement
When to Use. The Focus Statement should be started immediately after or in concurrence with the Customer Roadmap. Perhaps the most suitable time for this is at the beginning of the project in conjunction with the development of the detailed project scope with Work Breakdown Structure (WBS). (WBS is featured in the next chapter.) Most often, this tool is used in larger projects, particularly those that have strategic value to the organization. This does not preclude its application in smaller and medium-sized projects, where it is valued as a checklist rather than as a fully developed written statement. Perhaps the most frequent users are new product development projects.

Time to Develop. It takes a couple of hours of team time to prepare the Focus Statement for a larger project. By contrast, smaller projects that use it as a checklist may spend 10 to 15 minutes in practicing the focus in their contacts with customers.

Benefits. The value that the Focus Statement creates for the users is related to the customer, team, and management participation. As for the customer, it ensures that the team collaborates to create alignment on the need, purpose, and results for their involvement in order to generate a customer-centric project scope and Work Breakdown Structure. Instead of seeing the customer input as an ancillary or trivial event, as is so often the case, the statement gives customers a premier role. Further, the Focus Statement allows the team to perform the "due diligence" of planning a set of customer contacts prior to investing the time and money in the actual set of visits. Finally, the Focus Statement also helps engage management and other stakeholders by ensuring their participation in reading the customer voice. To fully understand the value of the statement, imagine that it doesn't exist. In this case, the probability of building a project scope that deviates from customer needs, reworking the scope, and delaying the project is significantly higher.

Advantages and Disadvantages. The Focus Statement's advantages are that it is

- *Simple.* Its intuitive nature and straightforward structure give it the appeal of simplicity, a great point when it comes to selling it to new users.
- *Educational.* It helps educate team members who have not previously contacted project customers.

Disadvantages are that the statement may be

- *Cryptic.* Because of the format of this tool, there is a risk of being too cryptic — that is, of shortening information to the point of limiting communication.
- *Trivialized.* The false assumption that everything is known may preclude full participation in the development of the Focus Statement, creating the danger of reducing it to triviality.

Variations. Similar to the Focus Statement the variations are the addition of the objective or purpose statement of your project, [2] and the purpose statement table.

Customize the Focus Statement. We think of this tool as a generic one, designed to meet the needs of across-industry users. To derive a greater value from it, we need to customize it per our project needs. Several ideas are offered in the following table to help you visualize possible customizations.

Customization Action	Examples of Customization Actions
Define limits of use.	Use the full-fledged, written statement for all larger projects.
	Use as a checklist for all smaller projects.
Adapt a feature.	Adapt the scope box to include special questions for new product development projects such as:
	—Is this a replacement of a previous product/product line? —Is the intent of the new product to reduce the manufacturing cost? —Is the new product a revolutionary design? —Is the new product aimed at a new market?
Add a feature.	Add the cost code to the budget section of the statement.
	Add to the budget section the Work Breakdown Structure element to which customer visits belong.
	Add the name of the executive sponsor for customer visits.

Summary

The Focus Statement, an action guide for customer visits, is central to this section. Perhaps the most suitable time to develop the statement is at the beginning of the project in conjunction with the development of the detailed project scope with WBS. Used in both larger and smaller projects, the statement gives customers a premier role. It allows the project team to plan for a set of customer contacts prior to investing the time and money in the actual set of visits. Customizing the statement for specific project needs increases the benefits the statement provides. The presentation highlights are recapped in the following box.

Focus Statement Check

Make sure that Focus Statement is appropriately constructed. It should include the following:

- Preliminary project scope
- Purpose, objectives, and scope of visits
- Key dates and budget
- Who approves the final report
- Team for visits with responsibilities.

Sample Selection

What Is the Sample Selection?

The Sample Selection tool identifies the right individuals in the customer organization that will provide valid and usable input information for the development of the customer-centric project scope (see Figure 4.4). Conversely, the tool's purpose is to help avoid contacting individuals that do not have information or, worse yet, that provide bogus or invalid input. The Sample Selection tool should not be confused with market segmentation tools, typically provided by marketing or sales organization, and are generally "quatitative" in nature.

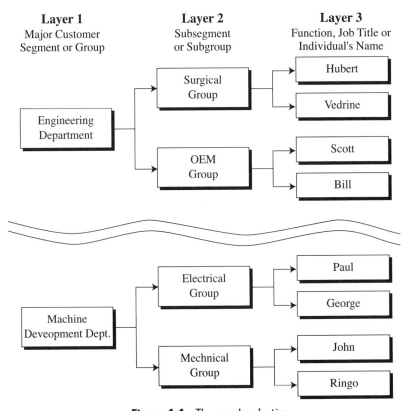

Figure 4.4 The sample selection.

Making the Sample Selection

Collect Necessary Information. Two information inputs dictate the quality of the sample selection:

- ▲ Focus Statement
- ▲ Customer organization chart

The Focus Statement's purpose, objectives, and scope of visits suggest the right target crowd to scan in order to select the right individuals who can provide customer input. Dissecting and understanding the target crowd is enabled by the information contained in the customer organization chart: organizational groupings, departments, job positions, and names.

Identify the Customer Segment or Group. The Sample Selection tool is divided into layers, with layer 1 as the highest level of segmentation. For internal customers, this could be major divisions, business units, departments, or any customer grouping that represents a major segment. For example, as shown in Figure 4.4, layer 1 includes engineering and machine development departments. They are major customer groupings in a project chartered to internally design and deploy the balanced scorecard for development projects. For external customers, it could represent customer segments per industry, business type, or geographic areas. It can also represent industries or even specific applications (for an alternative to customer visits, see box titled "Beyond Customer Visits" that follows). An example of the specific application is the high-tech fabs that may be a target customer for a new product project.

Select the Customer Sub Segment or Subgroup. Once layer 1 has been completed, selecting customer subgroupings in layer 2 is next. Here the rule is simple: All the customers in layer 2 must be a subset of those in layer 1. Refer again to Figure 4.4. The engineering department is divided into two groups: surgical and OEM (original equipment manufacturers). These two subgroupings work on projects that differ in almost any dimension— nature of projects, types of customers, life cycle, design techniques, and so forth. As a result, they have different needs with regard to the balanced scorecard, which the team in our example project has to internalize.

Choose Individuals. With layer 2 we have finished dealing with groupings. Now, in choosing layer 3 contacts, we face individuals—persons from the layer 2 groupings. They are specific people with specific functions, job titles, or, better yet, names, as indicated in Figure 4.4. This makes it simpler to track them down and contact them. At this point, we have made our Sample Selection.

Beyond Customer Visits

One disadvantage of customer visits is that they occur within a relatively short period of time, while the project may take months if not years to meet its objectives. During this time, the needs of the customers may have changed [2]. Then, there is the need to keep the project team engaged and morale high, which can be achieved through continuous input from the project customers.

An emerging type of customer contact that some firms have adopted is called the panel of experts, also known as customer councils [2]. The experts naturally are the project customers who provide ongoing feedback at a relatively low cost. The format is rather simple. We ask a small number of our project customers to agree to be in the panel for the duration of the project or for an acceptable period of time, say, about three months. To select the panel, we use the Sample Selection form. Our project customers agree to meet on a regular basis to provide input and answer specific questions about their needs. Clearly, we should ensure that the meetings do not become a burden to our busy customers by showing flexibility. For example, hold the meetings during the lunch hour or at the convenience of our customers.

The preparations for meeting with the panel are similar to those of a customer visit. That is, we need the Focus Statement, Sample Selection, and Discussion Guide. Apparently, there are many clever ways to obtain input from the customers; many require little or no investment. All it takes is a commitment to listen to the needs of our project customers.

Utilizing the Sample Selection

When to Use. Whether smaller or larger, projects typically make a Sample Selection of their customer contacts after the objectives of the customer contacts have been drafted in the Focus Statement. While more structure and details will characterize creating of a Sample Selection in larger projects, simplicity and informality are the crux of smaller projects' approach to the selection.

Time to Use. The Sample Selection tool can be completed in as short as one hour, although additional research may be necessary. For example, obtaining the contact information for individuals may require additional time.

Benefits. The Sample Selection allows for a rational selection of customer contacts that are capable of providing usable and valid information for customizing the project scope. This prevents a tendency to contact customers that do not have the information we require, which often results in wasted energy and higher risk for the project.

Customization Action	Examples of Customization Actions
Define limits of use.	Don't use the tool to accommodate very complex market or customer segmentation schemes (e.g., segmentation with many subsegments).
Adapt a feature.	Add more pages to represent additional customer segments and provide information about the project customers.
Add a feature.	If the Sample Selection is based on geography, add a map with labels calling out the customers.
	Add a sheet with quantification on each of the segments—for example, number of customers, percent of the market, volume of sales in dollars.

Advantages and Disadvantages. The Sample Selection is simple to use and visually effective in communicating its message. However, it is not always suitable for the selection of contacts in extended and large projects because it may grow big and complex

Customizing the Sample Selection. Tailoring this tool to account for specific project situations will create more value than using its generic, across-the-industry version that we have described here. Above are several ideas about how you can go about this customization.

Summary
In this section we detailed how to determine and utilize the Sample Selection tool. This tool creates the benefit of a rational selection of customer contacts that are capable of providing usable and valid information for customizing the project scope. More structured in larger projects and more flexible in smaller projects, the Sample Selection is typically prepared after the objectives of the customer contacts have been drafted in the Focus Statement. The following box briefly summarizes what the structure of the Sample Selection needs to include.

Sample Selection Check

Verify that sample selection has the required structure. It should include the following:

Layer 1—Major customer segment or group

Layer 2—Subsegment or subgroup

Layer 3—Function, job title or individual's name

Discussion Guide

What Is the Discussion Guide?

The Discussion Guide is a documented script or logical sequence of topics for discussion with project customers (see Figure 4.5). Its purpose is to help orchestrate each customer contact to make certain that it yields valuable information. This is accomplished by addressing the right topics in the right sequence and in the right priority during each customer interview. Also, with the same set of questions for each interview, the guide ensures repeatability across all the interviewed customers.

Constructing a Discussion Guide

Once the Sample Selection has been made, the customer visits team needs to have a selection of questions or topics for discussion with the project customers. The creation and selection of the questions occurs during the constructing the Discussion Guide.

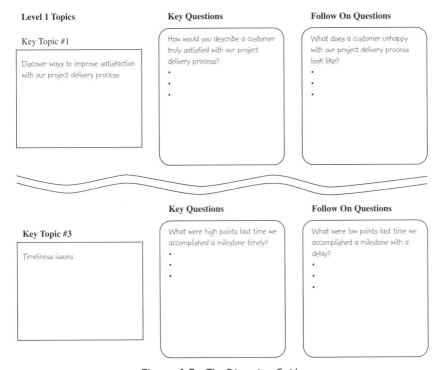

Figure 4.5 The Discussion Guide.

Collect Necessary Information. The major information inputs are the Focus Statement and the sample selection. Both will guide you selecting topics and phrasing questions for the discussion guide.

Frame the Discussion Guide. In the busy business world, odds are the customer contact may not have more than one hour for an interview. Generally, you can fit three or so topics or questions for discussion within this time slot. The philosophy here is that the objective is depth and not quantity; the idea is to get sufficient information about one topic and not small bits on many topics. Another objective is to have every interviewer use the same Discussion Guide filled with the same set of questions and in the same sequence to ensure consistency across interviews. Once all interviews have been completed, this enables you to consolidate all the information into a single report.

Identify Topics. What are key topics or important items for which you need customer input? The simplest and often most effective way to identify topics is to start with a brainstorming period, asking team members to list all the questions they believe should be discussed. Reminding the team of the reasons for contacting customers, outlined in the Focus Statement, ensures the necessary focus in the brainstorming session. When the questions have been captured, proceed to categorize them, organizing them into related topics or subjects. A popular tool to do this is the Affinity Diagram (see Chapter 8). You will likely end up with as many as seven or eight key topics.

Prioritize Topics. The next step is to prioritize the topics, not the individual ideas. Start by defining the prioritization criteria—always starting from the Focus Statement—and then use the criteria against each of the topics. The challenge now is to identify the top three topics that can fit into the framing of the discussion guide. Thus, assume that only the top three topics will be addressed. (We call them level 1 topics.) Admittedly, this will put pressure on the team, as all the topics will seem important. Some alternatives are to extend the duration of the interviews or to add more customer contacts to respond to topics four, five, six, and so on. Still, the most likely approach is three key topics for a one-hour interview.

Phrase Questions. With the three key topics identified, proceed to fill the Discussion Guide tool as in our example from Figure 4.5. Start by entering each of the three topics in the leftmost boxes. Then proceed to phrase questions that will enable the customer to address the topic under discussion. These questions are in two forms: key questions and follow-on questions. Like two sides of the same coin, they ensure that the corresponding level 1 topic is properly addressed. Together, they provide a holistic view of the issue.

Note that lots of such questions were already identified in the brainstorming session when identifying topics. Now that we know the top three topics, we may ask what question we missed in the brainstorming. This is helpful in developing a meaningful and comprehensive list. As a rule, we may have five to ten questions per key topic, including both key and follow-on ones. Finally, key questions should be entered in the middle and follow-up questions in the rightmost boxes in Figure 4.5. It is helpful to test the Discussion Guide with the team members, asking them to role-play the customer and with one team member asking all the questions in the form. Validating the choice of questions with experienced peers who are not involved in the project is also a beneficial tactic. This will help the team phrase the questions more appropriately and become familiar with the sequence of the topics.

Utilizing the Discussion Guide

When to Use. The most appropriate time to build a Discussion Guide is after the Sample Selection tool has been made and before any customer contact takes place. Logically, larger projects will strive for a more formal and structured guide, while smaller projects will demand an informal approach.

Time to Use. A team meeting is perhaps the most effective way of developing the discussion guide. This meeting should take about one hour, although for more challenging assignments, it may take up to two hours.

Benefits. Left to its own devices, a customer contact can become an exchange of ideas that are not relevant to the project. In this case, a customer interview or customer contact is not a dialogue, nor is it a discussion. That is what the Discussion Guide prevents from happening. Instead, it makes the interview a forum to enable customers to articulate their needs in their own language and address the topics of relevance (see box that follows titled "The Art of Asking Questions"). For the relevance to occur, the guide forces the team to prioritize topics and approach all customer contacts consistently, with the same questions. This allows for more productive customer contacts and the elimination of the proverbial "visit creep"; this is a colloquial term that describes conversations that move away from the topic, and end up covering subjects of little or no relevance. It may also refer to visits that take much longer than anticipated.

Advantages and Disadvantages. Designed for simple interviews where topics are relatively clearly defined, the Discussion Guide offers an advantage of simplicity. Its downside is in its inability to cover the note-taking aspect of the interview.

Variations. Several variations of Discussion Guides are available. One of them is conceived as a simple list of the topics and subtopics to be discussed, appearing like a script [2]. Market research organizations have their own formats, typically called topic guides.

The Art of Asking Questions

Many potential pitfalls stem from how we ask questions in customer interviews. Following are a few tips on the kind of questions to avoid:

- Don't let personal biases get into the questions.
- Don't use "leading the witness" questions. This is a term from the legal world where questions are asked that lead the customer to give a desired answer.
- Don't ask restrictive questions. These stifle the customer's ability to answer in full.

Three types of questions that are particularly useful for customer contacts include the following:

- *Open-ended questions.* These questions enable customers to articulate their needs. For example, "What are the top three problems that you face in our project delivery process?" As we can see, the customers are not constrained and are free to list the problems based on their perception and priorities.
- *Visualizing questions.* These are helpful in getting customers to visualize what is possible. For example, "What if your computer could also notify you that your project is exceeding the schedule and resource threshold of variance?" While this capability may not be readily available in a computer, it helps customers think in innovative ways.
- *Reversing questions.* These involve answering a question with another question. For example, if our customer says, "What technology will you use in your new project?" we respond with "What would the technology need to do to meet your needs?"

Customize the Discussion Guide. This generic guide will certainly create a significant value for the user, although we believe customizing the guide to match one's project situation can generate more value. Several ideas are provided in the following table to help you visualize the effort of customizing.

Customization Action	*Examples of Customization Actions*
Define limits of use.	Use discussion guides for face-to-face conversation (they are not a survey to be filled by the project customers).
Adapt a feature.	Add more questions under the main topics (if longer interviews are approved).
Add a feature.	Add more pages to help in the note-taking task.
	Include a diagram or a photograph to help the customer better visualize the question.

Summary

Covered in this section was the Discussion Guide, a script of topics for discussion with project customers. The most appropriate time to build a Discussion Guide is after the Sample Selection tool has been made and before any customer contact takes place. In so doing, the guide makes the interview a forum for customers to articulate their needs in their own language and address the topics of relevance. Consequently, the guide forces the team to prioritize topics and approach all customer contacts consistently, with same questions. The value of the guide can further be enhanced through its customization to specific project needs. In the box that follows, we summarize briefly how to structure the discussion guide.

Quality Function Deployment (QFD)
What Is QFD?

QFD is a customer-driven planning tool that builds the voice of the customer into the project. The intent is to ensure that customer requirements are integrated into each piece of the project—from the scope definition process to other project planning processes to project controlling processes to closing processes—all geared to produce the project product. For this to be possible, QFD focuses on identifying customer requirements and translating them into technical language related to the project delivery. When this is successfully done, every decision in the project management process helps meet the expressed requirements of project customers.

Constructing a QFD

QFD uses a set of matrices to relate the voice of the customer (i.e., customer requirements) to project requirements (see Figure 4.6). The figure makes obvious why QFD is often called the "House of Quality." The steps in building the House of Quality are as follows [12]:

- ▲ Prepare customer requirements
- ▲ Identify project requirements.
- ▲ Link the customer requirements with the project requirements.
- ▲ Benchmark.
- ▲ Develop target values.
- ▲ Stop or continue.

Discussion Guide Check

Verify that Discussion Guide is appropriately structured. It should contain the following elements:

- Three level 1 topics
- Key questions to address the topics
- Follow-on questions to address the topics.

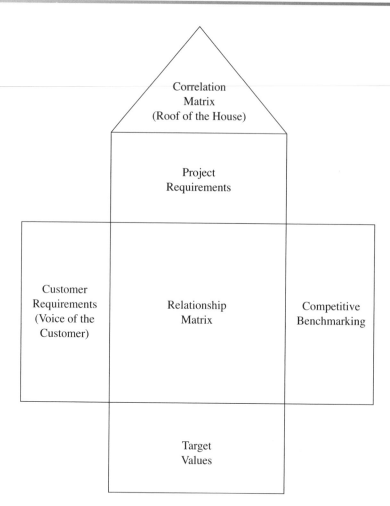

Figure 4.6 The House of Quality.

Building the House of Quality through these steps may come across as a very complex procedure, typically reserved for large projects. Actually, we hold the opposite view. It should be a simple tool: fast to deploy, formally or informally, with the purpose of serving primarily smaller projects that constitute the majority in the universe of projects. This is why our example is from a small project that took several months and some 200 person-hours to complete. The scope of the project included the development of the standard PM methodology document for a business unit in a technology corporation.

Prepare Customer Requirements. The voice of the customer is the primary input to building the House of Quality. Generally, this is the most critical and difficult step because it has to capture the essence of what the customers want. In an ideal case, the preferred scenario to identify customer requirements would follow the process of using the tools we have discussed so far: Customer Roadmap, Focus Statement, sample selection, and discussion guide. Their use would result in obtaining clear information concerning customer needs. In House of Quality terminology, customer requirements are also called the "whats"—meaning, "What do customers want?" or, if they represent prioritized wants, *value drivers.*

Informally, our example stuck with this process and identified the top five requirements (see Figure 4.7). In a nutshell, the customers—a group of seven managers and project managers from the business unit—wanted a short document that reflected their company culture, was clear to the target user, and related to the workgroup level of projects. Also, the methodology had to be based on the company's project life cycle.

These five requirements were extracted from a longer list of requirements that were prioritized by the sample of interviewees. The idea was to use a small number of requirements so that the House of Quality is simple and not time-consuming. A number of companies using the House of Quality follow this logic, limiting the list of "whats" to fewer than ten. The risk of having too many "whats" can be visualized from the "2500-Cell Nightmare" box that follows.

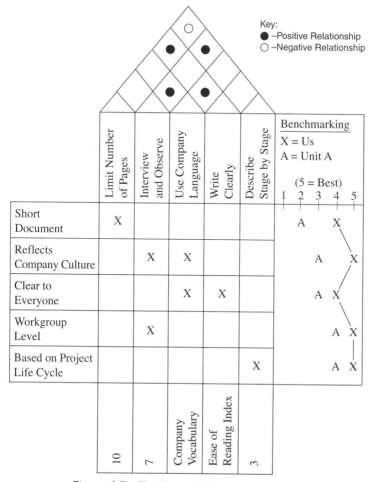

Figure 4.7 The House of Quality for a project.

2500-Cell Nightmare

Pat: What's happened with your QFD? It looks like it's far from being completed.

Jim: My team quit, and I resigned as the project manager. That's why it is incomplete.

Pat: All your team members have used QFD before, right? And then they quit. It doesn't make sense.

Jim: They just stopped coming to our meetings to develop QFD. They gave me all kind of excuses, but I know they believe it's too much work for nothing.

Pat: Too much work for nothing?

Jim: We spent 16 hours so far. We need perhaps 24 hours more. They say our QFD is too big, and, as such, it is a nightmare to use it. Plus, it takes too much time to build it.

Pat: Jim, it is a 50 (whats) X 50 (hows) cells QFD. Frankly, it is so cumbersome it is almost impossible to use in a beneficial way. And, yes, they are busy people who cannot afford to spend a whole week constructing a QFD. Why don't you reduce it?

Jim: I wanted to. I told them. But they didn't want to consider QFD anymore. That's when I resigned.

This is a case from the real world, where a misuse of QFD turned people against the project manager. The moral to the story: QFD is a great tool only when used properly.

Identify Project Requirements. Customer requirements are expressed in the language of customers. This language has to be translated into the language of the project team planning and implementing the project. Such translation is a list of project requirements. The key is that project requirements are measurable, since their output needs to be validated against the target values identified in the step "Develop Target Values."

Put it differently, project requirements are "hows" by which the project team acts to make "whats" happen.

In our example in Figure 4.7, there are a number of project requirements to consider. After a quick brainstorming, ten "hows" were listed and prioritized. The top five made it in the House of Quality. They are as follows:

- ▲ Limit the number of pages of the methodology document.
- ▲ Conduct interviews with project and other managers, observe their work, and review their project documents.
- ▲ Use company terminology when writing the methodology.
- ▲ Write the methodology document in a clear and readable fashion.
- ▲ Describe methodology stage by stage in their project life cycle.

The roof of the House of Quality deals with the correlation between any pair of project requirements. Various symbols may be used to describe the correlation. For simplicity, we use ● to denote a positive relationship and ○ for negative relationship. For example, through conducting interviews and observing, we learn about company language, which helps us in using company terminology when writing the methodology. Apparently, there is a positive relationship in this pair of "hows." By contrast, in describing the methodology stage by stage, there will be some repetition. For example, each stage will use progress reviews. Consequently, describing the progress review in each stage will lengthen the methodology, which is in conflict with our "how" of limiting the number of pages in the methodology. Apparently, the two "hows" have a negative relationship.

Why do we analyze the relationships and build the roof? The reason is that these relationships indicate how one change of project requirements affects others, as well as their trade-offs. We have seen this in the relationship of the two "hows" of limiting the number of pages and conducting interviews. Building the roof, then, enables us to view project requirements collectively rather than individually [12].

Link the Customer Requirements with the Project Requirements. The heart of the House of Quality is the relationship matrix. It uses a similar manner as used in roof of the house to indicate whether the "hows" (project requirements) adequately address the "whats" (customer requirements). Driven by the need for simplicity and lack of time, we use X, a simple way to show that there is a strong relationship between a "what" and a "how" (see Figure 4.7). For example, using company language will strongly help with both to reflect the company culture and to make the methodology clear.

The lack of strong relationship between a "what" and a "how" indicates that the customer requirement is not addressed, or that the project will have a problem meeting the requirement. On the other hand, if a "how" does not support any "what," chances are it is a redundancy. Most of QFDs applied in larger projects tend to use more sophisticated types of relationships. Instead of our simple X, they may use symbols for strong, medium, and weak relationships. These may be quantified, and when multiplied by the level of priority, they help us obtain importance weightings for each "how" [1].

Benchmark. This step takes care of two things: it assigns an importance rating for each customer requirement and benchmarks our project against others. We haven't rated the importance of "whats" in our example. Simply, we found them to be equally important, a valid choice for a small project coping with scarce resources. A larger project would benefit from having the ratings. In particular, they enable us to focus on requirements that customers value most.

Benchmarking highlights strengths and weaknesses against other existing or competitive projects. Since our project is internal to the business unit, it doesn't compete against other similar projects. However, management is expected to ask how our methodology stacks up against the methodology of the business unit A, which is viewed companywide as being the most mature in project management. The benchmarking shows, for example, that our methodology is much shorter. Also, it better reflects the company culture. In a technology company that highly appreciates short documents and its own cultural values, these are two major improvements over the unit A methodology. Then what is the purpose of benchmarking? It is to discover opportunities for improvement. This is, for example, very important in product development when facing very strong products of competitors.

Develop Target Values. Earlier we mentioned that it is of key importance for project requirements to be measurable. We come to address that in this step. Typically, a company performs in-house deliberations (or testing) and translates project requirements into target values. In our example in Figure 4.7, our target value for the "how" of limiting the number of pages for the methodology is ten pages. Simply our belief is that if the methodology is longer, our time-strapped project managers won't use it. Similarly, we rely on our experience in determining that interviews with seven managers and project managers will be sufficient to extract necessary information in a tight schedule. For writing clearly, we set our target at 45 percent Flesch-Kinkaid ease-of-reading index. In case of projects catering to external customers, the House of Quality often evaluates project requirements of competitive projects. That helps them establish target values for their project's "hows."

Stop or Continue. Many projects use only one House of Quality. This helps them build customer requirements into the project planning process. Going beyond this step are usually larger projects that see value in carrying the voice of the customer throughout the project implementation process. For this they build three more houses of quality (see Figure 4.8). In this process, "hows" from first house become "whats" of the second; "hows" of the second become "whats" of the third, and so on. The idea is to relate customer requirements to successively lower-level chunks of project work, all the way to the operational level. In a sense, this is the same logic we apply in Work Breakdown Structuring: get to the level where the work gets done—the work package level. Such an approach really ensures that customer voice is built into project basic building blocks. Once one or four houses of quality are finished, reviewing and refining them is a useful last step.

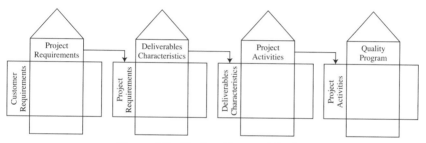

Figure 4.8 The four Houses of Quality.

Utilizing QFD

When to Use. It is not an overstatement to say that every project, large or small, would benefit from developing the House of Quality early in the project as part of the project scoping and planning. Such application is most meaningful when the development of the House of Quality follows a few guidelines. First, use a smaller House of Quality, with one level, because smaller houses tend to be more effective than the large ones. Second, build the House of Quality with the involvement of the team. Third, make sure the "whats" really come from customers. Finally, invest in training people on how to construct a House of Quality and mandate its use in each project. A disciplined emphasis on these guidelines can help build a customer-driven culture that looks at the House of Quality as a way of life.

Time to Use. The time necessary to complete a House of Quality depends on its size and the number of people involved. A small House of Quality such as one from our example in Figure 4.7, which involves three people, took about 45 minutes. Five or six project team members can easily spend five or six hours each to build a larger House of Quality, say, ten ("whats") times ten ("hows").

Benefits. There is strategic as well as tactical value in applying the House of Quality to projects. At the strategic level, the House of Quality offers a way out of the traditional narrow focus on project results that are measured at the project completion. This new way with the House of Quality concentrates on creating the process of how to accomplish these results, ensuring that the voice of the customer drives the project, rather than the desires and opinions of the project members. Tactically, such a new approach obliterates the tradition of the project planning process in which project teams assume to know what customers want, scoping and defining projects that will be later rescoped and redefined to meet actual customers needs. That waste of rescoping and redefining is gone when the House of

Quality is applied, because it necessitates the customer requirements to be identified first and serve as a starting point in project planning. This also provides a common forum for dialogue between functional groups involved in the project.

Advantages and Disadvantages. The House of Quality' advantages are in its

▲ *Conceptual simplicity.* The House of Quality is an easy-to-understand tool because its foundations are simple, logical, and grounded in common sense.

▲ *Visual power.* At a glance, you can fathom the message of the House of Quality, which makes it a useful communication tool to keep the project team in alignment.

Major disadvantages are related to the tendency of some project teams to grow a large House of Quality, which makes it

▲ *Cumbersome.* If the list of customer needs exceeds ten, the House of Quality may become difficult to follow and tedious.

▲ *Time-consuming.* Some team members may perceive it as too much work at the front end of the project.

Variations. Plenty of matrices are used in the industry to manage the process of translating customer requirements into the project scope. Names for them vary from customer matrix to decision-making matrix.

Customize the House of Quality. Our goal in describing the House of Quality was to provide details on how to build and utilize a generic House of Quality that can benefit projects across industries. We can get more value out of this tool if we tailor it to the needs of our own projects. Following are several ideas to point you in the proper direction for customization.

Customization Action	Examples of Customization Actions
Define limits of use.	Use the House of Quality in all large projects regardless of project type, either one- or four-level format will be suitable.
	Use one-level House of Quality in smaller projects, with maximum of five "whats" and five "hows."
Adapt a feature.	Compare target values with those of competitors (for product and software development projects).
Add a feature.	Generate a large-format (wall-size) version of the House of Quality for everyone in the team to view.

> **QFD (House of Quality) Check**
> Check to make sure that the House of Quality is appropriately structured. It should include the following:
> - Customer requirements (voice of the customer, "whats")
> - Project requirements ("hows")
> - Roof of the house (correlation matrix)
> - Relationship matrix
> - Competitive benchmarking
> - Target values.

Summary

In this section we presented QFD (the House of Quality), a tool that helps you build customer voice in the project scope. Large or small, each project benefits from identifying the customer requirements in QFD, serving as a starting point in project planning. When that is the case, QFD ensures that the voice of the customer drives the project, rather than the desires and opinions of the project members. This also provides a common forum for dialogue between functional groups involved in the project. Customizing QFD to account for specific project needs further augments its value. A summary of QFD structuring is given in the box above.

Concluding Remarks

Five tools that we covered in this chapter—Customer Roadmap, Focus Statement, Sample Selection, Discussion Guide, and QFD—are selected as a set of complementary tools, often referred to as customer intimacy tools. Together they can help you listen to the voice of the customer and then build it into the project product and process (see the summary comparison that follows). Although their primary purpose is to be deployed as a toolset, it is not unusual to see them employed individually.

Used individually, each tool serves a specific purpose. The Customer Roadmap provides a systematic approach to planning and organizing the voice-of-the-customer process. The reason for customer visits and what information to collect are at the heart of the Focus Statement. The Sample Selection tool identifies the right individuals in the customer organization that will provide valid information. The Discussion Guide allows you to document the script or logical sequence of discussion with customers. Finally, once the voice of the customer is heard, QFD helps incorporate that voice into the project. Overall, the use of the tools is formal and documented in large projects and informal in smaller projects.

Summary Comparison of Voice-of-the-Customer Tools

Situation	Favoring Customer Roadmap	Favoring Focus Statement	Favoring Sample Selection	Favoring Discussion Guide	Favoring QFD
Provide methodology for voice-of-the-customer process		√			
Provide focus in visiting the customer		√			
Know which individual customers to visit			√		
Document script for discussions with customers				√	
Build voice of customer into project product and process					√
Small and simple projects	√	√	√	√	√
Large and complex projects	√	√	√	√	√
Need to use the tool informally	√	√	√	√	√
Need to use the tool formally	√	√	√	√	√

References

1. Shillito, M. L. 2001. *Voice of the Customer.* Boca Raton, Fla.: St. Lucie Press.

2. McQuarrie, E. F. 1998. *Customer Visits.* Vol. 2. Thousand Oaks, Ca.: Sage Publications.

3. Goetsch, D. L. and S. B. Davis. 2000. *Introduction to Total Quality.* 3d ed. Upper Saddle River, NJ: Prentice Hall.

4. Hammer, M. and J. Champy. 1993. *Reengineering the Corporation.* New York: Harper Business.

5. McKenna, R. 1995. "Real-Time Marketing." *Harvard Business Review* 73(4): 87–95.

6. Scholtes, P. R. 1996. *The Team Handbook.* 2d ed.Madison,Wis.: Joiner Associates.

7. University of Michigan Business School and American Society for Quality. 1998. *American Customer Satisfaction Index: 1994–1998.* Ann Arbor: University of Michigan Press.

8. Thompson, A. T. and A. J. Strickland. 1995. *Crafting and Implementing Strategy.* Boston: Irwin.

9. Hoch, D. J., C. R. Roeding, G. Purkert, and K.S. Lindner. 2000. *The Secrets of Software Success.* Boston, Harvard Business School Press.

10. Norman, D. A. 1998. *The Invisible Computer.* Cambridge, Mass.: The MIT Press.

11. Shillito, M. L. 1994. *Advanced QFD.* New York: John Wiley & Sons.

12. Evans, R. J. and M. W. Lindsay. 1999. *The Management and Control of Quality.* St. Paul, Minn.: South-Western College Publishing.

Scope Planning

Be angry when you will, it shall have scope.

William Shakespeare

Major topics in this chapter are schedule development tools:

- ▲ Project Charter
- ▲ Project SWOT Analysis
- ▲ Scope Statement
- ▲ Work Breakdown Structure

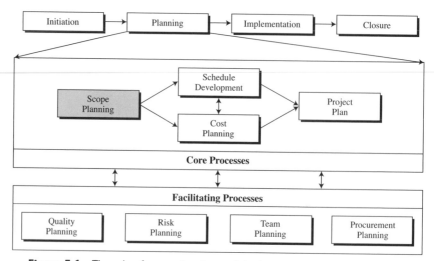

Figure 5.1 The role of scope planning tools in the project management process.

These tools aim at determining what work you need to do in a project in order to complete it successfully (see Figure 5.1). In particular, they form the baseline for further schedule and cost planning. Coordinated with tools of organizational, quality, and risk planning, they eventually lead to the project plan development. Later in the implementation, the scope baseline will be a vital foundation for the disciplined scope control and changes, a great barrier to scope creep.

This chapter prepares practicing and prospective project managers to achieve the following:

▲ Become familiar with various scope planning tools
▲ Select scope planning tools that match their project needs
▲ Customize the tools of their choice

These are prerequisite skills in successful project planning and building a standardized PM process.

Project Charter

What Is the Project Charter?

A Project Charter is a tool that formally authorizes a project (see Figure 5.2) [5]. Typically issued by a manager external to the project, it equips the project manager with the authority to deploy organizational resources in the project. This is especially important in environments where project managers have no direct authority over project team members and other resources but bear the responsibility for delivery of the project. In such a situation, for the charter to be effective, the issuing manager has to be on a level that has the control over the resources.

PROJECT CHARTER

Project Name: Metrics _____   **Date:** Feb. 13, 02 _____

Mission: Deploy new metrics for new product development (NPD) projects.

Business Purpose: Improve customer satisfaction regarding project timeliness to 4.5 / 5.0 within a year after the system launch (November 1, 2003).

Project Goals:
- Complete by November 1, 2002
- Budget: 600 hours
- Quality: Per specs
- Customer satisfaction: Level 4.0

Project Team Members: Berry Chilock (Project Manager), Chuck Ulster, Doug Kong, Doug Spitfire, Angi White

Project Sponsor: Jan Pallow, Vice President for NPD

Major Milestones	Timing	Resources (hours)
Needs Analyzed	April 1, 2002	80
Design Approved	July 1, 2002	300
Test Complete	October 1, 2002	100
System Launched	November 1, 2002	120

Vice President, NPD	Vice President, Marketing
Vice President, Finance & Administration	Vice President, Manufacturing

Figure 5.2 An example of the Project Charter.

Developing a Project Charter

When comparing the type of information described in the charter and Scope Statement, you will note a lot of similarities. Both contain the same elements—business purpose, project goals, and milestones, for example. Where these elements differ is their level of detail. More precisely, because it is an authorization tool, the charter tends to include fewer details, giving a mandate to the project team to proceed with a detailed planning loop, part of which is the Scope Statement. Naturally, then, the Scope Statement has more details about these elements than the charter. Therefore, we chose to describe the elements in more details in the Scope Statement section.

Collect Information Inputs. Issuing a Project Charter is a major decision because it commits resources to support organizational goals. For that reason, organizations tend to invest in generating information that would help make educated charter decisions. Crucial pieces of such information include the following:

- ▲ Strategic and tactical plans
- ▲ Voice of the customer
- ▲ Project proposal
- ▲ Project selection process.

That projects are vehicles for the delivery of organizational needs and goals cannot be overstated. Understanding which of the goals a project supports is therefore of crucial importance. This is typically described either in the organization's strategic plans for large projects or in the tactical plans when it comes to smaller projects. For a project to be successful, the voice of the customer needs to be heard, understood, and responded to. Also, to properly assess the viability of a project, you need to develop a project proposal or feasibility study. When this information is available, project selection criteria can be applied to rate, rank, and select projects (see Chapter 2). To issue a charter, then, all these information inputs are necessary.

Define Project Mission. Precision and clarity are two key words in the charter's definition of what the project should accomplish [2, 7], as indicated in Figure 5.2. Whether the charter is for a small product modification or a multibillion semiconductor fab, a few words can usually do the trick. The statement may identify major tasks such as design, prototyping, and programming, or it can be as simple and directive as "develop a new product platform."

To express the accomplishment expected of the project, we use the term *project mission*. *Project mission* has an aura of significance, which may be why it is often used in major projects. Alternative terms, such as project task or assignment, have less gravitas, but may nonetheless be appropriate. The selection of the term is often dictated by organizational lingo.

Identify Business Purpose. What drives the implementation of this project? Is its purpose to increase customer satisfaction (as is in the example in Figure 5.2), enabling their easier acquisition and retention and eventually leading to more repeat business that tends to be more profitable as well? Or you might be trying to crack a new market, increase market share, develop new competencies, and so on. On a strategic level, there may be several different reasons for the existence of the project. The point is, we should know it and spell it out.

Stretch Goals? Or Not?

How attainable should the project goals in the charter be? Is it okay to write charters using a stretch goal? Empirical evidence suggests that those who set stretch goals—that is, outline goals that are typically unattainable—outperform those with routine goals, which are typically attainable. If you are purely driven by performance, the choice is clear: go for stretch goals. Many of Intel's project managers do this because the corporate culture drives this behavior. What happens when they do not attain their stretch goals? According to one of them, "No managers around here will use this to catch people out. The idea is to always strive for more and do your best. If you do so, you won't have any problems when missing the stretch goals." Is this the case in all companies? "If you try stretch goals and miss, that is going to be held against you in performance evaluations. That's why everybody goes for routine goals in our company," says a project manager from a traditional company. The point, then, is that project managers tend to go with the flow.

Once we are out of the strategic territory and in the tactical world of small projects, many project teams struggle with what exactly is their project's purpose. They assume it is simply to put out the project's product. However, it is not. Like any other project, your small project exists to accomplish some tactical gains supporting your organization's business. When you are buying a piece of new equipment, the business purpose is not to buy and install the new equipment. Rather, the purpose may be to remove a bottleneck and increase the manufacturing line capacity, a coveted prize of manufacturing managers. Similarly, a project developing a standardized PM process is probably aiming at improving timeliness, repeatability, capacity, and so forth. Certainly, its business purpose is not to have the process itself.

Define Project Goals. By their nature, "project mission" and "business purpose" are broadly defined. To provide more specific guidelines to the project team, the charter needs to identify specific project goals (see the box on page 128, "Stretch Goals? Or Not?"). As a minimum, these usually include schedule, cost, and quality targets. Your schedule target is your desired project completion date, which in Figure 5.2 is November 1, 2002. To hit the date, we want to spend no more than 600 resource hours, aiming for the level of quality defined in the specifications. For example, one quality element of the specifications relates to the executive dashboard, a monthly report for executives about the performance of current 20+ NPD projects. The element requires that the report take no more than three minutes to be read and interpreted by an executive. Defining more than the three goals is the accepted practice, as shown in Figure 5.2 where a customer satisfaction goal is specified.

Appoint Project Team and Sponsor. One of the purposes of issuing a charter is to formally announce the names of the project managers and, possibly, team members. However, it is not important that all team members be immediately identified. The expectation here is that functional managers will nominate those members after the charter is issued.

In some organizations, the use of project sponsors is a regular practice for major projects. Sponsors provide guidance for the project team, making sure that the functional managers fulfill their resource commitments to projects and serve as a communication link with customers [8]. Typically, the sponsor is a senior manager who may be sponsoring several concurrent projects. In the case of less important projects, the role of a sponsor may go to a middle-level manager. Whatever the level of the project sponsor, issuing the charter is a convenient way to visibly announce the name of the sponsor. Still, some organizations do not use sponsors.

Specify Milestones and Resources. Major milestones include completion of certain deliverables by a specified date and are typically requested by those issuing the charter. The key word here is *major*; you should limit the number of major milestones to those that are absolutely vital. Specifying the number to three to five vital ones is a dominant practice. In other words, given the charter's purpose and the related level of detail, developing a long list of milestones is unnecessary. In addition, cost conscious organizations have a tendency to identify specifics of resource or cost budget for the major milestones. For example, 400 resource hours may be needed for the milestone called "Requirements definition complete."

Inform Resource Providers. All functional groups or departments in the organization that will be supporting the project need to be informed correctly and promptly about the start of the project [9]. For that reason, you need to put them on the distribution list for the charter. Why do the groups need to receive a copy of the charter? First, for some of them—the engineering department, for example—the charter is a signal to start the work. For others, receiving the charter means that your project is in existence and that you demand their support. For example, human resources may need to hire database programmers for your project or the information technology group may need to support your new intranet-based software for distributed teams.

In developing the distribution list, you may face some minor dilemmas. Should you go with a standard distribution list or not? Since standardization is a great timesaving strategy for repetitive activities, and issuing a charter is a repetitive activity, the choice is clear: standardize the distribution list. Also, is using job titles in the list more effective than using persons' names? While this may depend on the organizational culture, some organizations prefer job titles, since they are less likely to change.

Refer to the Supporting Detail. What is immediately evident in a Project Charter is the decision about bringing a project to life, stated in a laconic manner. What is not evident is the process that led to the decision. The decision was a result of the process of project selection that was based on information developed in strategic and tactical plans, the voice of the customer, the project proposal, and project selection methods. To make this visible and give credibility to the charter, refer to these documents in the charter.

Evaluate and Fine-Tune the Charter. You should check the elements of the Project Charter for completeness, making sure they are in tune with the information inputs and the information that your chosen format contains. Are the dates for milestones identical to those from the project proposal that served as a basis? Is that date consistent with the strategic and tactical plans, which identified the framework for this project? You should also ensure the consistency of other information in the charter; for example, does the distribution list include all functional groups from which team members come? Checks of this type and consequential changes to the charter should not be ignored.

Utilizing the Project Charter
When to Use. The Project Charter has been used in large projects since the beginning of formal project management in early 1950s. Because large projects engage substantial organizational resources originating in different

functional groups, this approach is quite logical.. For the same reason—resources that derive from various functional groups—the charter is popular with small, cross-functional projects as well. However, for other small projects that are not cross-functional, issuing a Project Charter is an infrequent practice—unless functional department members are not collocated, a growing phenomenon in our virtual world. For an example of the use of the project charter in different corporate situations, see the box that follows, "The Need for the Charter."

Time to Use. The Project Charter is no more than packing information that is already created in activities preceding it: strategic planning, project selection process, and signing up team members, for example. Therefore, it may take a skilled, smaller team little time, perhaps 30 to 60 minutes, to develop it. As the project grows in size and complexity, and when more is at stake, Project Chartering is likely to command more time.

Benefits. Projects often require organizational arrangements that span functional boundaries. Considering that in such cross-functional designs functional managers "own" resources, the charter is a practical way of telling functional groups to provide such resources to the project manager. To make it even clearer, many companies list functional groups (or their managers), providing resources and the names of the project team members. This practically defines specific resources, the amount and time of their use in the project, and who is responsible to provide them. Aside from this act of organizational legitimacy, the charter also helps a project get visibility by announcing its start and purpose, leaving the ball in the project manager's court.

The Need for the Charter

Do you need a charter for all projects? It takes a leading truck company months of work to issue a charter for the new truck development project. With millions of dollars involved, the company develops multiple scenarios of scope, cost, and time line, and evaluates them carefully before launching the effort. The launch begins with the issuance of a detailed charter, where typically the sponsor is a top vice president. In contrast, a major information technology upgrade project typically starts with a sentence-long charter, e-mailed to the functional mangers providing resources. No sponsor is identified, and not much of the charter-preceding planning is done. The rule for these projects is that any major project consuming major resources (over $10K) must issue a charter. Charters are not used for projects below $10K, usually performed within a functional group. Reason? It is considered an unnecessary paperwork step. This is a good example to review when deciding whether you need a charter for all projects. The need for the charter should be matched with the size, complexity, and the degree of the cross-functional nature of the project.

Project Charter Check

Check to make sure you developed a proper Project Charter. The charter should

- Be based on information inputs from strategic or tactical plans, voice of the customer, project proposal, and project selection process.
- Include all elements defined by its format.
- Provide consistency among the elements.

Advantages and Disadvantages. The Project Charter's major advantage is in its

- ▲ *Clarity.* With its typically laconic language, the charter cuts to the chase, mandating accomplishment of the project mission.
- ▲ *Simplicity.* The charter is focused on a few major elements of the mission, leaving out details and eliminating the possibility of blurring its message.

The Project Charter's major disadvantage is in its

- ▲ *Potential to be easily and frequently misused.* Rather than blaming this on its design, we see it as a flaw in basing the charter on an "order of magnitude" type of information. Specifying the mission, major deadlines, and necessary resources requires a significant amount of planning. Naturally, this is done in large projects, where a detailed proposal, business plan, or feasibility study is developed as the foundation of the charter. In contrast, charters for smaller projects are deprived of such serious planning. Rather, they are based on very rough resource estimates, usually called "order of magnitude estimates," which nominally signify a large tolerance for error. As a consequence, such charters are imprecise, error-prone, and not taken seriously, jeopardizing the very projects they bring to life.

Customization Action	Examples of Customization Actions
Define limits of use.	Use a Project Charter in major, cross-functional projects. Define what constitutes a major project.
	Use a sponsor only for corporate priority projects.
	Include major milestones, a rough estimate of resource hours.
	Review the charter in the project kickoff meeting.
Add a new feature.	Add information about major risks and risk responses (favored by high-tech projects).
Modify a feature.	Instead of resource estimates for major milestones, combine deliverables with milestones (favored by high-tech projects).

Variations. The practice of project chartering exhibits many variations and nuances, including its name, content, pattern of use, and formality. For instance, some organizations call it the "project authorization notice," others the "project birth certificate." In all cases, the charter is meant to bring a project into existence. As for the content, some organizations use charters that include specifics about budget and schedule for major milestones, as shown in Figure 5.2. Others, especially in smaller projects, find it sufficient to announce the purpose of the project, the start of the project, and the team composition.

Usually charters are issued once in the life of a project. Still, some organizations might find it beneficial to start with preliminary charters. Such charters authorize resources for the planning stage of the project, and when the planning is finished and detailed information about resource and other requirements is available, a new final charter is produced and disseminated. Typically, such charters are made formal, further enforcing their authorization message. Unlike the formality of these preliminary and final charters, charters in the fast-paced and distributed project environments are often reduced to informal e-mail messages.

Customize the Project Charter. Charters come in all sizes and shapes. Which one is for you? You should certainly be able to adapt the example from Figure 5.2 to fit needs of your projects. On page 132 are a few ideas to help you envision how to adapt it.

Summary
This section focused on the Project Charter, a tool that formally authorizes a project. Applied in both large and small cross-functional projects, the charter is a practical way of telling functional groups to provide such resources to the project manager. Aside from this act of organizational legitimacy, the charter also helps a project get visibility by announcing its start and purpose, leaving the ball in the project manager's court. Following are key points in the charter development.

Project SWOT Analysis
What Is Project SWOT Analysis?

This analysis is an extension of the classical SWOT (Strengths, Weaknesses, Opportunities, Threats) analysis on the project level (not the company level), performed to gain an understanding of and act on the project's capabilities and environment [10]. Project capabilities—expressed as the project's internal strengths and weaknesses—tell us what our project can and cannot do well. At the same time, our assessment of the project environment indicates what opportunities and threats we are presented with by the external world of the project. Information about the environment, combined with the knowledge of the project's capabilities, enables project teams to identify critical success factors (CSFs) to meet customer requirements (see Figure 5.3). Measurement of where the project stands regarding these factors provides clues about strategic gaps, prompting the team to consider strategies and actions to address the gaps. Awareness of the gaps and a clearly defined response allow the team to formulate realistic project scope and related strategies for attaining project goals. In short, Project SWOT Analysis should underlie scope planning and how to execute a project.

Performing a Project SWOT Analysis

Collect Information Inputs. For a good start in performing the Project SWOT Analysis, two information inputs are of vital importance:

 ▲ The Project Charter with its supporting detail.
 ▲ The voice of the customer.

While the Project Charter provides knowledge about the fundamental boundaries of the project, the supporting detail (strategic and tactical plans, project selection criteria and process, and project proposal) helps determine the context in which the boundaries were drawn. Why the customer voice is so relevant for the analysis is the subject of the next step.

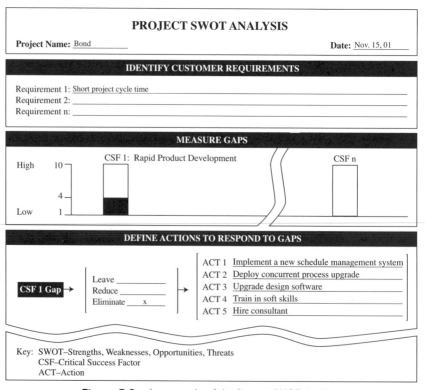

Figure 5.3 An example of the Project SWOT Analysis.

Identify Customer Requirements. Projects are implemented to help customers meet their requirements in creating value for their customers. Consequently, the voice-of-the-customer process is designed to furnish you information about the customer requirements (for more details on customer voice, see Chapter 4). When it comes to the Project SWOT Analysis, the requirements should center only on several major ones, those that can make or break the project. In our example in Figure 5.3, the customer made it clear that they were heavily involved in the time-to-market race with their competition and required that this project shave off 30 percent of what is considered a typical delivery time for this type of the project. This is a significant challenge for the implementing company and its project team, who have little experience in fast-track new product development projects. Because the company's management views this project as an opportunity to crack a new market of high-velocity clients, the project needs to be successful. But what does it take to be successful? The answer lies in CSFs.

Select CSFs. Fundamentally, CSFs are areas in which a company must do well in order to be successful [11]. So, what are these areas? The areas come from two major domains. The first one is called project capabilities and includes anything internal to the project. The other domain consists of anything surrounding the project, usually called project environment (see example in the box that follows, "How Bizarre: Even a Lettuce Vendor May Be a Critical Success Factor").

Which areas in the two domains will be CSFs is primarily dictated by customer requirements that we have identified as the voice of the customer. Begin, then, by asking what you need to do well within the project in order to meet or exceed customer requirements. Is it good design skills or a great prototype lab? In our example in Figure 5.3, rapid product development is a CSF. That is a very complex CSF, requiring the synchronization of several components, including concurrent engineering, collaborative design software, cross-functional teams with interpersonal skills, and scheduling. Of course, more may be necessary for rapid product development, but these four are good examples.

How Bizarre: Even a Lettuce Vendor May Be a Critical Success Factor
You should always consider the relationship your business has with local communities. In a fast-track project transferring manufacturing technology from Europe to a Middle Eastern country, this was not identified as a CSF in the project SWOT analysis. Just after the beginning of the project, the contractor's project manager got a call from a vendor claiming that all roads leading to his office were blocked by groups of violent protesters. As a result, some important computer equipment couldn't be delivered. A quick check proved this call correct. The people were local butter-head lettuce farmers, unhappy that the foreign contractor was buying the lettuce from Europe and not from them. The siege went on for days, and the deliveries were delayed, impacting the project schedule, and the protest was tacitly supported by the local authorities. Finally, the project manager figured out his mistake: he had not considered local communities with a big stake in the project. In fact, the communities turned out to be a CSF, prompting the project manager to start buying the local lettuce [4].

Screening Your Environment for Possible Critical Success Factors and Gaps

Here is a short general checklist of areas where you can find possible critical success factors and related strategic gaps [1] [3]:

- Stockholders
- Customers
- Governments
- Competitors
- The general public

- Creditors
- Suppliers/Vendors
- Union
- Local communities

Concurrent engineering is about overlapping project activities to speed up the project pace [12, 13]. Its crux is reciprocal dependencies between the activities [14] that exchange incomplete information [2], making it more difficult but faster to work. In the new product development context, that exchange is significantly more efficient if performed via collaborative design software, which command significant resources and skills. Rapid product development also demands agile cross-functional teams, fully equipped with soft, interpersonal skills to handle conflicts and negotiations germane to fast-track projects and the work with incomplete information [15]. On top of all of this, scheduling this project calls for the ability to juggle multiple critical paths, perhaps including 30 to 40 percent of all activities, with a constant need to use schedule crashing or fast-tracking techniques [5]. In short, we need a thorough knowledge of all components of a certain CSF.

When you are finished with project capabilities domain, look again at the customer requirements. In which areas in the project environment must you do well to meet or exceed customer requirements? You may need to include a first-tier vendor in your new product development team, because this has shown to be a proven technique to speed up projects [16]. Certainly, possibilities of external CSFs abound, but the checklist above, "Screening Your Environment for Possible Critical Success Factors and Gaps," offers a few ideas of where to look.

Again, possibilities are limitless, but note that these environment-related CSFs are more challenging to deal with because, unlike internal project capabilities, they are external to the project and less controllable. Brainstorm with the team to identify ten or so for each domain, internal and external. Rank them and focus on a few strategic ones; if the team picks too many, it may be overwhelmed. Once you have picked them, keep in mind that identifying a CSF without understanding its critical components, their dynamics, and interactions is useless.

Measure Gaps. After you have selected CSFs, the next step is measuring gaps. A gap is the difference between the ideal and actual level of a CSF. If the CSF in our example from Figure 5.3—rapid product development—were ideal, it would draw a perfect scoring on all of its four components, including concurrent engineering, collaborative design software, interpersonal skills, and scheduling. That would mean that the entire CSF and its four components are at a level ideal for meeting customer requirements. The actual level, on the other hand, is where you believe you currently stand with regard to that CSF and its components.

A measurement scale is necessary to identify the magnitude of the gaps. Depending on the degree of desired rigor and the time available for the Project SWOT Analysis, you can choose from among several alternatives. For example, smaller and simpler projects may do just fine by having a straightforward scale of small, medium, and large gaps. Other organizations categorize gaps by urgency, perhaps using colors — for instance, green for "take it easy," yellow for "caution," and red for "danger." This allows them to send an immediate visual message that identifies the gap. Consequently, the measurement of a gap would be easily performed and based on a quick group consensus.

In contrast, a perceptual scale spanning from 10 (the ideal level in Figure 5.3) to 1 (very poor performance) more precisely measures a gap. How do you identify a gap with this scale? One way is to assign a narrative description of the state of a CSF's component to each level in the scale. Then, after a team discussion, each team member assesses the actual level of a component. Further, the actual level of each component can be averaged out for the whole team. Assuming that all components are equally important, the team can average out the actual level of the CSF. For example, the actual level of rapid product development for Figure 5.3 would be 4. Comparing the actual level with the ideal one will yield the gap; in our case, a large gap of 6 units. In the most sophisticated option, a team can use the Analytic Hierarchy Process (AHP, see Chapter 2) to rank CSFs, establish their relative weights, select the top three, and then measure their gaps. Whatever gap measuring method you choose, keep in mind that measuring a gap is a subjective judgment, not an exact science.

Having a large gap on internal capabilities is an apparent weakness, while a small gap is a strength. Similarly, where we find a small gap in assessing the project environment, we can view it as an opportunity. Along the same line, an external sizable gap poses a threat. Expressing gaps as potential threats can help you better relate the idea to project sponsor, managers, and other stakeholders who may be used to more traditional SWOT analysis, but who are instrumental in getting the resources you need to address the gaps.

Decide How to Respond to Gaps. Identifying a gap brings you to another decision: what to do about it? Generally, you have three options: leave it as is, reduce the gap, or eliminate the gap [17].

As Figure 5.4 indicates, when your project has little or no gaps in CSFs in both internal capabilities and project environment (upper left-hand corner), your best option may be to leave things as is.

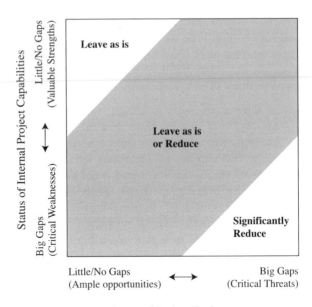

Figure 5.4 Project SWOT Analysis and strategies to act on gaps.

Why would anyone select this option? A lack of time or an insignificant impact of eliminating gaps may be reasons for opting to leave small gaps alone. On the other hand, those who decide to eliminate small gaps may find the motivation for doing so in the technical ease of closing them or in their drive for perfection. With larger gaps in either internal capabilities or project environment (the shaded area in Figure 5.4), the case may be more complex. First, acting on gaps usually requires resources, typically controlled by management. If your knowledge of the style of management and resource availability tells you that your action may not be fruitful, and therefore not worth the effort, leaving the gap as is may be understandable. However, it is often worth the case to management, especially if you use a language that management reveres: the bottom line language.

Take, for example, our case of the six-unit gap in CSF 1 in Figure 5.3. Not acting on this gap may prohibit the company's management goal of cracking a new market of high-velocity clients, resulting in loss of expected sales and profits. Given the strategic and profit significance of our case, management sees that our gap deserves an action of eliminating the gap. If it really was in the resource crunch, we might have ended up with the *leave as is* option.

Receiving an infusion of resources, though, is not an easy scenario in today's organization, where various projects often brutally compete for their share of the resource pie. Rather, it may take preparing well, presenting the case for the scarce resources, and fighting for them with a passion based on facts. Once the resources have been obtained, developing a plan to reduce the gap is important. Figure 5.4 also shows that in a situation burdened with large gaps in internal and external capabilities (lower right-hand corner), perhaps the most viable option is to act to significantly reduce gaps; eliminating them may be too challenging.

Define Specific Actions to Respond to Gaps. Reduce and eliminate are good strategic directions, requiring specific actions to deploy resources. Going back to the structure of a CSF, its components, and the actual state is helpful in identifying actions directed at gaps. In Figure 5.3, we list several actions, attacking gaps in all four components: concurrent engineering, collaborative design software, interpersonal skills, and scheduling. Some of them can start right away, for example, training on interpersonal skills and scheduling system. Others, like installing collaborative design software, will take a slower path. Because almost all of them will go in parallel with the project execution, help of an experienced consultant is enlisted. When the actions are identified and understood, they should be built into the Scope Statement and, possibly, into the Work Breakdown Structure.

Utilizing a Project SWOT Analysis

When to Use. Identifying a sound strategy for project implementation is not likely without a critical evaluation of the project's internal capabilities and surroundings, whether the project is large or small. Typically blessed with more resources than small projects, large projects should strive for a more comprehensive, systematic, and formal Project SWOT Analysis, preceding a detailed Scope Statement. Before such time, large projects should use the analysis in the project idea development, certainly, in an informal manner. That informal manner is realistic in applying the analysis in small projects. Faced with no time to be detailed, small-project managers should make the analysis part of their mental process, constantly questioning their project capabilities and surroundings. They should not be concerned that the analysis must be written; it does not have to be (see the box that follows: "10 Minutes Can Do It").

Time to Use. While a Project SWOT Analysis in a small project may take 10 to 15 minutes of informal time to a skilled project team, formal, detailed analyses of this type are known to exact hours in large projects, still with a skilled team. Not surprisingly, as project size, complexity, and team grow, so does the time to perform the analysis.

---

10 Minutes Can Do It

"When I first ran into the Project SWOT Analysis (PSA) tool, I was very pleasantly surprised and, frankly, very proud of myself. For a long time I was doing PSA in my projects without knowing there was a formal tool like that. As a manager of multiple small projects at a time, I never had time to do a formal, written PSA. Rather, I was doing it informally, verbally in ten or so minutes with my team members. We simply called it risk assessment. But it worked, and it worked very well." This story of a multiproject manager in a high-tech organization is not a solitary case. Many project managers do the same thing. Know your gaps before you venture into the project. Make them visible. Ask your manager to help reduce or close the gaps. And if it takes ten minutes, find them in your busy schedule.

Benefits. The most successful ventures are built on the ability to utilize one's strengths, moderate weaknesses, seize on the opportunities, and neutralize threats [18]. Pressured for time on many concurrent fronts, project managers all too often do not build their projects on this premise. Rather, they dive into detailed project planning without taking stock of their project's strengths, weaknesses, opportunities, threats, and related gaps. This is where the Project SWOT Analysis comes in; it takes the stock and offers a clear picture of project gaps. The value of the analysis, therefore, is in enabling the project to do the following:

▲ Position itself in a best possible way to take advantage of its particular strengths and opportunities, while minimizing weaknesses and threats.

▲ Uncover strengths that have not yet been fully utilized and in identifying weaknesses that can be corrected.

▲ Bring to the management attention significant gaps that can jeopardize the project and get their support to close the gaps and reduce the risk of failure.

▲ Have the customer focus as a focal point in identifying and closing the gaps.

Advantages and Disadvantages. Major advantages of the Project SWOT Analysis include the following:

▲ *Focus.* The analysis brings strategic gaps into the spotlight, prompting a behavior of "forewarned is forearmed."

▲ *Proactive nature.* It helps build the mentality that it is never too early to bring a defensive shield up. Facing strategic gaps early helps rehearse alternative project scenarios and prepare for major project danger zones.

As for its disadvantages, performing a Project SWOT Analysis may be

▲ *Time-consuming.* Taking time from their busy schedule to do the analysis is sometimes a challenge many project teams cannot handle.

▲ *Challenging.* Many project teams are skilled to operate in the tactical theater and horizons but find it challenging to walk the uncharted waters of strategic analysis.

▲ *Threatening.* Some teams may find it threatening to tell management that there are weaknesses in their projects. Also, some teams don't like to identify their own faults, which is where the analysis may take them.

Customization Action	Examples of Customization Actions
Define limits of use.	Use Project SWOT Analysis in all projects.
	In small projects, make it an informal tool that takes 10 to 15 minutes.
	In large projects, insist on sessions specifically dedicated to a formal Project SWOT Analysis.
Add a new feature.	Separate CSFs and pertinent gap analysis into traditional SWOT areas—strengths, weaknesses, opportunities, and strengths. This resonates well with resource providers used to the language of SWOT.
Modify a feature.	In addition to identifying specific actions, identify who is responsible for them, along with tentative deadlines (this can also be done in the detailed scope planning).

Customize Project SWOT Analysis. Because Project SWOT Analysis may be performed in many ways, you must consider which way would be of most value to you. As you ponder an answer, think of what we have covered in this section. Following are a few that may be helpful.

Summary

By positioning the project to take advantage of its particular strengths and opportunities while minimizing weaknesses and threats, Project SWOT Analysis helps identify a sound strategy for project implementation. Also, the analysis helps bring to the management's attention significant gaps that can jeopardize the project and to get their support to close the gaps and reduce the risk of failure. The summary of the analysis is highlighted in the following box.

Project SWOT Analysis Check

Check to make sure you performed a proper Project SWOT Analysis. The analysis should

- Be based on information inputs from Project Charter and the voice of the customer.
- Include strategic customer requirements.
- Identify key success factors, supporting the requirements.
- Determine strategic gaps in key success factors.
- Select a strategy to address the gaps: leave, reduce, or eliminate.
- Spell out actions to implement the strategy.
- Provide consistency among the elements.

Scope Statement

What Is a Scope Statement?

A Scope Statement is a written narrative of the goals, work, and products of the project (see Figure 5.5). Focusing on the fundamentals, the statement answers the crucial question of "What do we produce in this project?" The answer thus creates a big-picture view of what the project is all about, setting the scope baseline to follow in whatever is done in the project. In a sense, this baseline may be compared with project boundaries, mandating that stepping out of the boundaries is not allowed without the consent of approving managers. It also means that what is within these boundaries is the solution space within which the project team can operate. While there are many versions of the Scope Statement, the format described here is based on the concept that a project is a business venture.

Developing a Scope Statement

The fundamental premise in developing a Scope Statement is that the statement must be as much change-resistant as possible (see the box that follows, "Innovative Ways to Develop a Change-Resistant Scope Statement," for some of these principles). Note that this premise does not have to do with successful control of changes through change management systems such as change control plan and scope control. Rather, the premise here is very different and rooted in a set of principles learned in product development projects, which help minimize the impact of changes from the environment on your scope. These principles may help in non-product projects as well.

Collect Information Inputs. The quality of Scope Statement hinges in many ways on the quality of input information. Specifically, the following inputs carry great weight in developing a Scope Statement that has value:

- ▲ Voice of the customer
- ▲ Project Charter
- ▲ Project SWOT Analysis

Truly, a project's reason for existing is its customer's needs. Recognizing this, we reviewed earlier in the book several tools for listening to the voice of the customer with the purpose of internalizing the needs and incorporating them in the project. If you prepared any of the tools, the time is now to put them to use in developing the Scope Statement. When customer checklists are available, they should be used in conjunction with the voice-of-the-customer tools.

```
┌─────────────────────────────────────────────────────────────────────────┐
│                         SCOPE STATEMENT                                   │
│  Project Name: Heal                              Date: March 12, 02       │
├─────────────────────────────────────────────────────────────────────────┤
│  Business Purpose: Improve customer service for our products and services │
│  by 5% via deployment of a new customer relationship management software. │
├─────────────────────────────────────────────────────────────────────────┤
│  Project Goals:                                                           │
│  • Time: Finish by September 15, 2002                                     │
│  • Cost: $150,000                                                         │
│  • Quality: Per service level agreement                                   │
├─────────────────────────────────────────────────────────────────────────┤
│  Project Work Statement: Analyze workflow, configure software, develop    │
│  prototype, and release software.                                         │
├─────────────────────────────────────────────────────────────────────────┤
│  Key Deliverables:                                                        │
│  • Workflow analysis               • Configure settings                   │
│  • Prototype                       • Training                             │
│  • Release                                                                │
├─────────────────────────────────────────────────────────────────────────┤
│  Key Milestones:                                                          │
│  • Workflow analyzed by March 15, 2002   • Configure complete by April 15,│
│  • Prototype complete by August 15, 2002   2002                           │
│  • Release by September 15, 2002         • Training complete by August 15,│
│                                            2002                           │
├─────────────────────────────────────────────────────────────────────────┤
│  Major Constraints: Our key developers will not be available in June      │
│  because of their visit to our European ally.                             │
├─────────────────────────────────────────────────────────────────────────┤
│  Major Assumptions: Configure software to meet our workflow; we will not  │
│  change our workflow to meet software.                                    │
├─────────────────────────────────────────────────────────────────────────┤
│  Exclusions: This project does not include training on customer service   │
│  skills.                                                                  │
└─────────────────────────────────────────────────────────────────────────┘
```

Figure 5.5 A simple example of the Scope Statement.

Perhaps the tools were used to develop a better Project Charter as well. As a document authorizing the project, the Project Charter already looked at the business need and goals that the project was chosen to address. These are critically important to good Scope Statement with respect to its elements related to the business purpose and project goals. In addition, the charter probably includes a preliminary description of the project product, valuable information in further detailing the product, associated deliverables, milestones, and goals. Another useful piece of information is the Project SWOT Analysis. Clearly, the Scope Statement and its logic should be based on the project's SWOT (strengths, weaknesses, opportunities, and threats) that the gap analysis found. Before you delve into defining your scope, consider doing that in conjunction with WBS development, possibly the most effective way of answering and marrying the crucial questions of "What do we produce in this project?" (scope) and "How do we produce it?" (WBS).

Innovative Ways to Develop a Change-Resistant Scope Statement

The application of the following principles in scoping projects can help define projects with increased resistance to change later in the implementation:

- *Principle 1:* Reduce project complexity; go for smaller projects. Adding new work elements to a large project means the elements interact with more existing work elements than they would in a smaller project. Each new element adds complexity by increasing the number of interactions, so when a change hits, more elements need to be changed or redone. As a result, the more work elements, the more changes. In contrast, scoping projects with fewer work elements would reduce the number of interactions and increase the project's resistance to change [2]. It is easy to divide larger projects into smaller ones and scope them accordingly.

- *Principle 2:* Design robust project products. Some project products are designed to perform within a narrow range of conditions. Others, however, that are designed for a wider range of conditions are said to have a robust design. When a change in the range of conditions hits, because of the wide range, the more robust project products are more resistant and less likely to change. In contrast, even a slight change in the conditions can cause ripple-effect changes in a less robust project product. Think of the robustness and incorporate it in your scope whenever possible [6].

- *Principle 3:* Freeze the scope early. When you freeze the scope early, you reduce the likelihood of changes later in the project life cycle. Such late changes typically impact a large portion of the scope, impeding the progress and imposing delays and resource overruns. Therefore, an early freeze will enable a faster completion of the project, meaning that a shorter execution time makes it less likely that the changes in the environment can impact the scope. This circular reaction, early freeze–faster completion–fewer changes, offers you advantages in defining a change-resistant project scope.

Identify Business Purpose. Long gone are times when projects were viewed only as technical ventures whose major purpose was to provide a certain product or service. Today, in addition to producing a certain product or service, a business must support the delivery of business goals: desired profits, market share, competency building, customer satisfaction, productivity levels, and so forth. [19]. Therefore, we begin the development of the scope statement with the project's rationale—what is its business purpose? What business goals will the project accomplish? What business plans will it support? For a traditional project manager, this thinking of projects in business terms is not easy; rather, it is both challenging and demanding. But so are today's projects. The demand is for them to be faster, cheaper, and better than those of the previous year. This is so much the case that some experts believe the project business purpose should be called the "project passion statement" and answer the question: what unique and distinct value will your project create for the customer and the company's business? The German philosopher Hegel once said that nothing great in this world has ever been accomplished without passion.

Define Project Goals. As compelling and inspirational as the business purpose may be, it is still no more than a broad direction, lacking details about specific targets for the project. Those targets are defined by project goals for time, cost, and quality (see the "Project Goals" item in Figure 5.5). By specifying when you will finish the project—for example, by May 1, 2003—you set your time or schedule goal. To attain the schedule, you must determine what cost budget you need—for instance, "the budget for this fab upgrade is $400M." Projects in industries not using cost budgets may indicate the desired number of resource hours, such as 1200 resource hours. Unlike the schedule and cost/resources goals, expressing quality goals in a specific and measurable fashion may be a challenge. Since quality is about meeting or exceeding customer requirements, usually expressed by certain standards, a sound strategy is to define the project quality goal by referring to a certain standard agreed upon with the customer. For example, "this project manual will be in line with PMBOK" [5] (for more details about the standards, see Chapter 8, on quality planning tools).

With goals defined, we have entered the territory of the proverbial triple constraint, a convenient way to state that our goals are of competing nature [5]. If we try to shorten the target project schedule, we may easily increase the cost budget. Or, an increase in quality goal may impact the cost and schedule goal as well. Generally, the three goals are interrelated and to manage them means to make trade-offs. For this reason, some projects prioritize the three goals: the first priority is schedule, the second priority is quality, and the third priority is cost. The message it that if we find ourselves in a situation where we have to make a trade-off decision, our decision's first concern is the schedule goal; we don't care about priorities two and three. This issue becomes even more complex for new product developers who may add a fourth goal and constraint—product cost [2]. To make trade-off decisions with the quadruple constraint in hand, you need a very good understanding of the subtle nature of interactions between the project goals.

Describe Work Statement. What work exactly will you do in this project to deliver the project product and support the project goals? Can you express that very succinctly in a sentence or two, focusing on the big picture of the project? For example, the work statement for the erection of an optics plant may read: "Design an optics plant, procure it, erect it, and commission it." Or, as a software project described it, "Define workflow, configure software, develop training plan, develop prototype, train personnel, and release software." The idea here, again, is to identify *major* elements of project work, which will be described in greater detail in the supporting specifications and other documentation. Also, as you read the example statement, you might have noticed the way the work statements are structured: the verb (define) followed by the noun (workflow), the verb (configure) followed by the noun (software). Of course, you may choose to write work statements differently. Our way of writing has a very specific purpose of reducing the activity orientation of the work statement, and beefing up its goal orientation. Specifically, some organizational cultures believe that it doesn't matter whether the project team works (activity orientation).

Rather, what matters is whether or not the team delivers results (goal orientation). If you feel that our attempt to strengthen the goal orientation of the work statement does not work for you, the next part of the Scope Statement is all about results—specifically, deliverables and milestones.

Identify Deliverables. Performing the work described in the work statement must result in major deliverables. Take the preceding example of the work statement for the software project. Its *major* deliverables are workflow report, configuration settings, training plan, prototype, training complete, and software release. A closer look at this set of deliverables reveals several guidelines for identifying the deliverables. First, there is almost one-for-one correspondence between elements of work statement and deliverables. "Define workflow" (element of the statement) produces a "workflow report" (deliverable). Also, major elements of work lead to major deliverables. Focus your identification of deliverables in the Scope Statement on major ones that you can take and make level 1 deliverables in your WBS. It is in the WBS that you will identify *minor* deliverables (levels 2, 3, and so on), not here. Another guideline evident from the example set is that your deliverables may include interim deliverables—for example, products of early project stages (workflow report)—as well as end deliverables (software release). Finally, deliverables may be of the product (e.g., machine, facility, report, study, etc.) and service (e.g., training complete). Identifying deliverables should be followed by an additional step: define each one in terms of how much, how complete, and in what condition they will be delivered. This is typically done in the supporting specifications and other documentation. Should a WBS dictionary be available, it can take on this role.

Select Milestones. Milestones are major events and points in time, indicating the progress in implementing your work statement and producing deliverables. As a crucial part of the Scope Statement you need to identify major milestones in your project. Consider again the example of the preceding software project. The author of the project selected these milestones: Workflow report ready—February 15, 2002; configuration complete—February 28, 2002; training plan ready—February 28, 2002; prototype complete—March 30, 2002; training complete—April 16, 2002; and software release—April 30, 2002. Apparently, the author again used almost one-for-one correspondence between deliverables and milestones. This is a personal choice designed to provide full consistency from work statement to deliverables to milestones. Although there are many other ways of identifying major milestones, they share some common ground. They focus on several major milestones defined and understood well by project stakeholders, with a clear-cut date, and consistent with the list of deliverables.

Define Major Assumptions and Constraints. At the time of defining your scope you will have to take for granted some factors that are not entirely known or are uncertain. We call them assumptions. A few years ago a project team assumed that the product platform they were developing would serve only their domestic business unit and accordingly developed the Scope Statement. When their CEO saw the scope, he immediately changed the assumption, saying, "I want the platform to be used by our foreign subsidiaries as well." Consequently, the scope was changed. In another project, the assumption was that to hit the deadline date, the project would use all of the company's ten programmers for two months, day and night. They had to change their assumption and the scope when VP of engineering required the programmers to be shared by several projects during that period. Uncertainties like this, whether of technical, economic, resource, or any other nature, happen all the time in today's business arena.

In addition, all projects face serious constraints that may change the way they define project work, produce deliverables, and deliver milestones. These constraints may be physical, technical, resource, or any other limitation. Think of climbing Mount Everest as a project. The physical constraint here is the climate, which limits the time the project is workable to certain months. In a more relevant example, when a certain consulting company bid on a project manual development project, the owner required all consultants to have the designation of a certified project management professional (PMP) administered by the Project Management Institute. This technical requirement forced the consultancy to change the initial work statement and enter a joint venture with another consultancy who had more PMPs. In another case, management ordered a project manager to deliver a software package per a fast-track schedule. Because of the lack of resources for fast software quality testing, the project milestones had to be changed and pushed out by several months. These are a few examples with a crisp message: identify major constraints and build your scope and project around them, or face the derailment of your project.

To manage assumptions and constraint means to clearly identify them, size them up, and base the Scope Statement on them. As the project unfolds, you need to revisit them, verifying whether they still exist. Considering that they are a foundation stone of the scope, you need to rescope the project if the assumptions and constraints change. As long as you manage them like this they are controllable. Not managing them may mean that you are managed by them, certainly a poor practice.

Determine Exclusions. Habits die hard, as the following anecdote vividly paints. In the early 1990s, a contractor involved in the technology transfer to Africa scoped its projects to include computer centers with office furniture. It took the owners several years to realize that they would be better off buying the furniture locally than having the contractor import it from Europe. Consequently, the owners asked the contractor to leave the furniture out of the scope. The contractor's office was properly notified, but to no avail; the furniture kept pouring in the offices of the African owners. Why? The furniture used to be a habitual part of the scope, and the notification was neither strong nor visible enough. To avoid a situation like this, you need to raise a big red flag by using exclusions. State specifically what is often related to a similar project, but at this time is not part of the scope.

Supporting Detail. To stay clearly directive and crisp, the Scope Statement should be succinct, possibly written in the imperative mood (see the boxes "Simple Dos in Preparing the Scope Statement" and "Simple Don'ts in Preparing the Scope Statement" that follow). Technical and other details usually have no place in the statement; rather, they should be attached in the supporting documentation. Those accustomed to attaching technical specifications are welcome to do so.

Evaluate and Fine-Tune the Scope Statement. There are at least two levels of evaluation that deserve your attention. First, check the statement for completeness, comparing it against the information inputs and the information requirements that we have discussed here. Did you include all customer requirements? Did you define quality project goal? Did you identify major assumptions? Has any specifically excluded element of scope been identified? Second, you should the assess the quality of certain information, for example, schedule and cost aspects of the project goals. This clearly has to do with the method of project planning that we can express concisely as an iterative cycle of initial planning, detailed planning, and integration. In particular, initial planning is determining schedule and cost in your scope statement. It is then used for detailed planning—through work package planning explained in the next section on Work Breakdown Structure—of schedule and cost. These numbers are integrated and may be different from your schedule and cost numbers in the Scope Statement. What will you do? You can replace the numbers in the Scope Statement with those integrated numbers from detailed planning. Or you can reduce the scope and cut the integrated numbers to comply with the original numbers from the Scope Statement. Whatever you choose, it is obvious that the Scope Statement is only the first step in this iterative cycle of project planning, and for that reason, you have to fine-tune it as you go through the cycle. The following boxes, "Simple Dos in Preparing the Scope Statement" and "Simple Don'ts in Preparing the Scope Statement," offer some practical advice for defining the scope.

Utilizing a Scope Statement
When to Use. Not only is the Scope Statement a must-have tool, it is one of the most indispensable tools available. Every industry and every project family can use it. Except for highly repetitive projects that can utilize the same informal Scope Statement, every project regardless of the size and complexity would greatly benefit from having a formal written Scope Statement. Definitely, history has seen successful projects without formal Scope Statements. But researchers have found that having a well-defined idea of what you want to do in a project (the scope) is a critical success factor [20]. Therefore, if you strive for the heights of success, make sure you define the project scope and control its implementation properly.

Simple Dos in Preparing the Scope Statement
- If the statement is longer than a page, split it in two tiers: a one-page summary (as in Figure 5.5) and the supporting detail.
- Avoid repeating in one tier what is mentioned in the other tier.
- Use active rather than passive language, and spell out acronyms.
- Include functional groups providing resources in writing the statement.
- Have the statement approved by managers.

> **Simple Don'ts in Preparing the Scope Statement**
> - Don't write a Scope Statement without having some system to structure it.
> - Don't write it as a purely technical statement of the project.
> - Don't use ambiguous language (e.g., "nearly").
> - Don't mix major and minor goals, deliverables, and milestones.
> - Don't proceed without having an independent party review the Scope Statement.

Time to Use. Well equipped with information, a skilled project team may spend between 30 and 90 minutes developing a Scope Statement for a small or medium project. Growing size and complexity of the project will inevitably make this more time-consuming.

Benefits. Users value the Scope Statement for what it is—a first step/tool in planning for your project that directs all subsequent tools in the planning and control effort. It captures the fundamentals of a project, bonds them, and displays them for everyone to see. By setting the scope baseline along with the work breakdown structure, it helps project teams stay focused, allowing them to be guided by its big picture. Once this baseline is available, the project team should consider establishing the change control plan (see the box that follows, "Change Control Plan Helps Set the Direction for Controlling the Scope").

Advantages and Disadvantages. The Scope Statement offers some impeccable advantages:

- *It is comprehensive.* The Scope Statement looks at all major project dimensions, providing you with a comprehensive big picture of what you do in a project.
- *It is simple.* Its simplicity enables an easy grasp of versatile dimensions of project assignment.
- *It is easily adaptable.* With a little effort you can adapt it to your industry or company needs, taking out or adding new elements to it.

The disadvantages of the Scope Statement reveal risks of the following:

- *Temptation to grow.* To keep its advantages alive, you must fight the urge to add more elements to the Scope Statement and make it all-inclusive.
- *Being obsolete.* If not actively used as a scope baseline and rebaselined when necessary, the statement often becomes obsolete and useless.

Variations. The Scope Statement that we have described is designed to be an across-the-industry tool, including generics for serving as many project audiences as possible. Still, almost any industry or project family may find it useful to have its variation of the tool. Take, for example, product developers. Having a product statement included in the Scope Statement (after the project goals section) is a regular practice in their project business. Such a statement typically spells out the following [21]:

▲ Specification of the target market: exactly who the intended users are

▲ Description of the product concept and the benefits to be delivered

▲ Delineation of the positioning strategy

▲ A list of the product's features, attributes, requirements, and specifications (prioritized: "must have" versus "would like to have")

Other industries may add other elements to the Scope Statement. In some of Intel's groups, for example, adding an element describing major risks involved and response strategies is a requirement. These examples support our point: again, there are many variations of the Scope Statement.

Change Control Plan Helps Set the Direction for Controlling the Scope

Working on a significantly large project will almost inevitably call for some sort of the change control plan, typically part of the project plan. Although smaller projects usually cannot afford such a level of documentation, they still need to set a clear direction for controlling changes. Therefore, both smaller projects informally and larger projects formally need to address these issues:

● Who has the change approval authority? If the authority is to rest with the change review board, then chairman and board members need to be appointed for the project. Clearly define their responsibilities. In some cases, especially in smaller projects, project managers may have full authority to make changes and will not use a board.

● How is the scope of the change authority defined? For example, a board may be authorized to deal with major changes substantially affecting the scope. For the sake of responsiveness, the board chairman may have the approval authority, with other members reviewing (not approving!) their areas of expertise. Changes with small or no impact on scope may be in the purview of the project manager.

● What is the change request procedure like? The change control plan needs to describe the change request procedure and any forms or documentation to be used for submittal to the change authority.

● How is the management reserve managed? (For a definition of the reserve, see the "Risk Response Plan" section in Chapter 9.) To mitigate the risk of changes, you may need to establish the monetary amount of management reserve and define procedures for its use.

● How do we follow up the implementation of the approved changes? For example, nominating an administrator for this role is a possible solution.

● When in the project life cycle do we begin using and stop using the change request procedure? The change control should address this issue. For more details see the "Project Change Request" section in Chapter 11.

Customization Action	Examples of Customization Actions
Define limits of use.	Use a Scope Statement in every project.
	Small projects should include no more than business purpose, project goals, and milestones.
Add a new feature.	Include product definition element (works for hardware and software developers).
	Add information about major risks and risk responses (favored by high-tech projects).
Leave out a feature.	Leave out the work statement item (a frequent move by results-oriented companies).
Modify a feature.	Combine deliverables with milestones (favored by high-tech projects).

Among variations, one of the crucial points is length. In the majority of small and medium projects ranging from less than 1000 to several thousand resource hours, the statement is no more than a page long. In contrast, large projects' statements may be many pages long, supported by detailed technical specifications.

Customize the Scope Statement. The real value of the statement is in its ability to give you exactly what your projects need. For this to happen, you have to adapt the generic format we have provided. The following hints may help you visualize how to adapt it.

Summary

The Scope Statement is a written narrative of the goals, work, and products of the project. Except for highly repetitive projects that can use it informally, every project regardless of the size and complexity would greatly benefit from having a Scope Statement. It is valued for providing the first step in planning for a project, directing all subsequent planning and control effort. It captures the fundamentals of a project, bonds them, and displays them for everyone to see. By setting the scope baseline along with the Work Breakdown Structure, the statement helps project teams stay focused, allowing them to be guided by its big picture. Recapping the essence of the Scope Statement development is the following box.

Scope Statement Check

Check to make sure you developed a proper Scope Statement. The statement should

- Be based on information inputs from voice of the customer, Project Charter, and strategic gap analysis.
- Include all elements defined by its format.
- Make the elements consistent.
- Be fine-tuned in line with iterative cycle of project planning.

Work Breakdown Structure (WBS)

What Is WBS?

A WBS is deliverable-oriented grouping of project elements that organizes and defines the total scope of the project—work not in the WBS is outside the scope of the project [5]. When presented in a graphical format, it becomes obvious why WBS is often described as a project family tree, hierarchically displaying project deliverables (interim and end ones), which are further broken down into more detailed deliverables (see Figure 5.6). This same family tree analogy helps picture deliverables at one level as parents to those from the next lower level, who then become parents to the next-lower level's deliverables, and so on. Also, WBS can be presented in a "table of contents" format, where each next-lower level of deliverables is indented (see Figure 5.7).

Don't confuse WBS with an array of intimidating acronyms such as CWBS, BOM, or OBS. These tools are as logical and conceptually simple—as is WBS—but they have different purposes. For example, CWBS (contractual WBS), which is less detailed than WBS, is used to define the level of reporting that the project contractor will provide to the owner in larger, contractual projects. Widely applied in manufacturing industries, BOM (bill of materials) is a hierarchical representation of the physical assemblies, subassemblies, components, parts, and so forth that are necessary to produce the product. Finally, WBS is different from an OBS (organizational breakdown structure); an OBS indicates which organizational units are responsible for which work elements from a WBS (for basic terminology see the box that follows, "WBS Language").

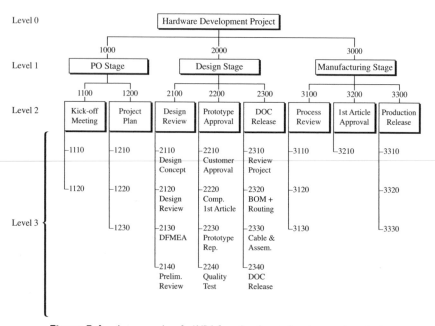

Figure 5.6 An example of a WBS for a hardware development project.

```
1000  PO Stage
       1100   Kickoff Meeting
              1110
              1120
       1200   Project Plan
              1210
              1220
              1230

2000  Design Stage
       2100   Design Review
              2110 Design Concept
              2120 Design Review
              .
              .
              .
```

Figure 5.7   A table of contents (TOC) format of the WBS.

Constructing a WBS: A Top-Down Approach

There are two basic ways of developing a WBS: top down and bottom up. Here are more details on the top-down approach, which is a very convenient way for project managers and teams with significant experience in project work and knowledge of project deliverables.

Collect Information Inputs. The development of WBS is likely to be an easier and more meaningful exercise if you are equipped with information about the following:

- ▲ Project Scope Statement
- ▲ Project workflow
- ▲ Voice of the customer (issues include customers wanting the product or service fast, the need to use rapid prototyping, the need for outsourcing)
- ▲ Pool of available resources
- ▲ Project situation

The Project Scope Statement, whether a sentence long or very detailed, is still no more than a broad direction for what you will do in the project. You first need to know "what you will produce" (scope) before you decide "how you will produce it" and depict it in WBS. Note the practice of some experienced project teams who prefer to develop Scope Statement and WBS in parallel rather than sequentially. In constructing a WBS, the knowledge of the project workflow is crucial. In particular, to develop a meaningful WBS for a software development project, you need to understand the process of software development. Knowledge of the process will indicate which activities are necessary to produce the required project deliverables. However, these activities and related deliverables may also depend on the voice of the customer. For example, the customer may require an extra fast-paced product modification conditioned on rapid prototyping. This will automatically create rapid prototyping deliverables in the structure of your WBS, which typically does not include such deliverables. Another powerful factor shaping the WBS is the amount of available resources. Using the same example of rapid prototyping, you may do it within your company, if such a resource is available, and include a branch of multiple levels of related deliverables in the WBS. Or, you may outsource it, because of the lack of such a resource in the company. In that case, you will end up with one deliverable in your WBS, while the vendor will have a whole WBS about this rapid prototyping. Finally, specifics of your project situation are known to influence the anatomy of WBS. Take, for example, the case of a product developer where each department had its own WBS when working on a project until it was acquired. The new owner immediately ordered all projects to develop an integrated WBS for each project, mandating that a large project's WBS add a project management branch. List these political and other factors influencing the WBS makeup, and include them into the process of its structuring.

WBS Language

- *Work elements.* Any deliverable in the WBS is called a work element, consisting of an item of hardware, software, service, or data. While some elements are the direct outcome of work, others are the aggregation of several logically grouped deliverables.

- *WBS level.* This is the hierarchical location of a work element in the WBS. Work elements at the same stage of structuring are on the same level. There is no universal system for numbering levels. We number the overall project level as 0 and subsequent levels as level 1, level 2, and so forth. Using level numbers, you can uniquely code each work element, providing a basis for cost control, for example.

- *Work package.* These are work elements at the lowest level of the WBS. We assign each of them to individuals (often called work package managers), who are responsible for managing tasks such as planning, scheduling, resource planning, budgeting, risk response, quality assurance, and, finally, the WBS's proactive cycle of project control.

- *Cost account.* This is a summary work element that is one level higher than the work package. Cost account includes one or several work packages and is often described as a management control point where actual performance data may be accumulated and reported.

- *Branch.* All work elements underneath a level 1 deliverable constitute a branch. Branches may vary in length.

- *WBS Dictionary.* As a minimum, a WBS may include brief descriptions of work packages, along with entry conditions (what inputs to a work package are necessary) and exit conditions (what outputs of a work package are required to call it complete). Adding more to it—for example, schedule dates, cost budgets, staff assignments for work packages, and descriptions of other work elements—may make sense in large projects.

Table 5.1: Methods for Structuring WBS

Level in WBS	Method of WBS Structuring		
	Life Cycle	System	Geographic area
0	Project	Project	Project
1	Phase	System	Area or region

Select the Type of WBS. After acquiring all necessary information about WBS shaping factors, you have more roadblocks to remove before being able to construct a WBS. Which method will you use to structure your WBS? Consider the three major ones: WBS by project life cycle, WBS by system, and WBS by geographic area (see Table 5.1).

The underlying principle of the WBS by project life cycle method is self-explanatory: You break the overall project into phases of its life cycle on level 1 of the WBS. Apparently, this principle of conveniently following the natural sequence of work over time is widely popular in some industries. A good example is a software development project consisting of phases such as requirements definition, high-level design, low-level design, coding, and testing. In contrast, those who prefer WBS by systems disregard sequential phases, tending to divide the project into its constituent physical systems and display them on level 1 of the WBS. For instance, in an airplane development project, systems on level 1 of its WBS may include fuselage, wings, and engine. This approach has been well established in some traditional manufacturing industries, where WBS resembles a BOM. Followers of the WBS by geographic area may also be found in construction, where, for example, level 1 of the WBS for a project consists of building A, building B, and so on. It is not unusual that literal geographic regions—for example, northwest site, southwest site, southeast site, and northeast site—are used as areas for level 1 of the WBS.

You have probably noticed that our discussion about the three structuring methods was limited to level 1 of the WBS. What about level 2 and other lower levels? Each of the three can continue with the underlying principle of further dividing the work into the lower levels. For instance, WBS by geographic area can have work elements on levels 2 and 3 that are areas as well. There is nothing wrong with using the same underlying principle on all levels of the WBS. Many practitioners, however, find hybrids more practical, combining the two or three methods in the same WBS. In particular, they may have WBS by project life cycle on level 1, systems on level 3, and so forth. Some even mix two structuring methods on the same level, such as systems and areas on level 1.

Which of the three methods of WBS structuring is the right one for you? Before answering this question, you should know that structuring a WBS is not an act of science. It is rather an act heavily influenced by a company's culture, shaped by top managers with the purpose of determining "how we get things done around here." When your peers are used to a certain way of getting WBS developed, that is probably the right approach for you. If there is no previous history of using WBS in your company, you should probably follow your industry culture. In both of these cases, we really suggest that you go with the flow of either your company's or industry's culture. This does not rule out the use of a different type of WBS structuring, but the bar of resistance to a new WBS method will be set higher than if you go with the flow. Should you decide to go against the grain, get ready for some serious change (resistance) management [22].

Establish the Level of Detail of the WBS. How many levels in the WBS? How many children per parent in the WBS? (See the box that follows, "Too Many Levels Can Create Turmoil.") Answers from these questions will determine the total number of work packages. Considering that the number of work packages influences the necessary time and cost you need to manage a project, you need this number to be on a level that your available time and cost budget can tolerate.

As explained earlier, the work package is the central point for managing WBS (see Figure 5.8). Simply said, work packages are discrete tasks that have definable end results—deliverables—that assigned organizational units "own" and need to produce. When using them for the integration of project planning and control in a very detailed WBS, for each work package, you will assign the responsibility, develop the schedule, estimate the cost or resources, develop plans for risk response and other planning functions, measure performance, and exercise proactive work package control. Obviously, as the number of work packages grows, so does the necessary time (and cost) to perform project planning and control. Reaching a point when there may be too many of them may render their management impractical and cost-prohibitive. Closely related to the number of work packages is their average size. Clearly, work packages need to be small deliverables that are manageable. But how small?

In summary, establishing the level of detail of the WBS includes determining the number of levels in the WBS, the number of work packages, and the average size of the work package that are compatible with your tolerance level and the industry practice. Table 5.2 shows data on WBS from some actual projects. From the table, you can derive the following guidelines for the majority of small and medium projects in the field of information technology and software and product development:

- 3 to 4 levels of WBS
- 15 to 40 work packages
- 20 to 50 hours per average work package
- 1 to 2 weeks' duration of the average work package
- 3 to 7 percent of the total hours budget per average work package.

Too Many Levels Can Create Turmoil

"How many WBS levels do we want our projects to have?" This was a question that designers of the PM process in an organization asked themselves. To prepare a good answer, they first looked at the projects they manage: 10 to 15 projects per year, ranging from $100K to $5M, mostly involving the design and construction of electrical substations. After some benchmarking, the designers made their decision: Each project shall have a five-level WBS. Soon after the deployment of the PM process began, a silent rebellion in smaller projects began, followed by outright refusal of their project managers to use the process. The justification was simple. The amount of work on scheduling, budgeting, and control of 250 work packages in a five-level WBS drove smaller projects to a halt. As a result, project managers went to the old, ad hoc way of management. The moral to the story: Match the size and structure of WBS with the size and structure of the project.

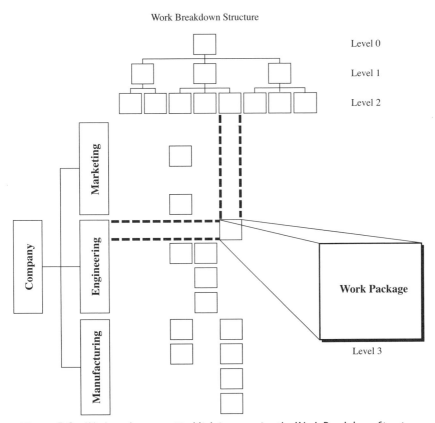

Figure 5.8   Work package, a critical link in managing the Work Breakdown Structure.

Table 5.2 Examples of the Level of Detail of the WBS

Project	Project Duration (days)	Project Budget (person-hours)	# of Levels in WBS	# of Work Packages	Mean # of Hours/ Work Package (3)/(5)	Mean # of Days/ Work Package* (2)/(5)	Mean % of Budget/ Work Package [(6)/(3)]×100
(1)	(2)	(3)	(4)	(5)	(6)	(7)	(8)
IT infrastructure	90	500	3	15	33	6	6.6
Selecting a billing platform	180	1200	4	36	33	5	2.7
S/W development	270	1200	3	25	48	11	4
Hardware development	365	500	4	29	17	13	3.4

* Assuming all work packages are sequential without any overlapping

For larger projects, however, the level of detail often quoted in the literature seems to include [23, 24]

- ▲ 5 and more levels of WBS
- ▲ 80 to 200 hours per average work package
- ▲ Less than 2 to 4 weeks per average work package
- ▲ 0.5 to 2.5 percent of the total project budget per average work package

Whether you manage a small or large project, these numbers should be adapted for your personal and possibly cultural preferences. For example, some individuals and cultures favor more detailed planning and control, and accordingly, more detailed WBS, while others have the opposite tendency [25].

Structure the WBS. Once you are equipped with necessary information, along with the type of WBS and its level of detail, you are ready to logically construct a WBS for your project (see the box that follows, "Golden Rules of WBS Structuring"). The steps are outlined in the following list:

- ▲ Start by identifying the major deliverables of the project. Depending on the type of WBS you selected, these may be phases, systems, geographic areas, or their combination. One useful approach here is called the *scope connection*. In particular, when developing the Scope Statement, you identify major deliverables, which could be borrowed to serve as major deliverables in WBS. This helps integrate the Scope Statement with WBS, linking business and project goals—via major deliverables—with lower-level deliverables all the way to work packages.

- ▲ Divide major deliverables into smaller, more manageable deliverables, level after level, until a point is reached where the deliverables are tangible, verifiable, and defined to the level of detail that enables them to be used for the integration of the project planning and control activities.

- ▲ How will you represent your WBS? In the case of smaller projects, drawing a WBS as a tree diagram is a very visual and preferential way (see the upper part of Figure 5.6). As your number of WBS levels grows, so does the complexity, making it more difficult to stick with the tree format. A way out may be in the TOC (table of contents) format. For example, Figure 5.7 shows the TOC format for the WBS from Figure 5.6. To some extent this WBS construction process looks like a random process. Bringing more orderliness to it is possible by following several guidelines (a shortened version appears in the "Golden Rules" box) [26, 27].

- ▲ Make sure WBS is deliverable-oriented. Since WBS is about deliverables, there is no place in it for activities [28].

▲ Be sure that WBS includes all project work. What is left out will not be resourced and scheduled, which is a risky proposition.

▲ Make each work element relatively independent of others on the same level.

▲ Keep breaking the work down into work elements until you reach a level at which there is a method in your organization capable of producing the element. Stop there. Considering that you may outsource deliverables to vendors, this practice may produce branches of unequal length, an acceptable practice.

▲ Produce a WBS that integrates work elements or separate levels to the point that their aggregate is an equivalent of the project completion.

Evaluate Whether the WBS Is Structured Properly. Since the WBS development lacks the rigor and discipline of the scientific approach, there is no single correct WBS. Rather, there can be many different but equally good WBSs. To make sure that your WBS is good enough, you should evaluate it against the preceding guidelines. If there is a need for some revisions, making them will ensure that you have created a WBS capable of acting as a framework for integration of project planning and control.

WBS Templates Increase Productivity. Having each project team develop a WBS from scratch can create several problems. First, WBS development consumes resources. Even if the resources are expended, there is no assurance that the WBS will be a good enough framework for the integration of the management of the project. Also, when each project uses a different WBS, the benefit of comparing various projects and drawing synergies is lost. All these problems have been successfully cured by the use of WBS templates.

Specifically, this means adopting templates for certain project families. Highway construction projects are an example of a project family. Other families may include software development projects, manufacturing process developments, and hardware development projects. Generally, a family of projects is a group of projects that share identical or sufficiently similar project assignments. When the templates are developed and adopted, the development of a WBS for a new project is reduced to adapting the template. This saves time, produces a quality WBS, and enables interproject comparability. In a nutshell, templates increase productivity.

Golden Rules of WBS Structuring
- Include deliverables only.
- Show all project work.
- Make deliverables relatively independent.
- Use (a)symmetrical branches when justified.
- Build WBS as an integrated effort.

Constructing a WBS: A Bottom-Up Approach

Brainstorming all project work that needs to be performed and then organizing it into a WBS hierarchy is at the heart of the bottom-up approach. This approach, essentially an application of the affinity diagramming method, is beneficial to those without much project experience and first-time users of a WBS. Projects developing or deploying novel technologies, typically fraught with high uncertainty and lacking precedents, can benefit from the approach as well, even if the project team is experienced. Also, when you are working to develop a template WBS from a bunch of competing templates used by various project managers, a bottom-up approach is helpful. Despite its brainstorming nature, the bottom-up approach may be preceded by collecting necessary information for the WBS development, selecting the type of WBS, and establishing the level of its detail — in other words, some of the steps taken in the top-down approach. Other steps in the bottom-up approach follow.

Generate a Detailed List of Deliverables. This step includes having each project team member brainstorm what is going to be delivered in the project. Each deliverable may be recorded on a sticky note and posted in a visible place, with the number of generated deliverables ranging from 40 to 60—a good number for a small or medium-size project. Higher numbers may be required for large projects. In the process, adhering to the brainstorming principle of not critiquing the ideas is crucial.

Sort Deliverables into Related Groupings. The result of this step is the grouping of related deliverables. The aim may be to create groupings of five or so deliverables. Screen them carefully for relationships and group them into new groupings, striving to have three or four levels of groupings for small and medium projects, possibly more for larger projects.

Create Duplicate Deliverables, If Necessary, and Consolidate Deliverables. Project team members may have different and conflicting ideas about grouping deliverables. If that occurs, create duplicates of the deliverables and post them in the groupings suggested by the team members. Develop the discussion to understand the rationale for the conflicting groupings, and try to reach a consensus. If this is not possible, use a preferred type of voting to decide on the final grouping. Also, consolidate similar deliverables and eliminate redundancies. This should give you the preliminary WBS hierarchy.

Create Names for the Groupings. The hierarchy needs names for groupings/deliverables on different levels. As much as possible, the idea here is one of having the consensus among team members. Spending enough time on naming the groupings/deliverables is useful for a good understanding of what is going to be delivered in the projects, as well as for building a buy-in.

Evaluate Whether WBS Is Structured Properly. Similarly to the top-down approach, bottom-up WBS development lacks the rigor and discipline of the scientific approach, leaving room for mistakes. Therefore, this is the time to evaluate the WBS developed against the guidelines for structuring WBS. Again, useful revisions and corrections are welcome, aiming at the improvement of the WBS to the level it needs to perform as a framework for integration of project planning and control.

The bottom-up approach is a good method for novices and unfamiliar projects. For full evaluation of its potential, we should recognize that it provides an easy start, encourages strong involvement, and downplays vocabulary issues. Its ease of use may give it an edge over the top-down approach, which requires more time to start and a shared vocabulary, and also limits the participation.

Utilizing WBS

Time to Use. A reasonably skilled and prepared project team of smaller size can develop a three-level, 15-work package WBS in 30 to 60 minutes. The necessary time will expand as the WBS and team size grow.

When to Use. WBS was initially used to bring order to the integration of management work faced with large and complex projects in the government domain. Logically, then, most of the "science" of WBS was hatched out in governmental agencies and is very well reflected and covered in the works of prominent PM books [27]. What is not well covered is how to adapt the science for what is the dominant stream of projects in today's business world—small and medium projects. For these, WBS is one of those few must-have tools. Whether you are in software or hardware development, marketing or accounting, manufacturing or construction, practically any area and industry, small and medium projects need WBS to glue all their pieces together. In fact, it is possible to run a successful project without using a WBS. We have seen and heard about such projects. The point here is that experience suggests the probability of success is higher when a good WBS is used, as opposed to having an inappropriate WBS or not having one at all (see the box the follows, "Tips for WBS Use").

Benefits. There are two reasons why the value of WBS cannot be overstated: It helps organize the project work and it creates a framework from which management of a project can be fully integrated. In particular, WBS enables a project team to organize the work into small deliverables that are manageable, making the assignment of responsibility for each of them easy. Because the deliverables are relatively independent, their interfacing with and dependence on other deliverables is reduced to a minimum. Still, they are integratable as you move up the WBS so the team can get a view of the total product. Finally, the progress of the deliverables is measurable. This extraordinary capability of WBS to help organize the project work is an enabler of what is an even higher value of the well-constructed WBS; it serves as the framework for integration of project planning and control functions. Some call it the single most important element in project management [23].

Tips for WBS Use

- Develop a WBS in each project, small or large, starting off a template WBS.
- Adopt a template WBS for each project family.
- When developing the template, begin with few levels. Add more if project team members ask for it.
- Allow smaller projects to use a fewer number of levels in the template.
- Build "blank" work elements in the template to be used by unusual projects.

At the heart of the significance of WBS is its ability to serve as a project planning and control framework, enabling the achievement of these fundamental project management actions (see Figure 5.9):

- ▲ Assign the responsibility for the project work.
- ▲ Schedule the project work.
- ▲ Estimate the cost or resources to complete the project work.
- ▲ Develop the response to risks associated with the project work, and perform other planning functions such as quality planning.
- ▲ Measure performance.
- ▲ Control the project work effort to accomplish project goals.

The center of these actions is the work element. For each work element, we assign an individual responsible for producing the deliverable listed under that work element. If, for example, we identify 20 elements in the WSB, each work element would have an "owner" responsible for the element, its schedule, cost, risk response, performance measurement, and project control. A convenient tool for assigning responsibilities in this manner is the *responsibility matrix*. Listed on its vertical axis are work elements, while across the top are persons involved in the project. In the cell at the intersection of a work element row and a person column different responsibilities are assigned for the accomplishment of the work element.

The second planning action made possible by WBS is scheduling the project work. Here the integration of the schedules begins at the work package level. In particular, once the schedule for each work package is developed, the schedule for a work element on the next level is no more than a summarization of schedules of its work packages. Similarly, the schedule for any higher-level work element is the summary of schedules of its constituent work elements. This summarization is enabled by a WBS as illustrated in Figure 5.9 and is explained in the section on hierarchical schedules in Chapter 6. A point to remember is that each work element can have only one schedule. For those practitioners who argue that WBS is static and does not display dependencies between its work elements, keep in mind that it is the network schedule's role, not the WBS, to show the dependencies.

In a way similar to schedule integration, WBS provides a formal structure for resource estimating. Again, resources are estimated for work packages, and their aggregation up the WBS leads to total resource requirements (as shown in Figure 5.9). When resources for a work element are spread over the element's schedule, you will obtain a time-phased resource plan, a great baseline against which to compare the actual performance and strategize corrective actions, if the negative variance occurs. The reason we emphasized resources is that the majority of small and medium projects prefer the use of resource-based estimates to the cost-based estimates. Should you prefer the latter, multiply resources by their cost rate to obtain cost estimates.

Planning for other management functions such as risk, quality, and change management, should also be performed around the skeleton provided by WBS. Take, for example, risk response—a must function in the majority of today's projects. The basic place for a risk plan including risk identification, quantification, impact, and response is the work package. Summing up risk plans for work packages that belong to the next higher level work element will yield a risk plan for the element. Continuing with this summation up the WBS hierarchy will produce a risk response plan for the total project (see Figure 5.9). Whether Monte Carlo analysis or a simple risk response worksheet is deployed for the summation purpose is not the primary issue; rather, it is the need to structure risk planning around the framework offered by WBS.

Performance measurement is another management action facilitated by WBS. Go back to the structure of WBS: a hierarchical tree of deliverables. The process of measuring performance again starts on the work package level, where completion of a certain deliverable is easily verifiable. Comparing the planned with the actual schedule, resources, and quality for the deliverable provides its status. Aggregating the status of all deliverables under a parent will give us the status for the parent. Following this pattern up the WBS hierarchy will help establish the status for the total project, as shown in Figure 5.9.

The ultimate purpose of performance measurement is proactive project control. More precisely, when performance status is known for each work element and the total project, we know the variance between the baseline and the actual project performance. Armed with this information, we are ready to seek answers to the questions of the proactive cycle of the project control (see the box titled "Be Proactive: Five Questions of the Proactive Cycle of Project Control" in the "Jogging Line" section in Chapter 12).

Note again, these questions relate to each work element and the total project. However, the ground work, which is the most time-consuming aspect, is done on the work package level, while other levels are a product of summarization. Using WBS to integrate the management of a project through six project planning and control actions may look overly complex for managers of small and medium projects. Such an assessment couldn't be further from the truth. With a three-level WBS and some ten work packages, a small or medium project team can easily and quickly exercise the six actions, providing their project with fast and sound management.

Advantages and Disadvantages. Advantages that WBS creates abound, but the most prominent among them are the following:

- ▲ *Projects a strong visual impact.* As a practicing program manager commented, WBS brings order to disorder in a visual way. At a glance, the WBS turns the disorder of the verbal and obscure scope statements into the order of a clearly structured family tree or TOC format.

- ▲ *Simplicity.* The old adage that simplicity drives perfection is rejuvenated when it comes to the WBS. Very limited training is usually sufficient in helping project participants read and construct a WBS.

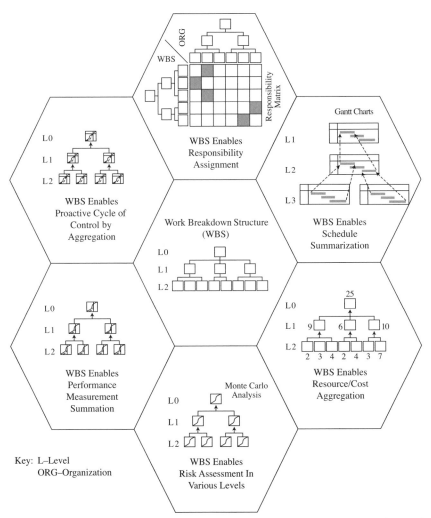

Figure 5.9 WBS serves as the framework for integration of project planning and control functions.

Customization Action	*Examples of Customization Actions*
Define limits of use.	Use WBS by project life cycle in all projects.
	Projects will have three or four levels in WBS, at the discretion of project managers.
	All four-level WBSs will have a project management branch.
	The number of work packages may vary between 10 and 50.
	Each work package must be broken down into several tasks that will not be shown in the project WBS. The work package manager will use the tasks for her planning and control of the work package.
	Develop templates that you can adapt for new projects.

In certain situations WBS may become a liability:

▲ *Excessively large WBS may be a time and productivity killer.* When composed of too many levels and work packages, the WBS's purpose as a framework of integrating project planning and control activities becomes meaningless, time-consuming, and resource-wasting.

Customize the WBS. The generic WBS that we have covered here offers ample opportunities for customization to match the specifics of your company and project. The examples that follow may give you ideas on how to tailor the WBS for your needs and get more value out of it.

Summary

WBS is a project family tree that hierarchically displays project deliverables. Practical and useful in each project, WBS is often viewed as the single most important element in PM. The reason lies in its ability to help organize the project work and create a framework from which management of a project can be fully integrated. Key highlights of WBS structuring are offered in the following box.

WBS Check

Verify your WBS against the golden rules of WBS structuring to make sure you developed a proper one. It should

- Be based on information inputs.
- Include deliverables only.
- Represent all project work.
- Contain deliverables that are relatively independent.
- Reflect an integrated effort.

Concluding Remarks

The four tools reviewed in this chapter go together, adding to each other's value (see a summary comparison that follows). The point is that they do not compete against each other for the project team's attention and time. The Project Charter's reason for existence is to bring the project to life, authorizing resources based on commitments from the functional departments and giving it organizational visibility. Logically, the charter should follow a sound dosage of planning, as is often not the case. Such planning should rely on Project SWOT Analysis to recognize the project's strengths and capability gaps and develop strategies to close the gaps. Building on the analysis, a broad project direction should be charted in the Scope Statement, providing the solution space within which the project team can operate. Emerging from the scope should be the WBS, a project work baseline and the skeleton for integration of all management actions in a project.

A Summary Comparison of Scope Planning Tools

Situation	Favoring Project Charter	Favoring Project SWOT Analysis	Favoring Scope Statement	Favoring WBS
Small functional projects		√	√	
Small cross-functional projects	√	√	√	
Large and complex projects	√	√	√	
Provide project visibility in the organization	√			
Make departmental resource commitments clear	√			
Need to use the tool informally		√		
Need to use the tool formally	√	√	√	√
Need strategic analysis of project		√		
Provide broad direction for project execution			√	
Provide the project baseline			√	√
Take little time in small projects, longer time in large projects	√	√	√	√
Provide skeleton for managing a project				√

References

1. Cleland, D.I. 1990. *Project Management, Strategic Design, and Implementation*. Blue Ridge Summit, Pa.: TAB Books.

2. Smith, P. G. and D. G. Reinertsen. 1991. *Developing Products in Half the Time*. 1st ed. New York: Van Nostrand Reinhold.

3. Cleland, D. I. and W. R. King. 1983. *Systems Analysis and Project Management*. 3d ed. New York: McGraw-Hill.

4. Milosevic, D. 1990. "Case Study: Integrating the Owner's and the Contractor's Project Organization." *Project Management Journal* 21(4): 23–32.

5. Project Management Institute. 2000. *A Guide to the Project Management Body of Knowledge*. Drexell Hill, Pa.: Project Management Institute.

6. Stevenson, W. J. 1993. *Production and Operations Management*. Boston: Irwin.

7. Katzenbach, J. R. and D. K. Smith. 1993. *The Wisdom of Teams*. Boston: Harvard Business School Press.

8. Kerzner, H. 2000. *Applied Project Management*. New York: John Wiley & Sons.

9. Lock, D. 1990. *Project Planner*. Aldershot, U.K.: Gower Publishing.

10. Harrison, J. S. and C. H. St. John. 1998. *Strategic Management of Organizations and Stakeholders*. 2d ed. Cincinnati: South-Western College Publishing.

11. Liedecker, J. K. and A. V. Bruno. 1984. "Identifying and Using Critical Success Factors" *Long Range Planning* 17(1): 23–32.

12. Handifield, R.B. 1994. Effects of Concurrent Engineering on Make-to-Order Products. *IEEE Transactions on Engineering Management* 41(4): 384–393.

13. Zirger, B. J. and J. L. Hartley. 1996. The Effect of Acceleration Techniques on Product Development Time. *IEEE Transactions on Engineering Management* 43(2): 143–152.

14. Thomson, J. 1967. *Organizations in Action*. New York: McGraw-Hill.

15. McDounough, E. F .I. and G. Barczak. 1991. "Speeding Up New Product Development: The Effects of Leadership Style and Source of Technology." *Journal of Product Innovation Management* 8(3): 203–211.

16. Handifiled, R. B., et al. 1999. "Involving Suppliers in New Product Development." *California Management Review* 42(1): 59–82.

17. Wright, P., C. D. Pringle, and M. J. Kroll. 1992. *Strategic Management*. Boston: Allyn and Bacon.

18. Thompson, A. T. and A. J. Strickland. 1995. *Crafting and Implementing Strategy*. Boston: Irwin.

19. Frame, J. D. 1999. *Building Project Management Competence*. San Francisco: Jossey-Bass.

20. Rosenau, M. D., C. Griffin, G. , and N. Anschuetz. 1996. *The PDMA Handbook of New Product Development*. New York: John Wiley & Sons.

21. Cooper, R. G. 1993. Winning at New Products. 2d ed. Reading, Mass.: Perseus Books.

22. Hammer, M. and J. Champy. 1993. *Reengineering the Corporation*. New York: Harper Business.

23. Kerzner, H., 2001. *Project Management: A Systems Approach to Planning, Scheduling, and Controlling*. 7th ed., New York: John Wiley & Sons.

24. Lavold, G. D. 1983. "Developing and Using the Work Breakdown Structure" in Project Management Handbook. Edited by D. Cleland and W. R. King. Van Nostrand Reinhold: New York, 283–302.

25. Schneider, A. 1995. "Project Management in International Teams: Instruments for Improving Cooperation." *International Journal of Project Management* 13(4): 247–251.

26. Department of Defense. 1998. *MIL HDBK-881; Department of Defense-Work Breakdown Structure*. Washington, D.C.: Department of Defense.

27. Department of Energy. 1996. *DOE G 120.1-5, Performance Measurement Systems Guidelines*. Washington, D.C.: Department of Energy.

28. Berg, C. and K. Colenso. 2000. "Work Breakdown Structure Practice Standard Project: WBS vs. Activities." *PM Network* 14(4): 69–71.

Schedule Development

You win battles by knowing the enemy's timing, and using a timing which the enemy does not expect.

Miyamoto Musashi

Major topics in this chapter are schedule development tools:

▲ Gantt Chart
▲ Milestone Chart
▲ Critical Path Method Diagram
▲ Time-scaled Arrow Diagram
▲ Critical Chain Schedule
▲ Hierarchical Schedule
▲ Line of Balance

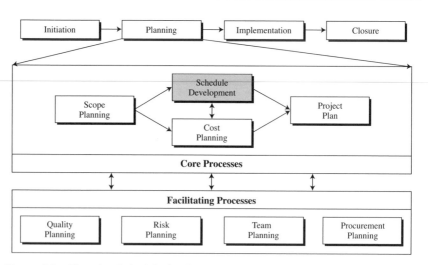

Figure 6.1 The role of schedule development tools in the standardized project management process.

These tools will help successfully develop a calendar-based project schedule. Schedule development tools are deployed in coordination with scope and cost planning tools to culminate in an integrated project plan (see Figure 6.1). A significant role in that effort belongs to tools of the facilitating processes such as team, quality, and procurement planning, and risk response development. This chapter seeks to help practicing and prospective project managers accomplish the following:

- ▲ Become familiar with various schedule development tools.
- ▲ Select a schedule development tool that fits their project situation.
- ▲ Customize the tool of their choice.

These are critical skills in project planning and building the standardized PM process.

Gantt Chart
What Is the Gantt Chart?

Using bars to represent project activities, the Gantt Chart shows when the project and each activity start and end against a horizontal timescale (see Figure 6.2). Although the Gantt was developed around 1917 and is the oldest formal scheduling tool, it is still widely used. It is also called the bar chart.

Constructing a Gantt Chart

Developing a Gantt Chart takes several steps. Though the first step—determine level of detail and identify activities—is normally part of scope planning, we include it here in order to provide an integrated procedure.

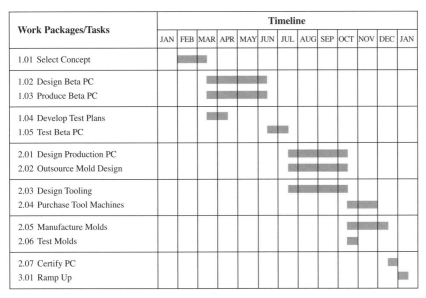

Work Packages/Tasks	Timeline												
	JAN	FEB	MAR	APR	MAY	JUN	JUL	AUG	SEP	OCT	NOV	DEC	JAN
1.01 Select Concept		▨											
1.02 Design Beta PC			▨▨▨										
1.03 Produce Beta PC			▨▨▨										
1.04 Develop Test Plans			▨										
1.05 Test Beta PC					▨								
2.01 Design Production PC							▨▨▨						
2.02 Outsource Mold Design							▨▨▨						
2.03 Design Tooling							▨▨▨						
2.04 Purchase Tool Machines									▨				
2.05 Manufacture Molds										▨▨			
2.06 Test Molds										▨			
2.07 Certify PC												▨	
3.01 Ramp Up													▨

Figure 6.2 An example of the Gantt Chart.

Prepare Information Inputs. The quality of the Gantt Chart is firmly rooted in quality inputs about the following:

- ▲ Project scope
- ▲ Responsibilities
- ▲ Available resources
- ▲ Schedule management system

Information about the scope helps schedulers analyze and understand the project activities that are being scheduled. Naturally, those responsible for the execution of the activities should be in the best position to schedule them, because they are the best source of knowledge about the activities. Part of their knowledge relates to priorities, activity sequencing, and duration estimating. Another part of their knowledge relates to what resources are available and in which time periods. Still another part of their knowledge is the schedule management system, a method leading companies deploy to ensure their schedules are systematically developed and used (see box that follows, "Schedule Management System).

Determine Level of Detail and Identify Activities. How many activities should there be in the Gantt Chart? Twenty-five? Fifty? Seventy-five? The answer will determine how big the individual activities will be. Consider, for example, a practice in a company. For a certain type of project, the team's decision is that a typical Gantt Chart will have around 25 activities, with no activity longer than three weeks and shorter than one week. This provides guidance for the next step and ensures that the chart is just about the right size, neither too cumbersome and time-consuming nor too small, lacking necessary information to manage. Specifically, the amount of detail should be enough for the intended user to monitor progress and for coordination. However, the amount of detail should not be so great that it cannot be easily statused using the project's progress reporting systems. Note that it may take some practice before you establish the right level of detail.

Brainstorming and breaking the project down into constituent activities that need to be performed in order to get the project done is next. Or, you can enlist the help of the Work Breakdown Structure (Chapter 5) for this purpose, identifying activities necessary to achieve work packages in WBS. At this time, you don't need to worry about how big the activities are; rather, the emphasis is on making sure that all necessary activities are identified. Once that point is reached, refer to the level of detail established in the previous step. If the level tells there are too few activities, breaking down some of them to hit the desired level of detail is a natural choice. Or, if there are too many activities, it is helpful to combine similar ones to arrive at the desired level.

Schedule Management System

Scheduling a project is rarely a matter of sitting down and developing a single schedule before the project starts. The process of developing and using schedules needs to be planned and organized [2]. To ensure such an approach, leading companies tend to resort to a schedule management system, determining the following:

- *What schedules are needed?* Depending on the size of the projects, you can use a hierarchy of schedules for larger projects (see the Hierarchical Schedule section later in this chapter) or a single schedule for small projects, as well as preliminary and implementation schedules.

- *How schedules will be used?* Some schedules, such as a summary one, can be used for management oversight, while detailed schedule can be used for organizing work.

- *What is schedule detail?* Limiting the number of activities in a schedule, such as up to ten in a milestone schedule, is an attempt to preempt unnecessary detail and waste of time.

- *What tools will be appropriate?* Any of the tools in this chapter are appropriate, if used for their design purpose.

- *When will the schedules be prepared?* Whenever possible, before the project starts. In uncertain projects, a rolling wave concept may be applied: Prepare a front-end schedule for the first 60 days of a project, then develop a more detailed schedule as the project unfolds.

- *How schedules will be monitored and updated?* Frequency and schedule control tools need to be defined in tune with needs of a company.

Scheduling can be overkill, unless an appropriate measure of scheduling is found. Developing a schedule that is more detailed than necessary is as useless as the schedule that management cannot follow or understand. Inappropriate scheduling emphasis like this may easily turn people against scheduling and create the attitude of "how about stopping this red-tape exercise and get down to real work?" It is the job of the schedule management system to prevent this.

Sequence Activities. Sequencing activities involves arranging them in a logical order of execution. This requires a good knowledge of the technology and priorities of the project and ensures that we first perform activities that need to produce outputs necessary for work on the subsequent activities. Illogical sequence of activities is destined to cause rework and slow down the project.

Estimate Activity Durations. Resources, human and material, drive the process of estimating activity durations. Begin the process by asking, "What resources do I need to successfully complete this activity?" The answer should provide the names of resources and work time for each one of them to complete the activity; for example, 100 hours of work from a programmer. Next, knowing the availability of a resource and the company's work calendar (e.g., no work on Sundays), convert the work time into calendar time. For instance, because she is involved in multiple projects, the programmer's 100-work time hours will have to be spread over 12 weeks to get her job done. This should be repeated for each activity.

Draft and Refine the Gantt Chart. Drawing a Gantt Chart requires a sheet or form with a horizontal timescale and the list of activities across vertical axis (see Figure 6.2). Adding up calendar times for project activities gives a rough idea of the total time the timescale needs to show. Adding to it a bit of spare time is helpful to improve the visual impact and have space for possible changes in the future. This is a good scale for a chart in which all activities are sequential. Should there be overlapping activities, you can reduce the duration of the timescale accordingly.

Next is listing all activities on the vertical axis per determined sequence, followed by these steps:

▲ Draw a barrepresenting each activity, with its length proportional to its duration on the timescale.

▲ For multiple activities that form a phase of work, add a summary bar, called a *hammock activity* or simply a *hammock*, just above the first activity. A hammock begins when the first of the activities begins and ends when the last of the activities ends. A reasonable measure is to have a hammock activity for every four to ten detailed activities that relate to each other. Because management needs a big-picture view of the project, the hammocks' summary level of detail makes them very convenient for that purpose.

▲ Look again at the whole chart. Are all necessary activities there? Logically sequenced? With appropriate timescale and reasonable durations? Make any changes necessary to finish refining the chart and get ready to use it.

Utilizing the Gantt Chart
When to Use. The Gantt Chart is an effective tool for smaller and simpler projects, where there is no need to show dependencies between activities,

since they are well known to all involved planners [11]. As the project size and complexity increase, the Gantt Chart becomes less applicable. Simply, the chart gradually looses the ability to handle the increasing number of activities, data, and dependencies between activities involved. In large and cross-functional projects, using the Gantt Chart as the primary scheduling tool is impractical and ineffective (see the box that follows, "Tips for Gantt Charts").

In contrast, in large and complex projects, the complementary use of the Gantt Chart and CPM chart may be a very smart strategy. The latter effectively copes with number of activities, data, and dependencies between activities involved. It, however, does a poor job of showing to project people in a simple and visual manner what activities to work on in the next week or the next two weeks. That is where the Gantt Chart may step in. Extracting from the sizable CPM chart activities due in the next week or two, presenting them in the Gantt Chart format, and handing to the people portions of the Gantt Chart they are responsible for provides clear and practical partial short-term outlook schedules. Those in management will still be responsible to coordinate the interfaces between the owners of the schedules.

Time to Develop. Depending on one's knowledge and experience, a 20-activity Gantt Chart can be developed in anywhere from 10 to 40 minutes. Some experienced project managers use the rule "activity per minute," meaning it should take one minute for each activity in the Gantt Chart. Note that the more people are involved in the Gantt Chart development, the more time may be needed.

Benefits. Having a Gantt Chart helps ensure that everyone understands the timetable for project activities. Then, project participants will have the necessary time allocated on their calendars and be available to perform their activities.

Advantages and Disadvantages. The chart has the advantages in that it is

- ▲ *Visual.* It creates a pictorial model of the project. This makes it a superb communication tool.
- ▲ *Simple.* With little or no instruction, almost anyone can read or construct a Gantt Chart, from a project team member to the executive sponsor.
- ▲ *Useful to show both the planned and actual status* (see the Jogging Line tool, Figure 12.4, in Chapter 12) of the project.
- ▲ *Useful in resource planning or allocation.* First marking on each activity the numbers of different human resources and adding them up for each time period helps get the total number of a certain resource over time for each activity and the whole project (see the Cost Baseline tool in Chapter 7).

Tips for Gantt Charts

- Rely on Gantt Chart as long as it has fewer than 25 activities.
- Use a single Gantt Chart as a primary scheduling tool in small, simple, functional projects.
- Don't use a single Gantt Chart as a primary scheduling tool in large, complex, cross-functional projects.
- Team-developed Gantt Charts lead to higher quality, better buy-in, and stronger commitment of team members.

Gantt Charts have disadvantages that may limit its relevance. In particular, they

▲ *Do not show dependencies between activities, making it impossible to clearly identify the sequence of project activities and, consequently, the critical path.* Without this information, Gantt Charts are not effective in large and cross-functional projects.

▲ *Cannot cope effectively with projects containing large number of activities, measured in hundreds, for example.* This can be overcome by using hierarchical Gantt Charts, in which an activity in a higher-level Gantt Chart is broken down into more detailed activities in the lower-level Gantt Chart (see the section "Hierarchical Schedule" later in this chapter).

Customize the Gantt Chart. The Gantt Chart we have presented here is no more than a generic one. You can expect to get maximum value out of it only if its format and features are adapted according to your own needs. Here are some examples of how this can be accomplished.

Customization Action	Examples of Customization Actions
Define limits of use.	Use Gantt Charts for functional projects or phases of a large project with fewer than 25 activities.
	Use Gantt Charts for cross-functional projects with fewer than 25 activities, only if team members have excellent knowledge of their interdependencies.
	Develop templates that can be used as a starting point in charting Gantts for new projects.
Add a new feature.	Add a column showing the owner of the activity.
	List resources under each activity.
	Add major milestones to the chart.

> **Gantt Chart Check**
>
> Check to make sure you developed a good Gantt Chart. The chart should have
> - All necessary activities to complete the project
> - A logical sequence of the activities
> - Activities with reasonable durations
> - An appropriate timescale.

Summary

The focus of this section was on the Gantt Chart, a tool that uses bars to represent project activities and shows when the project and each activity start and end against a horizontal timescale. The Gantt Chart is an effective tool for smaller and simpler projects, where there is no need to show dependencies between activities, since they are well known to all involved planners. Having a Gantt Chart helps ensure that project participants will have the necessary time allocated on their calendars and is available to perform their activities. Customizing the chart for specific project needs enhances its value. The following box recaps the key points in building the Gantt Chart.

Milestone Chart

What Is the Milestone Chart?

This chart shows milestones against the timescale in order to signify the key events and to draw management attention to them (see Figure 6.3). A *milestone* is defined as a point in time or event whose importance lies in it being the climax point for many converging dependencies. Hence, "Complete requirements document" is a distinctive milestone for software applications development projects, and "Complete market requirements document" is a characteristic milestone for product development projects. While these milestones relate to the completion of key deliverables, other types may include the start and finish of major project phases, major reviews, events external to the project (e.g., trade show date), and so forth.

Developing a Milestone Chart

A relatively simple procedure for developing a Milestone Chart includes several steps that build heavily on a schedule with activity dependencies that is constructed in a separate procedure. We nevertheless include it here in order to present an integrated picture of Milestone Chart development.

Prepare Information Inputs. A Milestone Chart is as good as its inputs are, specifically:

- ▲ Project scope
- ▲ Responsibilities

- ▲ Schedule management system
- ▲ Project schedule, possibly one that shows dependencies.

Having a solid definition of the project scope provides those who schedule with a good understanding of the milestones being scheduled. The quality of the chart is certainly expected to be higher when the owners of milestones are responsible for their scheduling (see the box that follows, "Who Owns Schedules?") and follow guidelines established in the schedule management system (see the "Schedule Management System" box earlier in this chapter). If scheduling of milestones is based on the previously developed detailed schedule, the quality of the chart is bound to further improve.

Prepare a Detailed Schedule Showing Activity Dependencies. This may be any of the network diagrams described in this chapter. Their value is in dependencies between activities that they show. The schedule should be used later in the step of sequencing milestones.

Who Owns Schedules?

Involvement of project participants in developing schedules hinges to a great extent on the organizational strategies for project management. In the matrix environment, for example, many players are involved—team members, project managers, functional managers, the project office, and executives.

Team members typically own work packages/tasks, reporting their completion and estimating how much more time will be necessary to complete each unfinished work package or task. While they have to know some scheduling terms, such as start date, finish date, data (reporting) date, resource availability, and so on, there is no need for their extensive knowledge of scheduling theory. As providers of resources to a project, functional managers care about the accuracy of the estimates and availability of resources when projects need them [6]. Like team members, their knowledge of scheduling theory is basic.

Project managers are the ultimate users and owners of the project schedule. They facilitate schedule development and monitor the data furnished by team members for completeness and feasibility. Then they run data and schedules on a computer or use the project office for this purpose and check out the results. Finally, they communicate with functional managers and make schedule modifications. They need a decent amount of knowledge about scheduling theory. The project office (or the scheduling group) should have theoretical scheduling experts who are capable of designing and maintaining a project scheduling system in which all other players have to fit. Also, their knowledge in running scheduling software and checking out time, cost, and resource estimates in order to support the system and individual projects in need is essential.

The job of executives in project scheduling is not about scheduling theory, tools, or software. Rather, their focus is on asking questions, reading reports, directing project-related personnel and providing overall support. Like a well-conducted orchestra of master musicians, these players need to synchronize their actions to produce a meaningful scheduling concert.

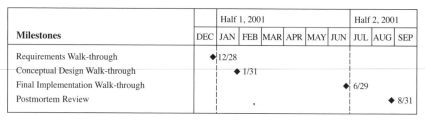

Milestones		Half 1, 2001							Half 2, 2001		
	DEC	JAN	FEB	MAR	APR	MAY	JUN	JUL	AUG	SEP
Requirements Walk-through		◆12/28									
Conceptual Design Walk-through			◆ 1/31								
Final Implementation Walk-through							◆ 6/29				
Postmortem Review										◆ 8/31

Figure 6.3 An example of the Milestone Chart.

Select the Type of the Milestone Chart. Again, you may develop a management type of the chart with only a few high profile milestones intended to entertain managers or outside stakeholders. Another option may be a working type of the chart designed to help manage work to get milestones accomplished. Which one is better? It depends on the situation. An example may help visualize this question. For a certain type of project, the company can use high-level Milestone Charts with five standard milestones identified as break points between project phases. At these milestones, senior management reviews the project and decides to either continue or kill the project. In addition, the company uses a more detailed chart with 14 milestones, at which the project team reviews major project deliverables. For another example of such a chart, see the "Management Milestone Chart" box that follows. Selecting the type of chart and clearly defining its purpose is an important step in building the Milestone Chart.

Select Milestones. In this step we let the type of the chart that is chosen guide the selection of the desired milestones. Consider all types of milestones—key deliverables, start and finish of the project and its major phases, major reviews, important events external to the project, and so forth. Which ones are key to the project progress? If a company uses standard milestones, the answer is simple: those same standard milestones. If this is not the case, consulting management may help in making the choice.

Sequence Milestones. Sequencing milestones is about studying dependencies between activities and comprehending how their outputs will converge to a culmination point—the selected milestone. Marking their position in the detailed schedule will provide their sequence, indicating which activities have to be started or completed to proclaim that this milestone is accomplished.

Management Milestone Chart

These milestones are eight sacred cows in one company's product development projects: product concept approval, requirements definition, plan and specs review, design complete, product evaluation review, launch plan, launch complete, and product release. Spanning from one to two years and costing millions of dollars, these projects are the company's engine of growth. The milestones signify the end of key phases and require upper management review. They are displayed in the management milestone chart that is used to report progress to senior management.

> **Tips for Milestone Charts**
> - Don't cram milestones. Space them out.
> - Use both charts for key events and detailed milestones.
> - Use the charts in both large and small projects for both the plan and actual project progress.
> - Use the chart in conjunction with another schedule showing activity dependencies.
> - Team-developed Milestone Charts lead to higher quality, better buy-in, and stronger commitment of team members.

Draft and Refine the Milestone Chart. Once milestones are marked on the detailed schedule with activity dependencies, the Milestone Chart is essentially drafted. This is the time to ask the following questions. Are all necessary milestones there? Are they logically sequenced? In appropriate positions on the schedule? It is important to ensure that enough milestones have been chosen so as not to have a prolonged period where there are no milestones. It is easy to select all milestones at the project beginning and end, where many activities start and complete. However, that would leave the middle void of milestones, reducing the ability to control project progress. As these questions get answered, the information will be created to make any changes and distill the chart.

With milestones marked on the schedule, two points should be discernible. The first is the laborious work that must be done. Second, the selected milestones that signify keys to the project progress must stand out so as to avoid getting lost in the details of the laborious work. In short, not only should the trees (details) be evident, but the forest (milestones) as well.

Finalize the Milestone Chart. Quite understandably, managers won't like to bother with detailed schedules. Rather, they will ask for a chart showing only milestones so they can glance and figure out key data on project progress. You should use information from the detailed schedule with milestones—timescale, milestone names, and time positions of milestones—in order to prepare the Milestone Chart (see Figure 6.3). To do so, list milestones on the vertical axis of a sheet, draw the timescale across the horizontal axis, select symbols for milestones (diamond, for example), and place the symbols across the timescale. (Also see the box above, "Tips for Milestone Charts.")

Utilizing the Milestone Chart

When to Use. Traditionally, the Milestone Chart has been used to focus management on highly important events whether projects are large or small [12]. As a result, when used for this purpose, the chart typically shows a few key milestones. In other words, use the chart to provide key data on the planned and actual project progress to higher management. When the WBS is used in the project, these highly important events and key data are usually related to level 1 in the WBS. Recent developments have seen the increased use of multiple Milestone Charts at a time, whereas each chart corresponds to a certain level of WBS. Consequently, a chart on level 4 of a five-level WBS, for example, may easily have a couple of hundred milestones, each one tied to a work package. The chart is then used in conjunction with a detailed network diagram chart so that the effect of dependencies on milestones is discernible. This practice is well accepted in technology organizations that compete on fast project cycle times (see the "Hierarchical Schedule" section later in this chapter).

Time to Develop. The development of a chart with several key milestones can take 20 to 30 minutes, while a Milestone Chart with 300 milestones may take several hours, provided that a more detailed schedule with dependencies was previously prepared. Should more people be involved, you can expect the time necessary for the development to grow because of more interface among participants.

Benefits. A chart with a few key milestones—related to level 1 of WBS, for example—is in a solid position to capture and enjoy management attention and time with high profile project events (see the box that follows, "The Lack of Milestones May Kill Thousands") [13]. Not so with charts including many milestones linked with work packages. The gain from such charts is the ability to beef up the emphasis on goal orientation ("milestone accomplished" or "milestone not accomplished"), while reducing focus on activity orientation ("I am working on it").

Advantages and Disadvantages. The chart has the advantages in that it is

- ▲ *Visual.* It creates a pictorial model of the project ideal for effective communication between project team and management, and within project team.
- ▲ *Simple.* Minimal or no training is necessary for project participants and their managers to develop or interpret a Milestone Chart.
- ▲ *Useful as both planning and tracking tool.*

The Lack of Milestones May Kill Thousands

Napoleon, a master of warfare, never made more than the sketchiest plans even for his great projects [1]. Those sketchiest plans probably resembled milestones to signify key events in a campaign. When he invaded Egypt, he did so making his expectations clear to each of his generals. Their effort proceeded very much per this high-level schedule, leading them to the zenith of glory. In contrast, when Napoleon invaded Russia with his 400,000+ troops in June of 1812, he chose not to communicate the plans. Despite his generals' desire to have the plans communicated because of the harsh Russian winter, Napoleon never shared them. In December of 1812, Napoleon's defeated army left Russian soil, this time numbering only 20,000+ troops. The lack of a clear milestone schedule might have contributed to the finale of Napoleon's Russian campaign—from the zenith of glory to ashes.

Milestone Chart Check

Check to make sure you developed an effective Milestone Chart. The chart should have the following:

- Key milestones
- A logical sequence of milestones based on activity dependencies
- Well-spaced-out milestones
- Milestones appropriately positioned on the timescale

The Milestone Chart's disadvantages include the following:

▲ *When used separately from a detailed schedule with activity dependencies, it is difficult to understand how to reach a milestone.* This is especially true when the chart includes many milestones.

▲ *As the number of milestones grows, the chart loses its appeal.* By being overcrowded, it may become ineffective in managing the work, thereby defeating its own purpose. Coupling it with a schedule with activity dependencies may be the best option to mitigate risks associated with detailed Milestone Charts.

Customize the Milestone Chart. The features of the Milestone Chart that we have described so far are of a generic nature. To get the real value out of the chart, the project team has to customize it to fit specific needs of their project. Here are some examples of how that can be done.

Summary

This section discussed the Milestone Chart—a tool that shows milestones against the timescale in order to signify the key events in the project. Traditionally, the Milestone Chart has been used to focus management on highly important events, whether the project is large or small. It can also help strengthen the emphasis on goal orientation while reducing focus on activity orientation. Its value can be additionally improved if the chart is customized to specific project needs. In the following box, we summarize the key points in the Milestone Chart.

Customization Action	*Examples of Customization*
Define limits of use.	Use the Milestone Chart only for events on level 1 of WBS, with no more than six milestones on the chart.
	Develop templates that can be used as a starting point in constructing milestones for new projects.
Add a new feature.	Add a column to show the owners of milestones.
	Link the milestones to indicate dependencies.

Critical Path Method (CPM) Diagram
What Is a CPM Diagram?

CPM is a network diagram technique for analyzing, planning, and scheduling projects. It provides a means of representing project activities as nodes (see Figure 6.4) or arrows, determining which of them are "critical" in their impact on project completion time and scheduling them in order to meet a target date at a minimum cost [14].

Building a CPM Diagram

Constructing a CPM schedule is an exercise in patience and discipline that involves proceeding through several major steps. In it, as with all schedule development tools, a crucial step is to determine the level of detail and to identify activities. Although the step usually belongs to the scope planning process, we include it here. That helps explain this tool development in an integrated way.

Prepare Necessary Information. The process of building a CPM schedule is destined to produce a better product if quality information about the following inputs is developed:

- ▲ Project scope
- ▲ Responsibilities
- ▲ Available resources
- ▲ Schedule management system.

In this context, the purpose of the information about the scope is to provide schedulers with the knowledge of the project activities that are being scheduled. Clear definition of responsibilities—who does what in the project—points to who has the best information about the individual activities and should therefore schedule them. To develop realistic schedules, these "owners" of the activities also need to know which resources are available and when. Finally, the schedule management system will direct schedulers in developing and using the CPM diagram (see the "Schedule Management System" box earlier in this chapter).

Activity	Description	Immediate Predecessor	Duration (days)
a	Start		0
b	Get materials for a	a	10
c	Get materials for b	a	20
d	Manufacture a	b, c	30
e	Manufacture b	b, c	20
f	Polish b	e	40
g	Assemble a and b	d, f	20
h	Finish	g	0

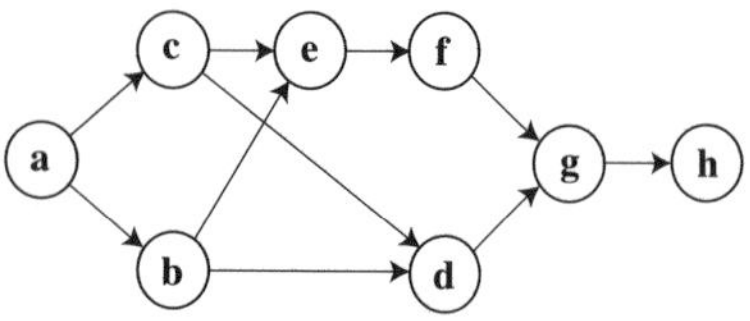

Figure 6.4 An example of a CPM Diagram.

Why a Team Approach to CPM Development?

Using a project team to build a CPM diagram is perhaps the most effective way of doing it. Here is why:

- Team members are usually the best source of knowledge about their piece of the schedule.
- Each team member can see where and why he or she is critical to the success of the project.
- The team can find creative ways to best sequence and shorten the duration of activities and the total project.
- As a unit, the team can focus its energy and mind on mission-critical activities.
- Involvement of team members enhances commitment and a sense of ownership of the project.

Determine the Level of Detail and Identify Activities. How can large or small individual activities that are identified influence the number of activities in the CPM? A rule at one company may help clarify this point. Large fab construction projects run at around 2000 activities, lasting from two to four weeks. This helps everyone realize what level of detail is acceptable and what is unacceptable. The goal for scheduling is to account for the complexity and size of the project in a way that gives the team enough information—not too little and not too much—to direct the daily work, identify interfaces between workgroups, and monitor progress at an effective level (see the box above, "Why a Team Approach to CPM Development?").

When the level of detail is set, you are ready to

- ▲ Brainstorm and identify activities that are necessary to complete in order to finish the project. This can be done by means of the Work Breakdown Structure, perhaps the most systematic and integrated way of activity identification process (see the "Work Breakdown Structure" section in Chapter 5).
- ▲ Refocus the attention on the established level of detail. If the number of activities is lower than the intended number, continue breaking down larger activities. If the number of activities is over the target number, combine similar activities to reach the target.

Sequence Activities. Sequencing is about identifying dependencies between activities by determining an activity's immediate prerequisite activities, called *predecessors*. A portion of the dependencies will be arranged in pure "technological order." These are termed *hard* or *logical* dependencies, meaning that the technology of work mandates such sequence. An example is that one must write the code before testing it; the other way around is not possible. Disregarding hard dependencies may lead to rework and project delay. But not all of dependencies are hard; some of them are soft or preferential ones. They are not required by the work logic but set by choice, reflecting one's experience and preferences in scheduling. For example, we may decide to write part of code, test it, write another part, test it, and so forth. Dependencies may also be dictated by availability of key resources. If two activities require the same resources, one will have to follow the other. Once the dependencies are established, they can be recorded, as we have done in the table in Figure 6.4.

Assign Resources and Estimate Activity Durations. The age-old rule of scheduling is that people and material resources get the project work done. As a result, it is logical to estimate an activity's duration by identifying resources necessary to successfully complete it. Consider, for example, 100 hours of work from a business analyst. This is the work (effort) time, which in the case of mature work technologies is calculated by dividing the amount of work by the production norms [15]. With the information that the analyst splits her work time between this and three more projects, and knowing the company's work calendar (50 hours per week only; no work on Saturdays and Sundays), she may need eight weeks to get it done. This is the calendar time. Repeat the cycle of identify resources; figure out work time, convert it to calendar time for all remaining activities, and write the calendar time in the fourth column of the table in Figure 6.4.

Draft a CPM Chart. Each activity is drawn on the network diagram as a circle or rectangle, with identifying symbols and duration within the circle or per convention chosen (Figure 6.4). This format is called AON (activity-on-node). Later in this chapter, we discuss another format of drawing network diagrams, AOA (activity-on-arrow). To pursue the AON format, indicate sequence dependencies by arrows connecting each circle (activity) with its immediate successors, with arrows pointing to the latter. For convenience, connect all circles without predecessors into a circle denoted "Start." Similarly, connect all circles without successors into a circle marked "Finish."

Identify the Critical Path (CP). Normally, the diagram shows a number of different paths from Start to Finish, defined as sequences of dependent activities. To calculate the time to pass through a path, add up the times for all activities in the path. The CP is the longest path (in time) from Start to Finish. It indicates the minimum time necessary to complete the entire project. Essentially, CP is the bottleneck route, the highest priority to manage.

There is another way to calculate CP: using the forward/backward pass procedure [16]. While adding up activity times is simpler for smaller projects, it is too cumbersome and difficult for larger projects. Rather, the large projects use the pass procedure [17]. Say, for example, you have the start date for a project. Then, for each activity there exists an earliest start time (ES). Assuming that the time to finish the activity is t, then its earliest finish time (EF) is ES + t. Figure 6.5 shows how to go through the forward pass to calculate ES and EF for each activity. The process, from left to right, is as follows:

- ES is the largest (or latest) EF of any immediate predecessors.
- EF is ES + time to complete the activity (t).

Suppose now that you want to finish the project by the time that is equal to the EF for the project. If so, you can define the concept of late finish (LF), or the latest time that the project can be finished, without delaying the total project beyond EF. Thus, LF is equal to EF. Similarly, you can define late start (LS) as LF – t, where t is the activity time. Building on these concepts, we can go through the backward pass, from right to left, to calculate for each activity (see Figure 6.5):

- LF is the smallest (or earliest) LS of any of immediate successors
- LS is LF – Time to complete the activity (t).

Now that the forward/backward pass is finished, note that Figure 6.5 indicates that in some activities, early start is equal to late start, while in some it is not. The difference between an activity's early and late start (or between early and late finish) is called *total float*. Total float is the maximum amount of time you can delay an activity beyond its early start without necessarily delaying the project completion time. *Free float* is another kind of float equal to the amount of time you can delay an activity without delaying the early start of any activity immediately following it. While an activity with the positive total float may or may not have the free float, the latter never exceeds the former. The formula to calculate free float is as follows:

The difference between the activity's early finish (EF) and the earliest of the early start times (ES) of all of its immediate successors.

In our example in Figure 6.5, activity b and d have free float of 5 days and 15, respectively, while all other activities have zero free float. The question is why one needs the two floats.

Activities on the CP have zero total float and are called *critical activities*. They are shown in Figure 6.5, with thick arrows connecting critical activities on the only CP. It is, however, legitimate to have multiple CPs, a common situation in fast-tracking projects. An activity with zero total float has a fixed scheduled start time, meaning that ES=LS. Consequently, to delay the start time is to delay the whole project, which is why such activities are called *critical*. In contrast, activities with positive total float offer some flexibility. For example, we can relieve peak loads in a project by shifting activities on the peak days to their late starts. That won't impact project completion time. But this flexibility may vanish quickly. Consider a path with a very small total float, which we call *near-critical path*, the second-highest priority to pay attention to. If we let an activity on the near-critical path slip, its small total float may be gone and it becomes critical path. In case of the free float, we can delay the activity start by an amount equal to (or less than) the free float without affecting the start times or float of succeeding activities.

Review and Refine. Look closely at the drafted diagram and ask the following questions:

- ▲ Has any important activity been left out or forgotten to be included in the chart?
- ▲ Is the activity sequencing logical?
- ▲ Are durations of activities reasonable?
- ▲ Is the project scheduled as time-constrained or resource-constrained? (See box that follows, "Time- or Resource-Constrained? Or Both?")

Time- or Resource-Constrained? Or Both?

Although Intel is in a constant time-to-market race with its competition, project managers in one of its divisions understand well the problem of the relationship between time and resources when developing their schedules. The problem is in that no matter how fast they want to complete their projects, the availability of resources is limited. This helps classify schedules into [4]:

- Schedules under a time-constrained situation. The project must be finished by a certain time (called the drop-dead date in Intel language), with as few resources as possible. Here, it is time, not resources, that is critical. These are typically projects of higher priority.

- Schedules under a resource-constrained situation. The project must be finished as fast as possible without exceeding a certain resource limit. Here it is resources, not time, that is critical. Projects like these are usually of lower priority.

Between these two extremes are resource-leveling projects with their medium priority. In them, once a schedule is developed, project tasks are shifted within their float allowances to provide smoother period-by-period resource utilization. As long as management clearly communicates in which category each project is, project managers face no problem—most of commercially available software they use already have algorithms to develop schedules for any of the three situations. The trouble arrives when project managers are told their projects are time-constrained but with limited and insufficient resources. Faced with such systems constraints [8], they know they have to find resources on their own. So, what do project managers do? They make do. They throw in overtime, work long hours, convince team members to do the same, and navigate their way through. Most of the time they succeed. After all, Intel's culture is all about performance.

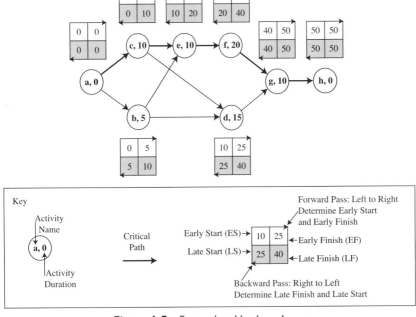

Figure 6.5 Forward and backward pass.

This is the time to answer the questions and, if the answers require, to make necessary corrections.

If the company competes on time, this is also the time to check if it is possible to reduce the project duration. The only avenue to do that is to find ways to shorten activities along CP. This is possible by fast tracking or crashing, or their combination [18]. Note that fast-tracking or crashing noncritical activities is irrelevant, because it does not reduce the duration of CP. Fast-tracking means changing the hard and soft dependencies—in other words, changing the logic of the diagram by obliterating previously established dependencies and creating the new ones. In the process, neither activity durations nor resource allocations will be changed. Simply, this can be done by taking certain activities from CP and overlapping them.

"Crashing" means shortening the duration of activities along CP without changing dependencies. The way to do this is by assigning more people to the activities, working overtime, using different equipment, and so on. The crucial question is whether the gains from the reduction of project duration exceed the costs of acceleration. For the majority in time-to-market projects the answer to the question is yes. For more details, see the "Schedule Crashing" section in Chapter 12.

Tips for CPM Diagrams

- If you need to accelerate the schedule, do it by fast-tracking and crashing.
- Watch out! Accelerating the schedule may increase the number of critical activities. While in earlier times 10 percent of all activities were critical, in today's fast schedules, you often see 40 percent to 50 percent of activities being critical.
- Sprinkle major milestones over your CPM chart. It helps you see woods (milestones) and the trees (activities).
- Color-code activities performed by various resource providers in order to identify their interfaces and provide their coordination.
- Develop template CPM charts. Then, use them consistently to develop the schedule for new projects.

Utilizing CPM Diagram

When to Use a CPM Schedule. The CPM tool was originally developed for large, complex, and cross-functional projects [19]. Even nowadays that is the major area of CPM application, because it can easily deal with large number of activities and their dependencies, directing our attention to most critical activities [20]. With the dissemination of PM and CPM knowledge, it is not unusual to see CPM used in smaller projects.

A fine application of CPM can be found in conjunction with the Gantt Chart. In short (see the "When to Use" discussion in the Gantt Chart section), extracting from the sizable CPM Diagram activities due in the next one or two weeks, presenting them in the Gantt Chart format, and handing to the people portions of the Gantt Chart they are responsible for provides clear and practical partial short-term outlook schedules.

Time to Develop. A skilled, smaller project team can take a day to a day and a half to build a 250-activity CPM schedule. As the team grows in size, so does the time it needs for communication and, consequently, CPM development time. Starting the development off a template schedule is a sure way to reduce the time.

Three Don'ts for CPM Diagrams

- Don't let the CPM chart control you. It is just a schedule and won't make a decision for you. You will.
- Don't consider it gospel. If there is a better way to schedule, go for it!
- Don't throw it aside when your project starts slipping. Review it, update, and improve, then use it again!

Benefits. Having a CPM schedule helps the project manager to see the total completion time, understand the sequencing of activities, ensure resources when necessary, monitor those that are critical, and measure progress (or lack of it). This is easier to accomplish if certain rules are followed (see the boxes on page 190, "Tips for CPM Diagrams" and "Three Don'ts for CPM Diagrams").

Advantages and Disadvantages. CPM offers

- ▲ *Graphic appeal.* CPM is easily explainable, even to the laypeople, by means of the project network diagram that clearly charts the technological order of work. Data calculations are not difficult and can be handled readily and quickly by personal computers.
- ▲ *Intuitive logic.* In a simple and direct way, it displays the dependencies in the complex of activities comprising a project. The logic reveals which activities have to be executed before others can proceed.
- ▲ *Focus on top priority.* It pinpoints attention on the small group of activities that are critical to project completion time. This focus greatly adds to higher accuracy and, later, precision of schedule control.

CPM's shortcomings include the following:

- ▲ *It looks convoluted and perplexing to first-time users.* The multitude of activities flowing into and branching out of the web of interrelated paths creates a sense of disorientation and complexity that is difficult to comprehend. As such a user put it, "When I used it for the first time, CPM appeared to me as mind-boggling crow tracks, almost impossible to decipher.
- ▲ *It is timeless, since it is a diagram without a timescale.* Certainly, the diagram is accompanied by a tabular report of schedule dates, but for today's project managers pressed for time, speed, and efficiency, the inability to quickly read dates and float off the CPM Diagram is frustrating.
- ▲ *It appears overwhelming* when it comes to maintaining it for a very dynamic project, where frequent changes are the order of the day. Consequently, updating and changing the schedule may be very time-consuming.

Variations. The CPM diagram taken here is one of the AON format. Other formats include CPM with AOA, PERT (Program Evaluation and Review Technique and precedence diagram (PDM). In AOA, activities are shown as arrows, and the arrows are connected by circles (or dots) that indicate sequence dependencies. In that way, all immediate predecessors of an activity lead to a circle at the beginning point of the activity arrow, while all immediate successors stem from the circle at the arrowhead. Thus, a circle becomes an event, where all activities leading to the circle are completed.

CPM is very similar to PERT [21], but while CPM's activity duration estimate is deterministic, PERT uses a weighted average to calculate expected time of activity duration as follows:

$$TE = (a+4m+b)/6$$

where

a = optimistic time estimate

b = pessimistic time estimate

m = most likely time estimate, the mode

PERT has been used primarily in research and development projects, while CPM, which was originally developed for construction projects, has spread across other industries.

PDM is an AON network that allows for leads and lags between two activities (see the "Time-Scaled Diagram" section in this chapter for more details on leads and lags). This makes it easier to portray rich and complex dependencies of the real-world projects, giving PDM a wider application across industries and an edge over CPM and PERT. These two methods allow for leads and lags only by splitting activities into subactivities, leading eventually to a significant increase in the number of activities in the network, and making it more complex and difficult to manage.

Customize the CPM Diagram. In this section we have described a generic type of the CPM. To benefit most from the diagram, you should adapt it to meet the specific project's need. Following we offer several examples of adaptation.

Customization Action	*Examples of Customization Actions*
Define limits of use.	Use CPM for cross-functional projects with more than 25 activities.
	Decide on the format of the diagram: AOA or AON.
	Develop templates that can be used as a starting point in building CPM schedules for new projects.
Amend an existing feature.	Amend the key to identify what information will be shown in the diagram. Some choices are activity name, duration, resources, owner, and so on.
	Decide on how to mark the critical path.
Add a new feature.	Add major milestones to the diagram and link them with activities to show dependencies.

CPM Diagram Check

Check to make sure you developed a good CPM Diagram. The diagram should show

- All necessary activities to complete the project
- A logical sequence of the activities
- No loose ends (beginning and ending points of all activities should be tied into other activities' beginning and ending points)
- Durations and resources assigned to all activities
- Critical paths

Summary

This section featured the Critical Path Method (CPM) Diagram. This tool provides a means of representing project activities as nodes or arrows, determining which of them are "critical" in their impact on project completion time. Originally developed for use in large, complex, and cross-functional projects, the CPM tool is nowadays employed in smaller projects as well. Having a CPM schedule helps the project manager to see the total completion time, understand the sequencing of activities, ensure resources when necessary, monitor those that are critical, and measure progress. A list of key points in building the CPM diagram is highlighted in the box above.

Time-Scaled Arrow Diagram (TAD)

What Is TAD?

TAD is the only Critical Path Method displayed against the timescale (see Figure 6.6). Its purpose is to analyze, plan, and schedule projects in order to meet a target date at a minimum cost. In the process, TAD determines which project activities are "critical" in their impact on project completion time so the project team can focus on them. As its nominal definition states, TAD is an activity on arrow tool.

Developing a TAD

Building a TAD is a workout that requires endurance and discipline, unfolding through several major steps. Like with all schedule development tools, the first thing to do is to figure out the level of detail and identify activities. Typically, this is part of the scope planning. We include it here in order to provide an integrated view of this tool's development.

Prepare Information Inputs. TAD's quality is heavily dependent on solid information about the following:

- Project scope
- Responsibilities
- Available resources
- Schedule management system

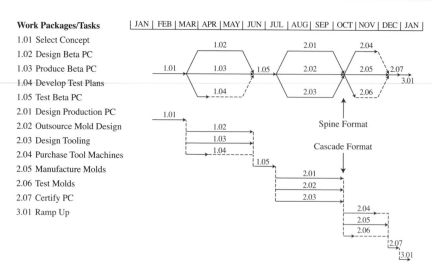

Figure 6.6 An example of a Time-scaled Arrow Diagram.

Clearly, you need to understand the project scope in order to schedule the project activities. The purpose of knowing who is responsible for certain activities is to indicate who will schedule and then manage the activities. For this to be possible, these schedulers need information on resource availability. Guidelines for how to develop and maintain TAD will come from the scheduling management system (see the box "Scheduling Management System" in the "Gantt Chart" section).

Determine the Level of Detail and Identify Activities. How large or small individual activities are influences the number of activities in a TAD. A company's example may shed more light on this issue. Large new-product-introduction projects typically include 300 to 500 activities, with durations between three and five weeks. The message is, only this level of detail is satisfactory. Its intention is to provide just enough information—neither more, nor less—than what one needs to direct and monitor project work of a certain size and complexity. Providing more or less information than necessary may either overload the team or deprive it of essential information, respectively.

Once the level of detail is chosen, the next actions are as follows:

▲ Identify activities that have to be performed to get the project done. As is the case with all types of scheduling tools, an excellent way for this is to employ the Work Breakdown Structure, perhaps the most systematic and integrated way of activity identification process. Activities necessary for developing TAD are those needed to produce work packages, the lowest-level elements of WBS.

▲ When the activities are identified, concentrate on the level of detail that was selected earlier. If the actual number of activities is smaller than the targeted number, resume breaking down larger activities. Should the actual number exceed the target number, some similar activities can be merged to reach the desired level of detail.

Sequence Activities. Sequencing is establishing dependencies among activities. This means putting activities into a specific order by determining an activity's immediate prerequisite activities, called predecessors, and leaving no loose ends (see the box that follows, "Loose Ends May Mislead the Team). As explained in the section on the CPM diagram, some dependencies will be hard or logical; others will be soft or preferential. Both types can be used to create overlapping activities, of course, for the purpose of fast-tracking TAD. For example, instead of writing the complete program and then testing it, you may decide to overlap the two activities by writing part of program, testing it, writing another part, testing it, and so on.

To overlaps like this and to other types of relationships between activities, TAD makes available another way to envisage dependencies: Finish-to-start (FS), start-to-start (SS), finish-to-finish (FF), and start-to-finish (SF) [22]. To each of those we may specify a lead/lag factor in order to accurately define the dependency. (See the examples in Figure 6.7.)

How much are these dependencies really used in practice? How much do we need them? Traditionally, FS has been used all the time. Generally, SS has gained huge popularity in businesses competing on faster project cycle times. If you are in that business, SS is what you need to fast-track a project [23]. Its major benefit is in allowing parallel work. Consider, for example, developing a new computer, where software development has an SS with lag dependency with hardware development. To start its work, the software team needs at least a hardware design, and once they get it, they can carry on with their work in parallel with the hardware development. FF and SF, on the other hand, haven't seen as much action.

Loose Ends May Mislead the Team

We frequently see TADs with loose ends, including arrow tails and arrowheads that are not connected to other activities. When we ask project managers why, we often get an answer like, "I only want to show the critical path and dependencies on it. Other paths and their dependencies are not important to me." This is a poor practice. To determine the critical path, the team has to evaluate all paths with properly connected activities. If there are loose ends, the team may not see the real critical path. Then the whole purpose of having a TAD—focus on really critical activities—is defeated.

FS—Activity B can start only when activity A is finished. When you add a two-day lead/lag, activity B can start only two days (lead/lag) after activity A is finished.

SS—Activity B must not start before activity A starts. Adding a lead/lag, activity B must not start before activity A has been in progress for at least two days.

FF—Activity B must be finished at the same time as activity A. With a lead/lag, activity A must be complete at least two days before activity B can be finished.

SF—Activity B cannot be completed before activity A starts. Inserting a lead/lag, activity B cannot be completed before seven days from the start of activity A.

Figure 6.7 Types of dependencies between activities.

Assign Resources and Estimate Activity Durations. The heart of schedule development is resource allocation and scheduling. Although it was touched on in the section on CPM, we repeat it here. The first rule of this allocation business is to identify resources necessary to successfully complete activities. For example, you may need a cost estimator and precisely 80 hours of his work (effort) time. With regard to mature work technologies, you can obtain this number by dividing the amount of work by the production norms [15]. Assume now that the estimator is shared by this one and two more projects. Look at the company's work calendar as well (50 hours week only; no work on Saturdays and Sundays). All this information tells that the estimator will need to spread his 80 hours' worth of work over 10 calendar weeks (calendar time). Reiterate steps of identify resources, figure out work time, and convert it to calendar time for all remaining activities. Estimating activity durations can be tricky, especially in multiproject environments (see the box that follows, "Switchover Time Adds to the Schedule Inaccuracy").

Draft the TAD. Draw activities as arrows, connecting them to one another—arrowhead to arrow tail to indicate the sequence of dependencies (see Figure 6.6). In that manner, all immediate predecessors of an activity lead to the beginning point of the arrow tail, while all immediate successors stem from the arrowhead. Thus, the beginning point of the arrow tail becomes an event, where all activities leading to the point are completed. Obviously, a TAD can be drawn in two different formats (see Figure 6.6 and the box that follows, "Cascade vs. Spine Formatted TAD").

Identify the Critical Path (CP). Normally, a TAD shows a number of different paths—that is, sequences of dependent activities. The paths can be used in two ways to find CP. First, you can visually find a path composed of activities without float—no complex calculations necessary. Among all network users, this convenience is available only to those who practice TAD. Adding up the times for all activities in a path (as we did with CPM) will tell how long it is. As a reminder, CP is the longest path in TAD, which indicates the minimum time necessary to complete the entire project. Second, you can find CP with the forward/backward pass procedure and calculate total and free float, as explained in the "Identify the Critical Path" part of the "Critical Path Method (CPM) Diagram" section.

Switchover Time Adds to the Schedule Inaccuracy

Some 90 percent of projects are implemented in a multiproject management environment. This means that a practice of having multiproject managers run multiple projects at a time, anywhere from two to ten, is widely accepted [3]. While such an approach provides outstanding benefits in terms of better management [7], it also generates a unique problem that calls for very meticulous scheduling of the projects. In particular, the problem here is the switchover time. When the multiproject manager switches from one project to another, she needs switchover time to align her thinking and get into the new project, physically and mentally [8]. Since team members also operate as members of multiple teams at a time, they suffer from the same problem. As the projects grow in complexity, so does the switchover time [10]. Clearly, this time is a loss in a busy day of a multiproject manager and team member. For example, some experts indicate that this loss may be up to 20 percent of the multiproject manager's or team member's time, when involved in four projects at a time. The real problem is then that this switchover time loss typically is not taken into account in scheduling multiple projects. As a consequence, project schedules are notoriously optimistic and inaccurate. There are at least two strategies to deal with the problem. One is to reduce the available monthly work hours of a multiproject person used for scheduling purposes by the corresponding switchover time loss. Another is to increase the estimate of a multiproject person's work hours for a specific project by the corresponding switchover time loss. These strategies are not attractive, but they are necessary for the realistic development of TAD or whichever tool is used.

Cascade vs. Spine Formatted TAD

Cascade

- *One zone, one activity.* A zone is a horizontal swath or strip across a TAD printout. The cascade format allows only one activity per zone.
- *Why called cascade?* A well-arranged succession of activities, one per zone, appears like a cascade.
- *Less complex.* The cascade resembles a Gantt Chart, a simple-looking tool, which creates a sense of lower complexity that is easier to apply.
- *Less practical.* Because of one activity per zone, a larger TAD may require lots of sheets to print it and large wall space to post it.

Spine

- *One zone, multiple activities.* TAD printout allows multiple activities per zone.
- *Why called spine?* Activities are symmetrically arranged around a central path, usually the critical path, resembling a spine of the network.
- *More complex.* The appearance of the spine is much like any other network, something that looks scarily complex to some project managers.
- *More practical.* Because of multiple activities per zone, you can print a larger TAD on a single sheet of paper and post on it on a small wall space.

Review and Refine. This is the time to review what has been developed. Directly off a TAD, you can

- ▲ Read the critical path, floats, the activity beginning and ending dates, along with their durations.
- ▲ Check hard and soft dependencies—leads and lags.
- ▲ Identify opportunities to speed up the schedule, add leads and lags.

In other words, when reviewing and refining TAD, we can modify to achieve a better schedule for our needs.

Utilizing a TAD

When to Use a TAD. Like any network diagram, TAD's original target were large, complex, and cross-functional projects. TAD is well suited to such projects because of the ease with which it handles a large number of activities and their intricate dependencies, directing our attention to the most critical activities. While it is still applied for this purpose, a large number of project managers have used TAD for medium- and small-size projects (see the box that follows, "Tips for TAD"). In this case, TAD is typically drawn in a cascade and often called a "Gantt Chart with links." Perhaps more than anything else, this format facilitated the growing popularity of TAD.

A sizeable TAD can be used along with Gantt Chart to provide focus on the day-to-day project work. In particular, we can take out from TAD those activities due in the next week or two, show them in the Gantt Chart format, and have their "owners" use them as short-term outlook schedules. This provides a balance in focusing on both the big picture of the project with TAD and daily work details with Gantt Charts.

Time to Develop. A skilled, smaller project team can take a day to day and a half to build a 250-activity TAD schedule. The larger the team, the more time it needs to construct a TAD. Developing a TAD off a template is a good strategy to reduce the development time.

Benefits. TAD offers a unique benefit not available to any other network diagram: the ability to read directly off the schedule's timescale when the project and each activity starts and ends, as well as the total float. Like other network diagrams, TAD helps identify the total completion time, understand the sequencing of activities, ensure resources when necessary, monitor those that are critical, and measure progress (or lack of it). Perhaps the highest value comes from TAD's focus on priorities; TAD directs our mind on the vital few activities of critical importance to the project completion date. The outcome is higher accuracy and, later, precision of schedule control.

Advantages and Disadvantages. TAD's advantages are as follows:

- ▲ *Lowered complexity.* TAD combines the best of two worlds: retaining the visuality and timescale of a Gantt Chart without losing dependencies of a network diagram. This makes its much more attractive to use than other networks.

- ▲ *Graphic appeal.* TAD's unambiguous sequence of work, supported by the timescale, is easier to clarify than any other network diagram. Data calculations are not difficult and can be handled readily and quickly by personal computers.

- ▲ *Intuitive logic.* TAD exhibits the dependencies between constituent activities of a project simply and directly. This helps fathom the order of activity execution.

Tips for TAD

- If you need to fast-track your schedule, use SS with or without lags. Be prepared to see and manage 40 to 50 percent critical activities in the schedule.

- Building on the similarity between a cascade-formatted TAD to a Gantt Chart, spread the use of the TAD in all smaller projects. This will significantly enhance the quality of scheduling.

- Add major milestones to TAD to help vital events serve as beacons in the sea of activities.

- Insist on the use of template TADs to boost quality and productivity of scheduling.

TAD Check

Check to make sure you developed good TAD. The diagram should show
- All necessary activities to complete the project
- A logical sequence of the activities
- No loose ends
- Durations and resources assigned to all activities
- Duration of each activity against a timescale
- Critical paths and total float

Undoubtedly, while the addition of the timescale simplified TAD's basic network diagram nature, it can still look to some potential users as

▲ *Complex and confusing.* The multitude of interconnected activities, even in the cascade format, may leave an inexperienced person at a loss.

▲ *Overpowering and time-consuming* in situations requiring frequent and significant TAD updates and changes.

Customize TAD. What we have covered up to this point is a standard, generic TAD. For a TAD chart to be more useful, you need to modify it to best match the project situation. The following examples illustrate our point.

Customization Action	*Examples of Customization Actions*
Define limits of use.	Use TAD for functional and cross-functional projects with more than 25 activities.
	Use the cascade format for up to 100 activities; the spine format for over 100 activities.
	Use templates as a starting point in developing TAD schedules for new projects.
Amend an existing feature.	Amend the key to identify what information to show in the chart. Choices include activity name, duration, resources, owner, and so on.
	Decide how to mark the critical path.
Add a new feature.	Add major milestones to the chart and link them with activities to show dependencies.

Summary

Presented in this section was the Time-scaled Arrow Diagram, or TAD. This is the only critical path method tool displayed against the timescale. It is applied in large-, medium-, and small-size projects. TAD offers a unique benefit not available to any other network diagram. Specifically, it enables reading directly off the schedule's timescale when the project and each activity start and end, as well as the total float. Like other network diagrams, TAD helps you identify the total completion time, understand the sequencing of activities, assure resources when necessary, monitor those that are critical, and measure progress. There is a special value of customizing TAD for your own project needs. Key points in constructing a TAD are highlighted in the following box.

Critical Chain Schedule
What Is a Critical Chain Schedule?

A Critical Chain Schedule (CCS) is a network diagram that strives for accomplishment of drastically faster schedules (see Figure 6.8). It uses several unique approaches. First, CSS focuses on the critical chain, the longest path of dependent events that prevents the project from completing in a shorter time. Unlike the critical path, the critical chain never changes. Second, its activity durations are estimates with 50 percent probability. For this reason, they are significantly shorter than those used in other scheduling tools, which are often with 95 percent probability. Third, in contrast to the critical path, the critical chain is defined by the resource dependencies. Fourth, buffers are built to protect the critical chain in the course of project implementation. Finally, CCS requires certain behaviors by the project team. Introduced in 1997, CCS is a relatively new tool to the world of project managers [24].

Building a CCS

Prepare Inputs. Because of the inherent challenge of faster schedules that it seeks to make possible and its newness, CCS's quality is even more dependent on the depth and degree of definition of the following inputs than other schedule development tools:

- Project scope
- Responsibilities
- Dedicated team resources
- Schedule management system.

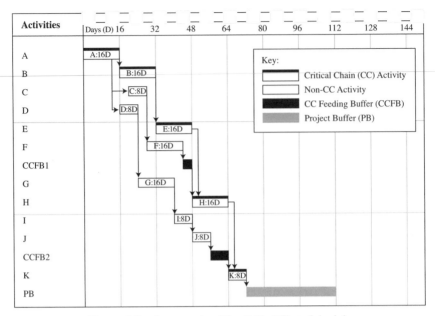

Figure 6.8 An example of the Critical Chain Schedule.

While the scope, responsibilities, and schedule management system will provide information about the what, and the who elements, as well as how to schedule project activities faster, the real emphasis is on CCS's requirement for dedicated team resources, meaning that team members work full-time on one project only. Because of this, the logic goes, members of the dedicated project teams are more productive than members who are shared by multiple project teams. A reason for this is the switching time cost created by one's work in multiple projects, as we discussed in the box "Switchover Time Adds to the Schedule Inaccuracy" in the TAD section. Although this is generally true, there are some exceptions. A study found that when a team member who is focused on a single project is assigned a second one, productivity often increases a bit because the team member no longer has to wait for the activities of other members working on that single project (see Figure 6.9). Rather, the team member can float back and forth between the two projects [25, 26]. When a third, fourth, and fifth project is added, however, the productivity plummets rapidly and the team member becomes a bottleneck of all projects he or she is involved in. This is why the CCS approach insists on using dedicated teams.

Determine the Level of Detail and Identify Activities. The number of activities in CCS is closely related with the size of activities in it. Therefore, choosing to have approximately 100 or 300 or 500 activities will help determine how large individual activities will be. To illustrate this point, look, for example, at one company's golden rule. Large product development projects, including between 5,000 and 10,000 person-hours, will have around 500 project activities that range in duration from two to four calendar weeks. Not only does this clearly tell everyone that neither180-activity schedule nor15-week activities are tolerated, it also spells out the company's belief about the right level of detail. Given the complexity and size of the project, such level of detail provides sufficient information to manage the project without making it unnecessarily burdensome and time demanding.

Once the decision has been made about the level of detail, these actions should be taken:

▲ Brainstorm and identify activities that are necessary to complete in order to finish the project. As with other scheduling tools, if necessary, resort to the Work Breakdown Structure for the activity identification (see the "Work Breakdown Structure" section in Chapter 5).

▲ In this process, disregard how large the activities are; rather, ensure that all necessary activities are on the list.

▲ Go back to the chosen level of detail. If below the intended number of activities, continue breaking down larger activities; if over the target number, combine similar activities to reach the desired goal.

Sequence Activities. Sequencing means arranging a logical network of activities and their dependencies. Deep knowledge of nature and flow of work are prerequisites here. The principle of sequencing is to know that a preceding activity produces outputs that become inputs to an activity that follows. If the obtained network diagram fails to observe the principle, the bets are we are missing the logic of the project work, resulting in rework and delays in project execution.

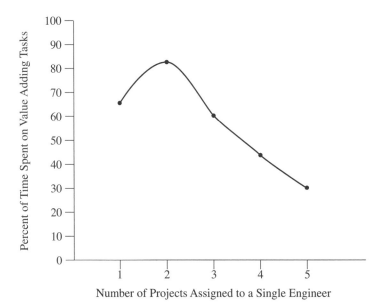

Figure 6.9 Productivity of team members.

Assign Resources and Estimate Activity Durations. Since people and material resources get the project work done, they dictate activity durations. Therefore, a natural starting point for estimating the durations is, "What resources do I need to successfully complete this activity?" The answer should provide the names of resources and work time for each one of them to complete the activity—for example, 100 hours of work from a programmer. The key point here is that CCS uses a unique technique of activity duration estimating that does not allow for contingencies (see the box that follows, "When Estimating Durations, No Contingency Safety Allowed"). Considering that the Critical Chain approach requires dedicated teams, and knowing the company's work calendar—5 days a week, 10 hours a day—those 100 hours turn into 10 workdays or 14 calendar days. Naturally, the estimation of each activity should undergo this process.

Identify the Critical Chain. The Critical Chain (CC) is the longest path in the network diagram, considering activity and resource dependencies. Or, it is the sequence of dependent events that keeps the project from completing in a shorter time.

Indicate a Resource Buffer. CCS always considers the resource constraints and includes the resource dependencies that define the overall longest path. Practically, this is handled by adding resource buffers to protect CC from unavailability of resources. Resource buffers are added to CC only, do not take any time in CC, and are termed *resource flags*. For example, any time a new resource will be used in a CC activity, we will add a resource buffer. This signals to the project manager and resource provider when to make the resource available to work on a CC activity ready to start. Since timely resource availability is critical to rapid execution that CCS advocates, some companies use incentives to reward behavior of early delivery of activity outputs and standby time of resources [5, 27].

Create a Project Buffer. Unlike other schedule development tools, CCS uses a novel concept of the project buffer. Its purpose is to protect the project due date by aggregating contingency time in the form of the project buffer at the end of CC (for management of the buffer, see the "Buffer Chart" section in Chapter 12). There are several methods to determine the buffer duration. One of them is to divide the duration of the critical chain by two (called the "50 percent buffer sizing rule"). The buffer is used to absorb uncertainty or disruptions that may occur on CC and has no work assigned to it (see Figure 6.8).

Create Feeding Buffers. Protecting the CC with the project buffer is not enough. There is a significant risk that activities that are not on the CC but feed into it may slip to the point of pushing out the CC. To protect the CC from the risk, we can aggregate contingency time at all points where non-CC activities feed into the CC (see Figure 6.8). These contingency times are termed *CC feeding buffers*. During the project implementation, these buffers will be used to absorb uncertainty or disruptions that may occur in non-CC activities. To determine these buffers, use one-half of the sum of the activity durations in the chain of activities preceding the buffer. No work is assigned to the buffers.

Review and Refine. The final step in developing a CCS is to review what has been developed:

- ▲ Take a close look at the CCS to read the critical chain, activity beginning and ending dates, their durations, and buffers.
- ▲ Do they make sense? Are there any opportunities to improve CCS?

When Estimating Durations, No Contingency Safety Allowed

Most project managers tend to include contingency time into each activity estimate without specifying it. The reason is simple: add the safety time. The Critical Chain Schedule strives to eliminate the safety. Here is what it means. Figure 6.10 shows a typical distribution of activity time performance. The solid line (the left ordinate) tells us the incremental probability of a given activity duration time on the x-axis. The dotted line indicates the cumulative probability (the right ordinate) that the activity will be finished in a time less than or equal to the activity duration time on the x-axis [6].

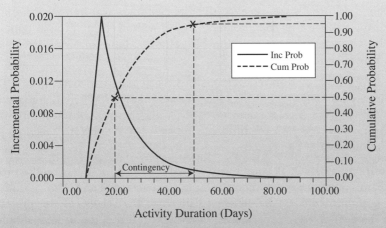

Figure 6.10 A typical distribution of activity time performance.

When project managers include contingency time within an activity, they really go for 95 percent probable estimate (cumulative probability). As the cumulative curve shows in the figure it is duration equal to or less than 50 days. Without the contingency time, the duration is less or equal to 20 days. That is 50 percent probable estimate. The difference between the 95 percent probable estimate and the 50 percent probable estimate is contingency time, 30 days in this example. To avoid excessive activity duration, and speed up the schedule, CCS eliminates the contingency time, using only 50 percent probable estimates.

> **Tips for Critical Chain Scheduling**
> - Use CCS in important projects that may afford a dedicated team.
> - Apply CCS in companies that are in a time-to-market race, always striving to shave off their cycle times.
> - Support CCS with performance measurements that promote behavior of transferring an activity's output to the succeeding activities as early as possible.
> - Deploy CCS where there exists a strong performance culture willing to take on 50 percent probable estimates.

The goal here is to make modifications that would lead to a more effective schedule for the project.

Utilizing the Critical Chain Schedule

When to Use a Critical Chain Schedule. The most appropriate application of CCS is for a dedicated project team seeking a significant reduction of the project cycle time in a company with an outstanding performance culture. The only job of this team is their project. Equipped with all necessary resources, the team operates in a company whose performance culture focuses on exceeding its customer expectations, creating maximum value for its shareholders, and providing strong growth opportunities to its employees (see the box above, "Tips for Critical Chain Scheduling").

Time to Build. A well-trained, small project team can develop a 250-activity CCS in anywhere from a day to day and a half. When a large team is involved in the development, more time may be needed for additional communication. Significant time can be saved if a template schedule is used at the start of the exercise.

Benefits and Costs. Beyond every schedule's purpose of having the project team understand the timetable for project activities and their personal time commitments, CCS intends to improve the results of the project team. Hence, as early experiences with CCS indicate, the project should see considerable improvements in schedule and cost performance, because CCS

- ▲ Is an important eye-opener. Simply, CCS recognizes that the interaction between activity durations, dependencies, resource requirements, and resource availabilities has a major impact on project duration [28].
- ▲ Protects a deterministic baseline schedule. This protection helps fight uncertainties by using feeding, resource, and project buffer to set a realistic project deadline. As a result, CCS offers a significant potential for radical acceleration of project completion times.

Project managers at 3M and Lucent reported up to a 25 percent reduction in project cycle times when using the CCS approach [5]. For those in rapid project cycle time business, this may be a tool worth trying.

▲ Makes a case for truth in activity duration estimation [29]. In contrast to other tools without a mechanism to prevent project managers from building contingency safety into activities, CCS's "no contingency allowed" mechanism eradicates such tendencies.

▲ Is a good fit for corporate and project achievers. By not allowing contingency safety, CCS drives the behavior of excellence, challenging even the most demanding performance cultures around.

▲ Enhances discipline. Greater and more open communication among internal project stakeholders results in enhanced discipline in project scheduling and control [30].

As a relatively new tool, CCS has been heavily scrutinized from both practitioners and researchers. Critics believe that the CCS approach includes assumptions, requirements and practices that may reduce its application and effectiveness. Specifically, CCS

▲ Needs more empirical testing. CCS is a relatively new tool, even though its concept has been around since 1964 [28]. Hence, there is little empirical evidence—beyond several case studies—about its true efficacy.

▲ May be based on an overly negative assumption. Some critics argue that the assumption that large contingency safety is "padded" into the activity duration estimation in critical path scheduling is overly negative and exaggerated [31].

▲ Requires a dedicated project team to apply it. In other words, when working at several projects sharing the same resources, project teams should not use the CCS approach. While the requirement makes economic sense based on a higher productivity of dedicated teams, it is not much in tune with political realities of today's organizations, where simultaneous work at multiple projects is almost a given.

▲ Excludes the use of project milestones. The lack of the milestones in CCS may severely hamper the ability of the project team to coordinate schedules with external vendors, often delivering critical components for the project [32].

▲ May overestimate the buffer. Experts testing the CCS capabilities found that its 50 percent buffer sizing rule might lead to a serious overestimation of the buffer protection. Also, updating the CC might provide shorter project durations than not updating as CCS argues [28].

▲ Requires extraordinary behaviors. CSS is well suited to strong performance corporate cultures, which are in relentless pursuit of excellence. Perhaps only such cultures may accept the challenge of 50 percent probable estimates on a routine basis without finger pointing when the estimates are not accomplished.

▲ May contribute to the burnout of project participants. As evidence about dedicated teams in accelerated product development projects indicates, people in such environments work long hours under high stress, resulting in physical and emotional burnout. Since work in such environments strongly resembles the CCS approach with its dedicated teams and 50 percent probable estimates, it is highly likely that an extended application of CCS with the same project players will have similar consequences.

Advantages. CCS's advantage is that its basic concept, including 50 percent estimates of activity duration, critical chain, and buffers is relatively simple. Although not evident upon first sight, CCS doesn't require any proficiency in statistics.

How to Customize CCS. We have described a standard, generic CSS. Adapting it to your needs is a good way to get the most out of it. Here are a few examples of such actions.

Summary

The focal point of this section was the Critical Chain Schedule. This is a network diagram that strives for accomplishment of drastically faster schedules. Its uniqueness lies in using project activity durations that are significantly shorter than those used in other scheduling tools. Its most appropriate application is in a dedicated project team, seeking a significant reduction of the project cycle time in a company with an outstanding performance culture. Although it is an eye-opener for many project managers, CCS is a relatively new tool that needs more empirical testing. The following box highlights key points in constructing this schedule.

Customization Action	Examples of Customization Actions
Define limits of use.	Use CCS for cross-functional projects of the most strategic nature.
	Use the cascade format as in TAD (because of its familiar look to managers).
Amend an existing feature.	Amend the key to identify what information is to be shown in the schedule. The choices are activity name, duration, and so on.
	Decide how to mark the critical chain and buffers.

> **Critical Chain Schedule Check**
> Check to make sure you developed a good Critical Chain Schedule. The chart should have:
> - All necessary activities to complete the project.
> - A logical sequence of the activities.
> - Activities with durations without contingency safety—that is, based on a 50 percent probable estimate.
> - Project buffer to protect the project due date.
> - Feeding buffers to protect the Critical Chain.
> - Resource buffers to protect the Critical Chain.

Hierarchical Schedule

What Is a Hierarchical Schedule?

This is a multilevel schedule with varying level of detail on each level (see Figure 6.11). Every activity in a higher-level schedule is broken down into several activities, sometimes entire schedules. Typically, the schedules from various levels are connected at major milestones or events. WBS offers a good skeleton for hierarchical scheduling.

Constructing a Hierarchical Schedule

Developing these schedules hinges on the size of project. A very large project can easily use three levels, while a medium-sized project cannot warrant more than two levels. To offer a sense of a more difficult situation, we will use a three-level hierarchy here. In developing the schedules, the rules for building the type of the chosen schedule—whether Gantt Chart, network diagram, or Milestone Chart—should be applied.

Prepare Information Inputs. Quality of the Hierarchical Schedule begins with the quality of its inputs:

- ▲ Project scope
- ▲ Responsibilities
- ▲ Available resources
- ▲ Schedule management system.

Project scope will provide information necessary for a good understanding of the project activities that are being scheduled. Such scope information will be furnished to or developed by those who are responsible for certain work packages of the project. When scheduling the work packages, they will rely on the information about availability of resources, seeking guidance from the roadmap for scheduling—the schedule management system (see the box titled "Schedule Management System" in the "Gantt Chart" section).

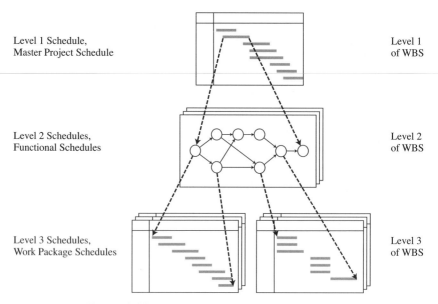

Figure 6.11 An example of the Hierarchical Schedule.

Develop Level 1 Schedule—Master Project Schedule. This is a summary schedule of the project, which is usually a Gantt or milestone chart format. It is an outline that will be used throughout the project as a tool to report progress to top management. Since it is developed in the phase of project planning, it is considered to be an initial plan as well. Included in the schedule are only principal activities and key milestones from level 1 or 2 of WBS (the project is level 0). Everything in the schedule is roughly estimated. For example, overall timing of phases, required resources, major dependencies, and major events in the schedule are all roughly estimated. It is important to highlight phases that need critical attention, such as material requirements, vital tests, and completion dates. Linking the development of the schedule with the definition of project objectives and strategies is a good strategy, because this is the time when the purpose and implementation methods for the project are shaped. In this linking, you should be able to easily chart, rechart, and evaluate multiple alternatives of the schedule in order to select the most viable one. Because the schedule is rough in its nature, it should not be used for the total integration of all project phases. That is why you need a level 2 schedule.

Figure 6.11 shows the master schedule for the product development project named OCI, consisting of eight work elements from level 1 of the WBS, each with a milestone at its end. It is in a Gantt Chart format and is used for the progress reporting to the product approval committee, an executive group responsible for the project selection.

For example, the first-level schedule (master project schedule) can be built of work elements from level 1 of the WBS. Work packages from level

2 of the WBS can be included into the second-level schedules (functional schedules because they are typically owned by functional units). Finally, constituent activities or task of the work packages (level 3) would be used to develop the third-level schedules (work package schedules).

Build Level 2 Schedule—Intermediate Schedule. The level 2 schedule will explode activities from the master project schedule, scheduling them in more detail. For this, a common choice is a Gantt or network chart, sprinkled with milestones. This schedule is a middle management planning and control tool, generally to assign responsibilities for work packages (level 3 of the WBS, for example). Clearly, activities in the schedule are not meant to provide daily or even weekly scheduling and directing of project work, except for the most critical activities. Still, it should be scheduled in sufficient detail to include major and minor milestones, crucial human resources, and sequencing and constraints in the project work. This enables you to scrutinize the structure of the project, dissect dependencies between various phases and milestones, and set boundaries within which shifting activities won't impact project delivery date. To deliver the punch, the level 2 schedule has to

- ▲ Allow the identification of the proper priorities.
- ▲ Determine critical and near critical paths.
- ▲ Get off to a quick start when the project receives a go ahead from management.

Project OCI had several level 2 schedules, each drawn as a Time-scaled Arrow Diagram in the cascade format. The largest of them had around ten work packages from level 2 of the WBS. Essentially, each level 2 schedule was the functional schedule for a certain discipline—marketing, electrical group, optoelectronic group, software hardware group, and so on.

Construct Level 3 Schedule—Detailed Schedule. This set of detailed schedules is intended to help lower levels of managers—work package managers, for example—in directing daily and weekly project work. Although it can be in the network format, a more frequent approach is to use a Gantt or Milestone Chart. Before getting to scheduling, you should size up available information, assess the size and complexity of the project, and weigh out experience and inclination of the involved project people. Then you should decide which of the following approaches to pursue for level 3 schedules:

- ▲ Create a fully integrated schedule for the entire project.
- ▲ Build a complete schedule for each activity from the level 2 schedule.
- ▲ Construct a separate detailed schedule for each phase as the project unfolds, and connect them via the level 2 schedule.
- ▲ Ask each project participant to develop detailed schedules for activities in the level 2 schedule that he or she is responsible for.

> **Milestone-CPM-Milestone Schedule Gets the Job Done**
>
> In a six-month-long, $70M project, a semiconductor company identified a few major milestones presented to the executive group (level 1 schedule). To direct the work and review progress weekly, the project management team relied on a CPM diagram with carefully weaved 400+ minor milestones (level 2 schedule). Minor milestones from the diagram were grouped into separate working milestone charts and handed to workgroups responsible for certain technical discipline (level 3 schedules). Including 40 to 50 milestones, each Milestone Chart was the key tool for doing work and reporting progress to the PM team. Rid of complex dependencies typical of the CPM diagram, the minor Milestone Chart provided clear and simple goals to go after.

Whatever the choice, the schedule must lay down the day-to-day, week-to-week work that an organization needs to successfully execute and control. It goes beyond saying that the schedule needs to be rooted in available resources, established dependencies, and time targets approved by management.

The choice of the OCI project was to use the Gantt Chart format to schedule in detail constituent activities of the individual work packages, keeping the number of activities per chart to less than ten (work package schedules). The total number of all activities in level 3 schedules was slightly below 500. While the OCI project offers one example of how to structure the Hierarchical Schedule, many other approaches are possible (see the box above, "Milestone-CPM-Milestone Schedule Gets the Job Done").

Utilizing the Hierarchical Schedules
When to Use. Hierarchical Schedules are used to confront challenges in two major project situations, including:

- ▲ Rolling wave scheduling. When starting some projects, we only have information about an early phase, while details about later phases emerge as the project progresses. In response, at the start of the project, we can develop a rough schedule encompassing the whole project and then build detailed schedules of the project's major phases as details become available. This approach is termed *rolling wave scheduling* and is implemented via the Hierarchical Schedule method.

- ▲ Multitiered schedule information for larger projects. Since different levels of management have different jobs in a project, to successfully get its job done, each level needs a different detail of schedule information. Different levels in Hierarchical Schedules provide those different details of schedule information. Tips for Hierarchical Schedules are described in the box that follows.

Time to Develop. A three-level Hierarchical Schedule with several hundred activities may take several days to develop in an organization that has a process for it. Also, the larger the group of people involved and the more inexperienced they are, the more time it takes. The good news is that that time is spread over project duration.

Benefits. The use of Hierarchical Schedules equips management of the project with the capability to integrate the scheduling of the earlier and later phases. Without the schedules, our scheduling and our attention would be focused on a piece of the project for which we have information, ignoring the whole project and its big picture. This would be tantamount to a runner who sees terrain just in front of her, having no idea how long the run is (a mile or 26 miles), what major milestones lie ahead (a steep hill, for example). Such a runner would have little chance to pace herself and successfully reach the finish line. In addition, Hierarchical Schedules enable managers and doers to perform their individual jobs.

Advantages and Disadvantages. Hierarchical Schedules offer advantages including:

▲ *Flexibility.* The project team is not forced to develop a schedule for activities for which they have no information. Rather, they can build a flexible big picture schedule and focus on first activities first, then schedule and deal later with activities that come later.

▲ *The right amount of information.* A frequent project tendency to inundate participants with both information they need and information they do not need is not an issue here. Rather, they get as much information as they need to do their job.

Hierarchical Schedules have disadvantages that may reduce their rate of use. Specifically, they are

▲ *Complex.* Multiple-level scheduling requires well-established process, skills, and involvement and coordination of many participants. This may make it a baffling and cumbersome exercise, raising the resistance level to its use.

▲ *Time-consuming.* Detailed scheduling of this type demands time, a quantity lacking in too many organizations. For this reason, some project participants may feel resentment toward it.

Tips for Hierarchical Schedules

● Go for a hierarchical schedule if you deal with uncertain projects, generating information as they unfold.

● Involve those responsible for work packages in the scheduling process, because they own the project work.

● If less experienced, use a combination of Gantt Charts and milestones in hierarchical scheduling to increase buy-in.

● Remember to use good links (mortar) to link schedules from various levels (bricks).

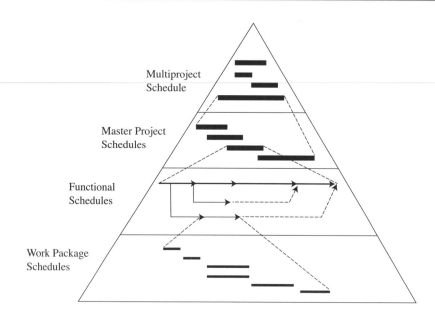

Figure 6.12 An example of the Hierarchical Schedule for multiple projects.

Variation. If we define multiproject management as an environment where multiple projects are managed concurrently [33], our Hierarchical Schedule (see Figure 6.12) can be adapted to help coordinate and integrate the multiple projects [25]. This requires an additional level of schedule, sitting on top of the three-level schedule from Figure 6.11 The additional level includes time lines for each of the multiple projects. Accordingly, we get a four-level schedule linking multiple projects, functional departments, and individual team members [34]. Product development and construction management are major users of this variation.

Customization Action	Examples of Customization Actions
Define limits of use.	Use a three-level hierarchical "one –Gantt-one TAD-multiple –Gantts" schedule for large and uncertain projects.
	Use a two-level hierarchical "one Milestone-multiple Gantts" schedule for smaller and uncertain projects.
	Develop templates that can be used as a starting point in developing Hierarchical Schedules for new projects.
Amend an existing feature.	Decide what information to show in the chart. The choices are activity name, duration, resources, owner, and so on.
Add a new feature.	Assign responsibilities by adding the name of the owner to each activity or milestone.

> **Hierarchical Schedule Check**
>
> Check to make sure you developed a good hierarchical schedule. The schedule should have
>
> - More than one level
> - Activities from higher-level schedules exploded into several activities or entire schedules in a lower level
> - Schedules linked together at major milestones or events
> - Proper level of detail on each level.

Customize the Hierarchical Schedule. To get the most out of the schedule, there is a need to modify it so it fits specific project needs. The following examples will help demonstrate this modification.

Summary

The Hierarchical Schedule—a multilevel schedule with varying level of details on each level—was the topic of this section. It ⚙ an effective tool when applied in rolling wave scheduling because it helps integrate the schedules of the earlier and later phases. Also, it works well in multitiered scheduling for larger projects. Customizing this schedule for specific project needs can generate possible extra value. In summary, the following box offers key points about the Hierarchical Schedule.

Line of Balance (LOB)
What Is the LOB?

LOB is a tool for scheduling and tracking progress of repetitive projects (see Figure 6.13). More specifically, LOB displays the cumulative number or percentage of components or units that must have been finished by a certain point in time for the schedule to be accomplished.

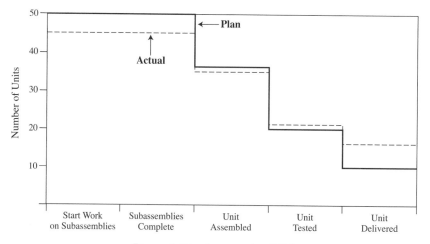

Figure 6.13 An example of LOB.

> **A Poor Bill of Materials Can Hurt Really Bad**
> A company received a request for special delivery of XYZ product according to a very tight schedule. An executive, who promised the customer a proposal the same day, asked for an LOB to highlight bottlenecks and assess the likelihood of meeting the schedule. The first step in preparing LOB calculation was getting BOM for a person to understand the parts buildup and define the program. Immediately a manufacturing representative discerned that the BOM was obsolete and not including all necessary parts. Updating BOM and checking missing lead times took a few days. As a result, executives could not keep their commitment and lost this business to a faster competitor.

Developing the LOB

Prepare Information Inputs. Developing a quality LOB requires the following information:

- ▲ Project scope
- ▲ Responsibilities
- ▲ Available resources
- ▲ Bill of materials, lead times for procurement, and production norms
- ▲ Schedule management system.

As with other scheduling tools, quality information about the scope, responsibilities, available resources, and schedule management system will help users of the tool understand the what and who issues, along with resources and scheduling requirements. What makes LOB different from other scheduling tools is the need for a bill of materials (for production projects, see the box above, "A Poor Bill of Materials Can Hurt Really Bad"), lead times for procurement, and production norms. This information is used to determine the program and then the time line for production/construction, as shown in the upcoming discussion on developing an S chart for multiunit programs.

Set the Objective. Picture a production or construction project aiming at producing a quantity of the end deliverables. A production example might include a series of special orders for 50 connection cables per agreed delivery commitments. To make the project more complex, the cables cannot be completed in one batch, because of the insufficient production capacity. In another case, the project's purpose is to construct a 15-house division. To start developing a LOB, you must set the objective of accomplishing the required delivery or completion schedule for 50 cables or 15 houses.

Define the Program. This is a production or construction plan that may be formatted as a network diagram, Gantt Chart, or Milestone Chart (see Figure 6.14). In the plan, we set the control points, which are key points in the production or construction process. Depending on the chosen format, these points are events in the network diagram, or the end of bars in the Gantt Chart, or milestones in a milestone chart. They are used to measure the progress of the project. An example that follows (Figure 6.14) shows such a program for one unit only. Having a project with multiple units to produce or construct requires the program for all units. There are two possible scenarios for this. In the first scenario, you can produce all multiple units in one batch, assuming sufficient production capacity. Here, all quantities for one unit are multiplied by the number of multiple units to arrive at the program for all units. The lead time for every unit would remain unaltered.

The fundamental premise of the second scenario is that the multiple units can be produced in several batches to accommodate for the insufficient production capacity. Considering that the lead time for each batch is equal to the duration of the one unit program, these batches can be fit into the production schedule. As a result, we obtain the program for producing multiple units.

Develop an S Chart for the Program. In a further step, draw the program for multiple units on a graph called an S chart, showing cumulative deliveries against timescale (see Figure 6.15) and also cumulative completion schedule for control points. Because of the linear production rate, S-curves are straight lines in our example. Essentially, this signifies the planned completion of final units and their intermediate parts or phases. This is for the second scenario—several batches to accommodate for the insufficient production capacity.

Draft and Refine the LOB. Draw a vertical line through the S chart. This is the LOB, a snapshot at a certain point in time (day 30, for example) that indicates the cumulative number of components or units that are planned to be complete by that date in order to comply with the schedule. To track the progress against the plan, you can draw another line that depicts the cumulative number of actually completed components or units (the dashed line in Fig 6.13). Figure 6.13, for instance, shows that the first two events are five units behind the plan, the third event is two units behind the plan, the fourth event is two units ahead of the plan, the fifth event is six units ahead of plan, and there is no real bottleneck. All of this is fine and helpful, but most of the time, the first draft may need refinement, including changes necessary to clean up the LOB.

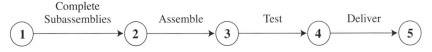

Events in L.O.B.:
1. Start work on subassemblies
2. Subassemblies complete
3. Unit assembled
4. Unit tested
5. Unit delivered

Figure 6.14 Program for a single unit in a multiunit project.

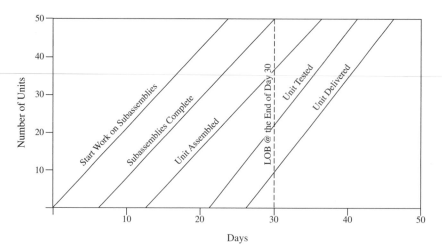

Figure 6.15 S chart for a multiunit program.

Utilizing the LOB

Time to Develop. If the prerequisites listed previously are prepared and an adequate software program is available, building an LOB diagram with 20+ units/components can be completed in between one and two hours. This time will increase when more people are involved, because more time is necessary for their communication.

When to Use the LOB. Although different versions have been developed since LOB's introduction in the 1950s, more than anything else, LOB is a tool for low-volume, new-production situations requiring coordination of the design and small-scale manufacturing or construction. Its main areas of application are as follows:

- ▲ Projects including the design and then the manufacturing of a limited number of units per that design (for example, a pilot run for microprocessors)
- ▲ Projects consisting of a number of identical units manufactured or constructed in sequence (for example, construction of a multiunit housing project)
- ▲ Projects of one-off nature (for example, shipbuilding)

Benefits. By presenting the planned completion status at each phase versus actual completion status at these phases, LOB shows to the project team in a very visual manner whether the project is ahead or behind the plan [35]. For executive use, it highlights the potential showstoppers and bottlenecks, urging an action to remove the problem areas (see the "Focus on Critical Components" and "Tips for LOB" that follow).

Focus on Critical Components

In a company, the prevailing view is that all components are not created equal. From previous experience, some components are known as potential show-stoppers in the delivery schedule, others are not. To respond to different level of risks related to the components, the company uses LOB only for monitoring the scheduling of circuit board components, long known as prone to delay. The point here is this: Keep your eye on the critical few, don't bother with trivial many.

Advantages and Disadvantages. LOB's major advantages are that it is

- ▲ *Visual.* It provides a visual display of both the planned and actual progress of repetitive activities.
- ▲ *Concise.* Its ability to report progress at a glance is what every reporting tool strives for.
- ▲ *Focused.* LOB highlights what requires critical attention.

When using LOB, you should watch out for the following shortcomings:

- ▲ *One-day use.* LOB is a snapshot in one point of time. For the next point of time, a new LOB needs to be developed. This is a time-consuming and laborious exercise, which in today's tight schedules may prove to be a discouragement for this tool's use.
- ▲ *Loss of focus.* Displaying the status of myriad of components may create a feeling that we don't see the forest for the trees. The remedy is in considering only those components that our experience marks as potential sources of risk to the delivery schedule.
- ▲ *Tricky scale.* Even in computer times, the vertical scale that has to accommodate a wide range of quantities may suffer from insufficient accuracy.

Customize LOB. To really exploit LOB in a beneficial way, you should adapt it to address specific project needs. The following examples illustrate how that can be done.

Tips for LOB

- ● Focus on critical components or units.
- ● Use LOB to detect bottlenecks.
- ● If you have an MRP (material requirements planning) system, use its power to define the program and develop S chart and LOB .
- ● Show LOB to executives when you need their support to solve a problem.

Customization Action	Examples of Customization Actions
Define limits of use.	Use LOB in conjunction with the Material Requirements Planning system.
	Develop templates that can be used as a starting point in building LOB for new projects.
Amend an existing feature.	For the y-axis, use a logarithmic scale instead of the linear scale to accommodate a wide range of quantities and increase accuracy.

Summary

Featured in this section was LOB, a tool for low-volume, new-production situations requiring coordination of the design and small-scale manufacturing or construction. By presenting the planned completion status at each phase versus actual completion status at these phases, LOB shows to the project team in a very visual manner whether the project is ahead or behind the plan. For executive use, it highlights the potential showstoppers and bottlenecks, urging an action to remove the problem areas. As a summary, the following box lists key points to an effective LOB.

Concluding Remarks

There are seven tools in this chapter. That begs a question, which one or ones are the most appropriate to select and use? Such a decision, of course, depends on your project situation. To help narrow down the field, in the table that follows we list a set of project situations and indicate how each situation favors the use of the tools. Identifying the situations that correspond to your own project is the first step. If that set of situations does not describe the project well, brainstorm more situations in addition to those listed, marking how each favors the tools. The tool that has the highest number of marks for identified situations becomes the tool of primary choice. Note, however, that more than one tool can be used, since some of them complement each other rather than exclude each other. A careful study of the material covered in this chapter will help you determine when this is practical.

The LOB Check

Check to make sure you developed an effective LOB. The chart should have

- Number of units or components
- Phases of operation
- Planned completion of components or units at a certain date of the delivery schedule
- Actual completion of components or units at the same date of the delivery schedule, if it is to be used for tracking progress

A Summary Comparison of Schedule Development Tools

Situation	Favoring Gantt Chart	Favoring Milestone Chart	Favoring Critical Path Method (CPM)	Favoring Time-scaled Arrow Diagram	Favoring Critical Chain Schedule	Favoring Hierarchical Schedule	Favoring Line of Balance
Small and simple projects	✓			✓			
Short time to train how to use the tool	✓	✓					
Focus on highly important events		✓					
Increase goal orientation		✓					
Large, complex and cross-functional projects			✓	✓			
Focus on top priority activities			✓	✓	✓		✓
Strong interface coordination needed			✓	✓	✓		
Need timescale in complex projects				✓	✓		
Very fast schedule for important projects					✓		
Multitiered schedule for large projects						✓	
Rolling wave scheduling needed						✓	
Short-term outlook schedule in large projects	✓						
Need tool to support resource planning	✓			✓			
Scheduling, tracking of repetitive projects							✓
Use of templates desired	✓	✓	✓	✓	✓	✓	✓

References

1. Segur, P. 1958. *Napolean's Russian Campaign.* Alexandria, Va.: Time-Life.

2. Powers, J. R., 1988. "A Structured Approach to Schedule Development and Use." *Project Management Journal.* 19(5): 39–46.

3. Adler, P. S., et al. 1996. "Getting the Most out of Your Product Development Process." *Harvard Business Review* 74(2): 134–152.

4. Mohanty, R. P. and M. K. Siddiq. 1989. "Multiple Projects-Multiple Resources Constrained Scheduling: A Multiobjective Analysis." *Engineerings Costs and Production Economics* 18(1): 83–92.

5. Leach, L. P. 1999. "Critical Chain Project Management Improves Project Performance." *Project Management Journal* 30(2): 39–51.

6. Sipos, A., 1990. "Multiproject Scheduling." *Cost Engineering.* 32(11): 13–17.

7. Ireland, L. R. 1997. "Managing Multiple Projects in the Twenty-First Century" at Project Management Institute 28th Annual Seminars and Symposium. Chicago.

8. Fricke, S. E. and A. J. Shenhar. 2000. "Managing Multiple Engineering Projects in a Manufacturing Support Environment." *IEEE Transactions on Engineering Management* 47(2): 258–268.

9. Meredith, J. R. and S. J. Mantel. 2000.Project Management: A Managerial Approach. 4th ed. New York: John Wiley & Sons.

10. Rubenstein, A. M., et al. 1979. "Factors Influencing Success at the Project Level." *Resource Management.* 16: 15–20.

11. Belanger, T.C. 1995. *How to Plan Any Project.* 2d ed. Sterling, Mass.: The Sterling Planning Group.

12. Bennett, J. 1981. *Construction Project Management.* London: Butterworth.

13. Harrison, F. L. 1983. *Advanced Project Management.* Hunts, U.K.: Gower Publishing Company.

14. Bowen, H. K., 1997. Project Management Manual. Boston,: Harvard Business School Press.

15. Project Management Institute. 2000. *A Guide to the Project Management Body of Knowledge.* Drexell Hill, Pa.: Project Management Institute.

16. James, K. E., Jr. 1961. "Critical Path Planning and Scheduling: Mathematical Basis." *Operations Research.* 10(3): 296–320.

17. Wiest, J. D. and F. K. Levy. 1977. *A Management Guide to PERT/CPM.* Englewood Cliffs, N.J.: Prentice Hall.

18. Crawford, M. 1992. "The Hidden Costs of Accelerated Product Development." *Journal of Product Innovation Management* 9(3): 188–199.

19. Levy, F. K., G. L. Thompson, and J. D. Wiest, 1963. "The ABCs of the Critical Path Method." *Harvard Business Review* 41(5): 98–108.

20. Snowdown, M. 1981. *Management of Engineering Projects.* London: Butterworth.

21. Miller, R. W. "How to Plan and Control with PERT." *Harvard Business Review.* 1962. 40(2): 92–102.

22. Crandall, K. 1973. "Project Planning with Precedence Lead/Lag Factors." *Project Management Quarterly* 4(3): 18–27.

23. Kent, B. H. 1997. *Project Management Manual.* Boston: Harvard Business School Press.

24. Goldratt, E. 1997. *Critical Chain Project Management.* Croton-on-Hudson, N.Y.: North River Press.

25. Wheelwright, S. C. and K. B. Clark. 1992 "Creating Project Plans to Focus Product *Development." Harvard Business Review*: 70–82.

26. Wheelwright, S. C., S. C., and C. K. B. 1992. *Revolution Product Development: Quantum Leaps in Speed, Efficiency, and Quality.* New York: Free Press.

27. Kania, E. 2000. "Measurements for Product Development Organizations." *Visions* 24(2): 17–20.

28. Herroelen, W. and R. Leus. 2001. "On the Merits and Pitfalls of Critical Chain Scheduling." *Journal of Operations Management.* 19(5): 559–577.

29. Steyn, H. "2000. An Investigation into the Fundamentals of Critical Chain Project Scheduling." *International Journal of Project Management.* 19(6): 363–369.

30. Elton, J. and J. Roe. 1998. "Bringing Discipline to Project Management." *Harvard Business Review* 76(2): 78–83.

31. Pinto, J.K. 2002. "Project Management 2002." *Research Technology Management* 45(2): 22–37.

32. Zalmenson, E. 2001. "PMBOK and the Critical Chain." PM Network. 15(1): 4.

33. Kuprenas, A. J. 2001. "Project Management Workload-Assessment of Values and Influences." *Project Management Journal* 31(4): 44–51.

34. Cusumano, M. A. and K. Nobeoka. 1998. *Thinking Beyond Lean: How Multi-Project Management Is Transforming Product Development at Toyota and Other Companies.* New York: The Free Press.

35. Lock, D. 1977. *Project Management.* 2d ed. Westmead, U.K.: Gower Publishing Company.

Cost Planning

Major topics in this chapter are cost planning tools:

- ▲ Cost Planning Map
- ▲ Analogous Estimate
- ▲ Parametric Estimate
- ▲ Bottom-up Estimate
- ▲ Cost Baseline

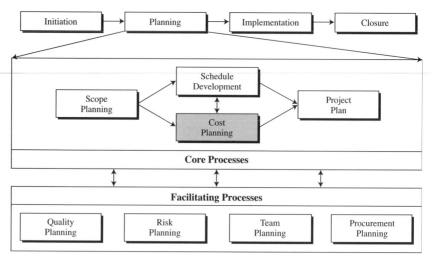

Figure 7.1 The role of cost planning tools in standardized project management process.

These tools will focus on developing an estimate of costs and the Cost Baseline (see Figure 7.1). Performing this task requires information about the scope of project activities, required resources, and a schedule of project activities. The Cost Baseline is integrated with the scope and schedule baselines to form a project performance baseline, the core of the project plan and a foundation for all other project planning from organizational to quality to risk to communications planning. In the course of the project implementation, the project performance baseline will become a corner stone of cost control and your ability to deliver the project on budget. This chapter's intent is to help practicing and prospective project managers accomplish the following:

- ▲ Learn how to use various cost planning tools.
- ▲ Choose cost planning tools that correspond to their project situation.
- ▲ Customize the tools of their choice.

These skills are of vital importance in project planning and building a standardized PM process.

Cost Planning Map
What Is a Cost Planning Map?

The Cost Planning Map (CPLM) is a tool for establishing a systematic approach to cost planning in projects (see Figure 7.2). To accomplish the approach, CPLM spells out steps and substeps a project team needs to go through in order to make choices necessary to develop basic definitions, terminology, estimate types, estimating tools, and the process for cost planning. When such choices are seamlessly integrated, CPLM can help establish a culture of cost consciousness that is proactive as well.

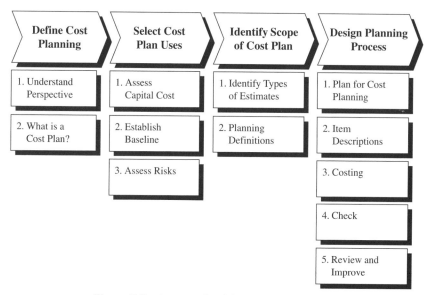

Figure 7.2 An example of the Cost Planning Map.

Deploying a Cost Planning Map

Project cost planning is an effort fraught with risks, which if not addressed may result in serious consequences. To tone down the risks and avoid the consequences, the deployment of a CPLM should boil down to a well-synchronized and integrated sequence of steps and substeps that are described in continuation.

Prepare Information Inputs. Two inputs have a significant impact on successfully deploying a CPLM:

▲ Financial policies

▲ Organizational policies of the organization performing cost planning

Financial policies dictate the design of the hardcore elements of the CPLM. For example, answers to questions such as which types of cost estimates will be used and with what purpose in the project process depend on the financial policies of the organization. Similarly, when performing resource planning, you must consider organizational policies regarding staffing and outsourcing, a key element of the costing substep in CPLM [11]. Clear understanding of the two policies is, thus, an important piece of designing the map.

Define Cost Planning. When configuring a CPLM, you should start with two simple questions: "Who is the cost planner?" and "What is a cost plan?" The former really asks about the perspective of the planner, for example, whether the planner is the owner of the project, or perhaps a contractor, and also whether the planner is experienced or just a novice. The approach to CPLM will greatly depend on the perspective, experience, and organizational culture in which the planner plans. Think, for example, of a company developing a new product. In developing a manufacturing cost estimate for the product, the company will have to consider its manufacturing process, plant, materials procurement strategy, design for manufacturability approach, and so forth. Each of these will contribute to risks and cost. If manufacturing were outsourced, the contractor is likely to have a different manufacturing process, plant, and materials procurement strategy, with no concern for manufacturability, all resulting in a different CPLM approach.

Regarding the second question, the project cost plan typically includes a cost estimate and a Cost Baseline (Time-phased Budget). Take, for example, the estimate. It is an assessment, based on specific facts and assumptions, of the final cost of a project [4]. This assessment and its results will heavily hinge on factors such as accuracy of scope, quality of available estimating data, stage of the project, time allowed for the estimate, estimator's perspective and experience, desired accuracy, available estimating tools, and so on. Consequently, by defining these factors, you will define the makeup of your CPLM. In summary, the relevance of asking the two questions is in getting answers that will help shape the CPLM for specific situation.

Select Cost Plan Uses. Once you have defined cost planning, you need to determine for what purposes you want to use the cost plan (see Step 2 in Figure 7.2). Three uses are available: to assess the capital cost, to establish a baseline, and to evaluate risks and productivity. As an assessment tool, a cost plan may serve multiple purposes — for instance, substantiating a request for capital appropriations or borrowing funds. In other situations, a cost plan can act as a basis for a proposal, bid, or contract document. Still in other situations, cost estimate, a part of the cost plan, is compared with other cost plans to validate their accuracy and increase the confidence level. Typically, this type of estimate is called an *independent cost estimate* [4], or in some companies' parlance, a *shadow estimate.*

The second capability of a cost plan, and also the second substep in this step, is to help establish two baselines—a schedule and a Cost Baseline. As we discuss later in this section, part of developing a cost estimate is identifying necessary resources such as hours of effort that are necessary to complete project activities. This is typically performed hand in hand with scheduling, so that the resource hours of effort may be turned into activity durations, thus setting a schedule baseline. By combining the scheduled activities with their budgets, you can develop a Cost Baseline, also known as a Time-phased Budget or cash flow curve. The Cost Baseline is described later in this chapter as a separate tool.

The third substep, evaluating risks, aims at establishing a reasonable amount of cost estimate to allow for changes that are likely to occur. This amount, called *contingency*, practically reflects project risks in the cost plan. The motivation for careful analysis of risks and subsequent identification of the contingency amount is usually driven by attempts to lower costs as much as possible. Risks, as we discuss in Chapter 9, may be related to any technical, schedule, and cost uncertainties in work packages. Once cost planning has been defined and the purpose of the cost plan selected, the stage is set for the next step to decide what will be in the cost plan.

Identify the Scope of the Cost Plan. Vital to this step is that you determine the types of cost estimates you want to use in your practice of project cost planning, as well as what exactly the estimates will include. There are many types available, but three of them are most often used across many industries, from software to construction to manufacturing: order-of-magnitude, budget, and definitive estimates [12–14]. As Figure 7.3 indicates, they differ in many attributes, from their purpose to accuracy to cost of preparation to information they require to type of the estimating tool they employ [15]. Each of the estimates can be used as a basis to develop the second element of a cost plan—Cost Baseline or Time-phased Budget. Naturally, the attributes (purpose, accuracy, etc.) of a so developed Cost Baseline would match those of the type of estimate they are developed from.

Cost estimates are typically expressed in currency units, such as dollars, enabling an easy comparison across and within projects [11]. In contrast, projects in some industries favor estimates in labor hours, lumping together labor hours of different types of expertise. This is an acceptable practice as long as it doesn't prevent comparison across and within projects. Also, the practice of providing estimates in multiple units of measure when management control requires so is acceptable.

Understanding what exactly an estimate and Cost Baseline mean calls for concrete definitions of their components and other cost planning terms. This is why we suggested that one substep should be developing cost planning definitions, such as those in the "Examples of Cost Estimating Definitions" box that follows. They will provide a shared language for all participants to communicate in the cost planning process that will be designed in the next step.

Order-of-Magnitude Estimates	Budget Estimates	Definitive Estimates
End Use: Feasibility Study, Project Screening, Budgeting/Forcasting	**End Use:** Budgeting/Forcasting, Authorization - Partial or Full Funds	**End Use:** Authorization - Full Funds, Bids/Proposals, Change Order
Accuracy: -30%, +50% Before Contingency	**Accuracy:** -15%, +30% Before Contingency	**Accuracy:** -5%, +15% Before Contingency
Cost Of Preparation (Typically): 0.04% - 0.15% of Total Project Cost	**Cost Of Preparation (Typically):** 0.15% - 0.60% of Total Project Cost	**Cost Of Preparation (Typically):** 0.45% - 2% of Total Project Cost
Information Required: Size, Capacity, Location, Completion Date, Similar Projects	**Information Required:** Partial Design, Vendor Quotes	**Information Required:** Specs, Drawings, Execution Plan
Estimating Tool: Analogous, Parametric	**Estimating Tool:** Parametric, Bottom-up	**Estimating Tool:** Bottom-up, Minor Parametric
Also Called: Global, Conceptual, Quickie, Ballpark, Guesstimate, Judgement	**Also Called:** Scope, Sanction, Authorization, Preliminary, Semidetailed	**Also Called:** Detailed, Control

Figure 7.3 Types of cost estimates.

Examples of Cost Estimating Definitions

- *Direct cost.* An item of cost, or the aggregate of items, that may be identified specifically with the project. These include direct labor, materials, and equipment costs.

- *Indirect cost.* The cost of labor, services, or supplies not easily or readily allocable directly to a project. It is accrued and charged to overhead accounts, the sum of which is applied as burden [3].

- *Most probable cost.* This is the cost most likely to occur, which is made up of all the itemized known items and a contingency estimate that together invoke a 50 percent degree of confidence [4].

- *Range of accuracy.* This is a prediction of the least expected and highest expected cost relative to the most probable cost. Higher quality of estimate, better scope definition, lower project risks, fewer unknowns, and more accurate estimate pricing will lead to a better range of accuracy.

- *Contingency.* An allowance added to an estimate to cover future changes that are likely to occur for unknown causes or unforeseen conditions. Contingency can be determined through statistical analysis of past project costs or from experience in similar projects [9].

Design Cost Planning Process. This is the process that you will use to develop any of the estimates and related Cost Baselines discussed later in this chapter—analogous, parametric, and bottom-up. Of course, the process for each one of them will differ in terms of level of detail, but the principle steps will be the same. To enable this, you need to design a proper cost planning process, consisting of several substeps (see Figure 7.2). The beginning one is to preplan how you intend to perform cost planning. Although this may sound like an excessive dose of paperwork, in reality it can reduce the total effort for cost planning while minimizing rework [4]. Several specific items are the focus of the preplanning. First, thinking through who the end users of estimates are and with what purpose may help you select appropriate estimating format and forms (such as one in Figure 7.9) and identify necessary contributors and their roles, as well as the resources for the cost planning. Getting to know the due date for the estimate and details of the estimate review are crucial to scheduling cost planning work and submitting an estimate of the desired quality. Also, as part of the preplanning, you may have to figure out the cost of preparing the cost plan and inform the end users.

Developing item descriptions is the second substep, one that may have more impact on the estimate quality than any other factor (other than contingency unknowns). But what exactly are item descriptions? These are descriptions of scope items that we want to develop an estimate for (for example, see the item in Figure 7.9). Typically, a complete item description should include several elements. Begin with a quantity and applicable measurement unit, followed by a physical description of the item in as much detail as possible. Continue by stating item scope boundaries that clarify any ambiguities or assumptions, and document any diversions from conventions and standards. Add sources of estimating data (e.g., standard production rates or cost index published by a trade association).

Costing is the third substep, computing an estimate for an item. Within the core of costing is an estimating algorithm or formula that processes project information—for example, item description and a source of cost data for both direct and indirect cost—into costs. Typically, these formulas or algorithms are called *cost estimating relationships* (CERs). Each cost estimating tool relies on a different CER. For instance, when quantities to produce, unit production rates, and an hourly rate of labor are available, a bottom-up estimate may use the following CER to calculate labor cost:

$$\$Labor = Quantity \times (hrs/unit\ of\ quantity) \times (\$/hr) = 200\ articles \times (5\ hrs/article) \times (\$80/hr) = \$80K$$

Again, which CER for labor cost calculation is used will depend on the type of estimate being prepared.

Labor, materials, and equipment can be estimated by CERs using ratios, parameters, cost chunks, or multiplication methods, as illustrated later in this chapter. The estimated amounts should be documented and entered on the estimating form you have developed earlier. When costing for individual items is complete, direct costs for all items are often totaled separately from indirect costs, and possibly by categories of work. If requested, this is the time to translate the estimate into a cost baseline.

Tips for Cost Planning

- *Know your customer.* Ask questions to clarify their needs, item descriptions, and project scope, and assume nothing.
- *Follow the cost planning process.* Don't skip process steps. If the process doesn't work, change it.
- *Go beyond a "number-cruncher" mentality.* Understand the big picture and philosophy of the project and its customer.
- *Document everything.* Include assumptions, references, sources, scope exclusions and so on.
- *Leave an audit trail.* The audits enhance quality of the estimate and your level of confidence.
- *Document changes.* The estimate you have originally planned for is almost certain to change. Record the change. Maintain document revision control record.
- *Get buy-in.* Make experts from performing functional departments part of estimate preparation; after all, they have to live with it.
- *Keep it in mind.* Estimate accuracy depends on the amount of information, available time and resources, and estimate type.

Costing needs to be checked, the fourth substep. This involves validation of calculations, verification of estimating data sources, and peer reviews [4]. With checking done, you can move to the fifth substep—review and improve. Management needs to review the estimate because they are responsible for supervision of the estimate preparation and typically can smell major problems. Then, the estimate can be issued following the principles of sound document management. The cost planning process does not end here. Rather, it ends when the project is complete. At that time, all actual costs are collected, analyzed, and compared with the cost plan, and historical data is updated. The essence of the cost planning process is summarized in the box on page 231, "Tips for Cost Planning."

Utilizing the Cost Planning Map

When to Use. While any project can find value in using it, organizations with large projects and organizations with a constant stream of small and medium projects may benefit most from CPLM. Consistency and discipline in cost planning that CPLM can generate is of vital importance to these users involved in the game of high financial stakes in their projects.

Time to Use. Building a CPLM is a significant time commitment. In organizations with large projects that are also complex and resource-intensive, heavy involvement of experts from various functions—technical, financial, accounting, for example—is typical, often resulting in hundreds of resource hours to construct a solid CPLM. Developing a CPLM in organizations with a constant stream of small and medium projects is less time-consuming for lower complexity and resource requirements in such projects, although it may easily take tens of resource hours.

Benefits. The value of a CPLM is in the clarity of the direction that it provides to project teams. Through a careful scripting and orchestrating of cost planning tasks, CPLM leaves no ambiguity as to what a certain cost estimate type and Cost Baseline are and how to develop them. This significantly reduces risks of poor cost planning and misusing the company's resources, increases cost consciousness, and leads behavior of cost planners in the right direction.

Advantages and Disadvantages. A major advantage of CPLM is its linear and clear structure, indicating visually steps to get to a good cost plan. The time-consuming nature of developing a CPLM and the high level of expertise required to build it are certainly its major disadvantages.

Cost Planning Map Check

Check to make sure that the cost planning map is appropriately structured. It should

- Be based on necessary information inputs.
- Include basic definitions, terminology, estimate types, and tools for estimating and cost baselining (time-phased budgeting).
- Include crucial steps and substeps in the cost planning process.

Variations. The CPLM described here is just one of many variations used in the industry. Some of them are more comprehensive, others less so. Sequencing of cost planning process steps may vary, as well as the number of estimates types and cost planning tools.

Customize CPLM. CPLM that we have described is a tool of general purpose, designed to match needs of a variety of organizations managing projects. You may find it more valuable to adapt the tool to your specific needs. Following are a few ideas about potential directions for such adaptation.

Customization Action	Examples of Customization Actions
Define limits of use.	Use CPLM for all organization's projects, large and small.
Add a new feature.	Include more types of estimates, such as the detailed estimate and more estimating tools.
Leave out a feature.	Eliminate definitive estimates (for organizations with small projects).

Summary

Presented in this section was the Cost Planning Map (CPLM), a tool for establishing a systematic approach to cost planning in projects. While any project can find value in using it, organizations with large projects and organizations with a constant stream of small and medium projects may benefit most from CPLM. The value of a CPLM is in a careful scripting and orchestrating of cost planning tasks that leaves no ambiguity as to what a certain cost estimate type and Cost Baseline are and how to develop them. This significantly reduces risks of poor cost planning and misusing of the company's resources. In sum, the key points in structuring the map are listed in the box that follows.

Analogous Estimate

What Is an Analogous Estimate?

An Analogous Estimate is the derivation of an estimate for a project that is being estimated—called the *target project*—based on the actual cost of a previous project or projects (*analogous* or *source project*) of similar size, complexity, and scope [16]. The estimators may use "gut feel," historical data, or rules of thumb that are modified to account for any differences between the estimated project and analogous project(s). An example of the Analogous Estimate is illustrated in Figure 7.4, while basic features of the estimates are shown in Figure 7.5.

Developing an Analogous Estimate

In general, the process of developing an Analogous Estimate will follow steps that we have defined in the cost planning process, part of the Cost Planning Map. Specifics will vary to reflect the nature of the analogous estimating.

Item	Analogous Size (KLOC)	Analogous Productivity Factor (LOC/Person-Month)	Analogous Effort (Person-Month) 2/3	Target Size (KLOC)	Target Productivity Factor (LOC/Person-Month)	Target Effort (Person-Month) 5/6
1	2	3	4	5	6	7
1	1	100	10	0.8	80	10.0
2	2	50	40	2.5	40	62.5
3	2	200	10	2.5	160	15.6
4	1	100	10	1.0	80	12.5
5	1	50	20	1.0	40	25.0
Totals	7		90	7.8		125.6

Key: KLOC–Thousand Lines of Code
     LOC–Lines of Code

Figure 7.4　An example of the Analogous Estimate for a software project.

Collect Necessary Information. Quality information inputs are a prerequisite to the development of a quality analogous estimate. They include the following:

- ▲ Project scope
- ▲ Historical information about analogous projects
- ▲ Resource requirements
- ▲ Resource rates.

Identifying the target project and analyzing its scope ensures that the project being estimated is understood. For the estimate to develop, however, we need an analogous project, which will be extracted from a historical database of previous projects with similar features. Resource requirements and rates are necessary to express the estimate in appropriate units.

Prepare the Estimate. The starting step is figuring out the preplanning specifics such as who the end users of the estimate are, the purpose of the estimate, estimating format, list of contributors and their roles, and available resources for the estimating. What follows is studying the target project's scope, size, and complexity features. In our example in Figure 7.4, the scope of the target project is broken down into five major items (column 1), each one with the estimated size (column 5). Now we can go to the database of your previous projects with similar features to evaluate them. The most appropriate project (or projects) is selected as the analog. The mapping of analogous features to the target project is fairly straightforward because the two projects share a common set of features [13]. Our example has chosen one analogous project with the same five items. Analyzing actual data about the analogous project indicates size and productivity (columns 2 and 3, respectively), as well as the effort for the completion of each of its items in column 4, essentially an analogous CER (cost estimating relationship). Then we transfer the solution that achieved the goal in the analogous project to the target, adjusting it for analogical elements that are not in correspondence with the target project. Specifically, for item 1 in our example, the project team is less experienced and their productivity (column 6) is judged to be

Use Combinations of Estimating Tools

None of the available estimating tools— Bottom-up, Parametric, and Analogous Estimate—is better than the others from all aspects. Rather, all of them have advantages and disadvantages that are complementary. This is especially important in the world of software development, where practitioners are concerned over their inability to estimate accurately project cost [2]. Therefore, in practice, Boehm suggests, we should rely on combinations of the estimating tools, comparing their results, and where differences occur, repeating the exercise [5]. While this can exact more estimating resources resulting in a higher cost, benefits related to higher accuracy of the estimates are obvious.

0.8 (judgmental factor) of that for the analogous project team (column 3). Applying a CER that divides an item size value (column 5) by the productivity factor (column 6) yields an item estimate (column 7) expressed in resource hours. To turn them into monetary terms, we can multiple the hours by the resource rates.

A sum of the all estimated items is equal to the total project estimate. Crucial in this effort is the ability of the estimators to identify subtle differences in the source and target items and estimate the cost of a target item based on the source item that is really similar or analogous [16]. Checking, reviewing, and improving the estimate, as explained in the section about the Cost Planning Map, is the final step in developing an Analogous Estimate.

Utilizing an Analogous Estimate

When to Use. An Analogous Estimate is a tool of choice when there is a lack of detailed information about the project. Typically, this is the case early in the project life cycle. Because other estimating tools have disadvantages of their own as well, an Analogous Estimate can be used in combination with the bottom-up and parametric estimates (see details in box that follows, "Use Combinations of Estimating Tools").

Time to Prepare. An Analogous Estimate operates on the assumption of a limited amount of information about the target project and a very summary type of information about the analogous project. Put together, these two facts mean that just a few hours may be enough for almost any project's analogous estimate. A smaller project would take less than that.

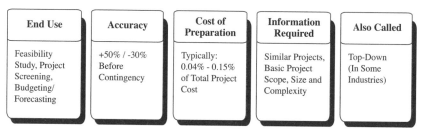

End Use	Accuracy	Cost of Preparation	Information Required	Also Called
Feasibility Study, Project Screening, Budgeting/ Forecasting	+50% / -30% Before Contingency	Typically: 0.04% - 0.15% of Total Project Cost	Similar Projects, Basic Project Scope, Size and Complexity	Top-Down (In Some Industries)

Figure 7.5 Basic features of Analogous Estimates.

Customization Action	Examples of Customization Actions
Define limits of use.	Use Analogous Estimates in combination with Parametric Estimates in decision making for project screening and budgeting.
Modify a feature.	Include separate items for labor and material (this is for projects with a significant amount of materials).

Benefits. The value an Analogous Estimate brings to a user is in the little time it takes to develop, while operating with limited available information about the project.

Advantages and Disadvantages. The Analogous Estimate is based on representative past experience so that the developed estimate can be substantiated [17]. Also, it is generally less costly than other tools [11]. These same advantages lead to its major disadvantages in that the estimate is as good as is past experience and is generally less accurate than other estimates [17, 18].

Variations. In analogous estimating, an estimator may choose to estimate only the total target project without breaking it down into items as we have done. He or she may judge, for example, that the target project may take twice the resource hours as the analogous one. This judgmental factor of 2 would then be multiplied by the resources deployed in the analogous project to obtain the estimate for the new target project.

Customize the Analogous Estimate. The relatively simple nature and process of the estimate offer limited opportunities to customize it to better account for your project needs. Still, some customization is possible, as the following examples indicate.

Summary

This section's focus was on the Analogous Estimate for a project that is based on the actual cost of a previous project or projects of similar size, complexity, and scope. An Analogous Estimate is applied when there is a lack of detailed information about the project. Typically, this is the case early in the project life cycle. The value an Analogous Estimate brings to a user is in the little time it takes to develop, while operating with limited available information about the project. Below we summarize key points in developing the Analogous Estimate.

Analogous Estimate Check

Check to make sure you developed a proper Analogous Estimate. This should be based on the criteria:

- Scope, size, and productivity of the analogous project
- Actual cost of the analogous project

Derive a judgmental factor and apply it to obtain one of the following:

- Analogous Estimates for items of the project scope, which when summed up give a total project cost estimate
- Just one figure for the analogous total project cost estimate

Parametric Estimate

What Is a Parametric Estimate?

A Parametric Estimate uses mathematical models to relate cost to one or more physical or performance characteristics (parameters) of a project that is being estimated [3]. Typically, the models provide CER that relates cost of the project being estimated to its physical or performance parameters, such as production capacity, size, volume, weight, power requirements, and so forth. Determining the estimate for a new power plant may be as simple as multiplying two parameters—the number of kilowatts of a new power plant by the anticipated dollars per kilowatt. Or it may be very complex, for instance, involving 32 parameters (also called factors or cost drivers) into an equation to estimate the cost of a new software development project. Values of the parameters can be entered into the CER, and the results can be plotted on a graph (see Figure 7.6) or tabular format. Basic features of the parametric estimates are described in Figure 7.7.

Developing Parametric Estimates

Collect Necessary Inputs. To develop a proper Parametric Estimate, you may need to collect quality information inputs that include the following:

- ▲ Basic project scope
- ▲ Selected project parameters
- ▲ Historical information.

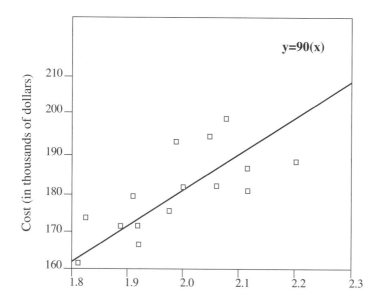

Figure 7.6 Typical CER in a Parametric Estimate model.

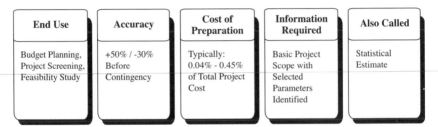

End Use	Accuracy	Cost of Preparation	Information Required	Also Called
Budget Planning, Project Screening, Feasibility Study	+50% / -30% Before Contingency	Typically: 0.04% - 0.45% of Total Project Cost	Basic Project Scope with Selected Parameters Identified	Statistical Estimate

Figure 7.7   Basic features of Parametric Estimates.

Basic project scope description provides understanding of what is being estimated. Its parameters are identified on the basis of the nature of CER model that will be used to collect and organize historical information, which will be related to the project being estimated.

Prepare the Estimate. As explained in the section about the Cost Planning Map, the general process of developing any type of estimate is pretty much the same. Per such process the parametric estimating starts with preplanning and continues with developing item descriptions and costing. The last step is very uniquely performed. Specifically, it includes database development, model building, and model application.

The purpose of database development is to collect and organize cost data from past projects in such a manner that the data can be used to build a model, which will be employed to estimate the cost of a new project. The first step is to select the framework to organize costs of past projects methodically. That framework is called the *basic work element structure form* and corresponds to that of the desired CER. For example, the structure may include project management cost (e.g., planning, controls), nonrecurring costs (e.g., design and engineering, software, facility), and recurring costs (e.g., production, operations). The necessary cost data from outside and within an organization are collected against the structure and normalized to established ground rules and assumptions. Typically, these databases are maintained in some constant-year price levels and are updated periodically to provide consistency in estimating from project to project.

Building a model aims at determining which equation type would best fit a data sample organized in the database and deriving a mathematical model for the CER that describes the project we want to estimate [3]. While many mathematical models in the form of equations can be used for the CER, in practice, lots of cost data can be fit empirically using one of the following forms—linear, power, exponential, or logarithmic curves. When graphed, all of them will look like either straight lines or well-behaved curves. As shown in Figure 7.6, the simplest CERs are as simple as the dollars per square foot relationship, a linear relationship of the form $y=ax$, where y is the estimated project cost (dependent variable) that is a function of x, the area in square feet (parameter and independent variable), and a is the parameter based on historical cost data relating the dependent variable to the independent variable (parameter). If this type of CER is used for rough cost estimating for a new home, and assuming that a number of homes between 1800 and 2300 square feet had costs of $90 per square foot, then the corresponding CER can be expressed as:

$$y = 90(x)$$

This simple linear model assumes that there is such a relationship between the independent variable (parameter) and cost that as the independent variable changes by one unit, the cost changes by some relatively constant number of dollars. Often, life is not that linear and simple, which leads to the use of nonlinear CERs (see an example in the box that follows, "Parametric Software Estimating"), as well as CERs with multiple independent variables and multiple regression analysis.

How do we determine which equation type would best fit a data sample organized in our database? If we enter all data points from the past projects into a graph, generally the best fit would be the equation type that can be drawn through the data points such that the sum of vertical distances from the CER curve to the data points above the line are about equal to the sum of vertical distances from the CER curve to the data points below the line. Or more sophisticatedly, the best mathematical fit is the equation type that minimizes the absolute value of total cost deviation between the data points and the CER curve [3].

Once the best-fit equation type is determined, model building continues with deriving a mathematical model for the CER. Among many statistical techniques available for this, the method of least squares appears to be the most frequently used curve-fitting method. Although linear in nature, the method can be applied to both linear and nonlinear CER equation types, to the latter only when they are transformed into linear forms.

Parametric Software Estimating

Many parametric software effort models are based on key software parameters as cost drivers. They are usually based on the statistical analysis of the results of previous software development projects [5–8]. These analyses included key parameters such as system size (e.g., line of code), complexity (e.g., degree of difficulty), type of application (e.g., real time), and development productivity (e.g., productivity). An expert suggested 59 parameters (factors) that can impact the outcomes of these cost models [10]. A simple model can take the form of:

- $Z = CY^L$ Where   Z = estimated project effort (person-months)

where:

- Y = Estimated project size (KLOC- thousands of lines of code)
- C = Regression coefficient
- L = Regression exponent

You can apply this model to estimate the effort for a new software development project by assuming the following values: C = 3.8, L = 1.4, Y = 2.

- $Z = CY^L = 3.8 \ (2)^{1.4} = 10.03$ person-months

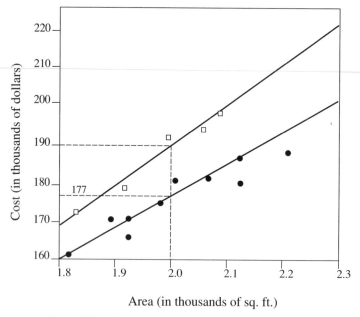

Area (in thousands of sq. ft.)

Figure 7.8 *Stratified cost estimating relationships (CERs).*

When database development and model building are complete, you can proceed with the next step—model application, using CER derived from past cost experience to estimate the cost of a new project. Clearly, the assumption in applying CER is that future projects will be performed as past projects. What if, as is often the case, the future project that is being estimated differs from the past projects in some details? This can be resolved by CER stratification and cost adjustments. Through stratification our historic database is divided into layers, each one a "family" of data points similar to each other in some respect. Then, a separate curve for each family is fitted. For example, five data points in Figure 7.6 have higher costs than the other ten. A close look reveals that these five data points are for luxurious homes with features such as central vacuum cleaner, surround system, stainless steel appliances, marble countertops, hardwood floor, stucco work, and so forth, while the remaining ten were ordinary homes with much simpler and less expensive features. Logically, we could stratify our database into two families of homes and fit curves through each of the two subsets of the database, thus obtaining two CERs, as illustrated in Figure 7.8. If we have a square footage of the home being estimated (for example, 2000 square feet), we can easily read off Figure 7.8 the parametrically estimated cost of either the luxurious ($190K) or the ordinary home ($177K).

Cost adjustments or complexity factors are also used in parametric estimating to adjust the CER estimated cost to mirror a difference between the project or its item being estimated and the database from which the CER was built [3]. For example, product development projects from the database typically included an average number of ten prototypes. If a project being estimated includes 40 prototypes, two methods can help resolve the difference. One is to stratify the database and develop separate CERs for different numbers of prototypes. The other is to separately compute the cost of the prototypes. Whatever the case, when cost adjustment is done, the Parametric Estimate needs to be checked, reviewed, and improved as discussed in the section on the Cost Planning Map.

Utilizing Parametric Estimates

When to Use. Parametric Estimates are most often used in the project definition stage as well as in the early design stages when there is not enough information to develop a Bottom-up Estimate. Considering that CERs typically relate project cost to high-level measurement of capacity or performance, it is exactly this information that is available early [3]. Naturally, such summary information makes parametric estimates very appropriate for calculating comparative cost assessments of alternate project approaches and providing a cross-check to other estimating tools, but not for developing a detailed competitive cost proposal [4]. To be used for such purposes, the Parametric Estimate must be based on accurate historic information, quantifiable parameters, and scalable model (applicable in both small and large projects).

Time to Develop. The most difficult and time-consuming part of parametric estimating is the methodology development, including database development and formulation of the CER. Depending on the complexity of the database, it may take anywhere from tens of hours to hundreds of hours to develop the database and CER. Once that's done, the estimating act can be measured in minutes or hours, having to do with the nature of the assignment.

Benefits. Parametric cost estimating tends to be faster and less resource-consuming than bottom-up estimating. Focused on the need to establish good CERs that properly relate project cost and cost-driving parameters, parametric estimates are bent on including only truly driving cost parameters, disregarding what is less important. This concentration on cost-driving parameters—coupled with greater speed and lower resource consumption—enables parametric estimates to be applied in estimating situations in which the Bottom-up Estimate is neither practical not possible. Think, for example, about a house cost estimating. To develop a Bottom-up Estimate of the house, you need detailed house blueprints, bill of materials, labor trades and rates information, and so forth. A lot of effort and cost is needed to prepare all of this. For estimating the cost of the same house using dollars per square feet parameter, you only need knowledge of the house structure, making it much faster and easier to estimate. Apparently, Parametric Estimates can be produced even though little is known about the project except its physical parameters.

Advantages and Disadvantages. Major advantages of parametric estimates are in that they are

▲ *Easy to use and repeatable*. The reason for these is that the estimates are based on mathematical formulas, which correlate the present estimate with the past history of resource utilization on similar project items. Still, to enjoy these advantages, you must rely on judgment and experience.

On the negative side, this calibration to the past that is not necessarily applicable to the future may

▲ *Incorrectly present cost trends.*

▲ *Be subjective.* Adjusting the CER calculated cost to account for identified differences between the item being estimated and the database from which the CER was derived is often subjective and difficult. This may be a reason why the accuracy of parametric estimates varies widely.

Variations. Parametric Estimates are extensively used in manufacturing, construction, and software industries, ranging from very simple ones to those with many parameters. Although many companies developed their own models, there are also commercially available models, such as three aerospace-oriented hardware models—PRICE, CAAMS, and FAST [16].

Customize the Parametric Estimate. To develop a parametric estimating model of a higher value to your projects, one that fits your needs, you may adapt the generic model we have described. The following examples provide some hints on how to accomplish this.

Customization Action	Examples of Customization Actions
Define limits of use.	Use Parametric Estimates for budget planning and project screening.
Modify a feature.	Adjust the CER calculated cost for design complexity of new projects caused by fast technology changes (this is for high-tech projects).
Leave out a feature.	Eliminate cost adjustments for projects with stable technology.

Summary

This section centered on the Parametric Estimate, which uses mathematical models to relate cost to one or more physical or performance parameters of a project. The Parametric Estimate is most often used in the project definition stage, as well as in the early design stages, when there is not enough information to develop a Bottom-up Estimate. The value of Parametric Estimates is in that they can be produced even though little is known about the project except its physical parameters.

Parametric Estimate Check

Check to make sure you developed a proper parametric estimate that

- Is based on a well-structured database of historic cost information.
- Is developed from best-fit CER that is derived from the database.
- Shows total project cost and its items, if these are required.

That value can be further improved if the estimate is customized for specific project needs. The box on page 242 summarizes the key points in developing the Parametric Estimate.

Bottom-up Estimate
What Is a Bottom-up Estimate?

A Bottom-up Estimate relies on estimating the cost of individual work items, then adding them up to obtain a project total cost [11]. Typically, an in-depth analysis of all project tasks, components, and processes is performed to estimate requirements for the items including labor and materials. The application of labor rates, material prices, and overhead to the requirements turns the estimate into monetary units [3]. Fig. 7.9 is a generic version of the Bottom-up Estimate for simpler projects, with more complex projects having more details and documentation. Basic features of the Bottom-up Estimates are summarized in Figure 7.10.

Developing a Bottom-up Estimate

Collect Necessary Information. The process of developing a Bottom-up Estimate and its accuracy heavily depend on the quality of the information inputs such as the following:

▲ Project scope (WBS)

▲ Resource requirements

▲ Resource rates

▲ Historical information

▲ Project schedule.

COST ESTIMATE

Project Name: Cablus   Estimate #: Cage - 010/1   Page #: 1 of 1
Compiled By: E. Shaw   Estimate Date: Aug. 5, 02

1	2	3	4	5	6	7	8	9	10	11
Code	Item	Quantity	Labor				Overhead 25%	Materials		Total $
			Unit Hours	Total Hours	Rate $/Hour	Amount $ 5x6		Unit Price	Amount $	7+8+10
3210	1st Article	10	0.5	5	60	300	75	45	450	825
010	Project Total	1	291.5	291.5	65	18,947.5	4,737		900	24,584

Figure 7.9 An example of the Bottom-up Estimate.

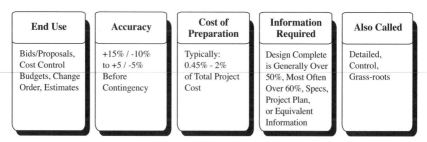

End Use	Accuracy	Cost of Preparation	Information Required	Also Called
Bids/Proposals, Cost Control Budgets, Change Order, Estimates	+15% / -10% to +5 / -5% Before Contingency	Typically: 0.45% - 2% of Total Project Cost	Design Complete is Generally Over 50%, Most Often Over 60%, Specs, Project Plan, or Equivalent Information	Detailed, Control, Grass-roots

Figure 7.10 Basic features of Bottom-up Estimates.

Project scope in the form of WBS provides a framework to organize an estimate and ensure that all work identified in the project is estimated [11]. For this to happen, resource requirements that define types and quantities of resources necessary to complete the work are multiplied by resource rates to obtain a cost estimate. Typically, the rates come from historic records of previous project results, commercial databases, or personal experience of team members. Considering that some estimates contain an allowance for cost of financing such as interest charges, which are time-dependent, the durations of activities as defined in the project schedule are an important input.

Establish the Estimate Format. Once all information inputs are available, work on a Bottom-up Estimate begins. Normally, the format of the estimate is established in the Cost Planning Map. A sound practice in that sense is to adopt a format that is based on a code of accounts, where a cost code is allocated to each item. In our example in Figure 7.9, the code in column 1 is actually built on WBS coding. This simplifies the analysis of the project while serving as the basis for cost reporting, cost control, and even information retrieval [12].

Prepare the Estimate. When the estimate format is set, take several steps to prepare the estimate. Since these steps are discussed in general in the section on the Cost Planning Map, here we will shed light on their specifics related to Bottom-up Estimates. First, you need to identify an item that is being estimated, then determine its quantity, along with the cost of labor (human resources), overhead, and materials. One good way to accomplish this is to proceed area by area or category by category in the project, before adding them up to arrive at the total project estimate. We take that approach in our example in Figure 7.9, selecting an item to be a work package in WBS. Specifically, our work package is called 1st article approval (column 2). By estimating work package by work package, our intent is to eventually aggregate them up the WBS to obtain total project cost.

The item of 1st article approval requires that the quantity (column 3) of ten prototypes of a high-tech cable are produced with equipment, tooling, fixtures, and materials that will be used later in the course of the regular production. Should you show the estimated labor cost for a single unit or a whole project batch of ten items? When project tasks are single and nonrepetitive, the question of quantity is irrelevant. When there are multiple identical items, as in our example, the cost for the whole batch needs to be estimated. Accordingly, our CER will multiply half an hour per unit (column 4) by the ten prototypes (column 3), which is a total of 5 resource hours (column 5), by a rate of $60 per resource hour (column 6) to obtain the cost of $300 (column 7).

Columns 4 to 7 indicate the labor times and cost for each estimated work package item. While we use monetary units to record cost in columns 6 and 7, we do recognize that many a project manager will not do so, but instead will only record labor or resource hours in columns 4 and 5. This is an acceptable practice in many industries, where project managers are not expected to manage dollars but resource hours only. Actually, when these labor times estimates (along with actually achieved times) are used for estimating future projects, the category of resource hours is much more relevant than the one of cost. With the passage of time, the accuracy of cost is eroded by inflation and other factors, while the resource (labor) times estimates should remain valid.

Involved in the project may be different departments and their specialists (for how not to exclude the departments, see box that follows, "How the Courthouse Disaster Was Courted"). Since they typically have different hourly rates consisting of their salaries or wages, plus employee-related expenses, an adequate estimate should be based on labor times and hourly rates of all contributing specialists. When labor cost for all these specialists is totaled in columns 4 to 6, it will yield total labor cost in column 7, the direct labor cost for each work package item. When quantities to produce, unit production rates, and an hourly rate of labor are available, labor cost may be calculated as explained in the section about the Cost Planning Map.

When the direct labor cost is calculated, you can move to the labor overhead (column 8). There are no hard-and-fast rules here, since company policies widely vary. While some companies zealously include labor overhead into the estimate, other companies do not factor overhead labor into the estimate at all. Those who do often have different overhead rates in different parts of the company, and even from a project to a project. Very frequently, this rate is based on a CER, calculating it as a percentage of the direct labor cost in column 7. In our example (Figure 7.9), the rate is 25 percent, for most industries a lean and competitive approach. Typically, the overhead labor relates to the wages and salaries of employees who are not directly connected with the project, such as supervisors.

So far, the estimate produced direct and overhead labor cost for an item, in our case a work package. Now, we will estimate a net cost of materials required for the item completion (column 10), using a CER that multiplies the cost per unit (column 9) by the number of units (column 3). Materials cost typically comprises costs of components, raw materials, or services for each item. It can include the cost of larger capital equipment as well, which is left out here for the sake of simplification. While our example for unit prices is based on catalog prices of materials, it is also possible to base it on vendor quotations or standard unit costs for stock items [19].

How the Courthouse Disaster Was Courted

Halfway through its construction, the courthouse project looked like a sure winner for the contractor. The project was on schedule, contract payments were made in a timely manner, and the owner was happy with project performance. Then Greg, the contractor's project manager who also developed the Bottom-up Project Cost Estimate that was the basis for the project contract, left the company. A month later, Pete, the new project manager, figured out that the whole project budget was spent although lots of work remained to be done. A quick audit commissioned by management revealed the following:

- A significant monetary loss was to be expected at the end of the project.
- Greg's project cost estimate was never reviewed by peers or managers.

When completed a few months later, the courthouse became one of the biggest losers in the company's history, ending $500K in the red, almost one-third over the original budget. In the postmortem session, the following improvements were adopted for future cost estimating:

- All major estimates will be reviewed by peers and management.
- All major estimates developed under time pressure will be compared to a shadow cost estimate (a cost estimate developed by an independent firm).

With direct and overhead labor cost already available, the materials cost is the last cost piece necessary to figure out the total estimated cost for the item. Adding them up in column 11, everything comes together. Repeating this exercise for each item (i.e., work package) and summing up estimates for all items will lead to a total project cost estimate. If this were a project for an external customer, this would be the time to add markup for profit. The work on a Bottom-up Estimate ends with checking, reviewing, and improving it. Details about this are discussed in the section on the Cost Planning Map.

Utilizing Bottom-Up Estimate

When to Use. Both small and large projects, whether simple or complex, are good candidates to apply Bottom-up Estimates. Typically, the application occurs just before project execution, or even in earlier phases if the required information inputs are available. This generally means that a substantial amount of design work is completed, often exceeding 60 percent. For their detailed nature, they are primarily used for cost control budgets, bids/proposals, and change order estimates (see the box that follows, "No Bottom-up Estimate, No Job!").

Time to Develop. The time to develop a Bottom-up Estimate varies with the size and complexity of a project that is being estimated. A 500-resource hour project without materials and equipment may take an hour or two to Bottom-up Estimate. In contrast, a team of estimators may spend thousands of resource hours preparing a bottom-up estimate for a $400M project.

Benefits. Bottom-up Estimates' value lies in their capacity to produce estimates of good accuracy, which is higher than that of any other estimating tools. Subsequently, they are the best basis for cost control [4].

Advantages and Disadvantages. Bottom-up Estimates' advantages are that

▲ They can easily be applied—in manual form or computerized form.

▲ Because of their involvement in the estimating process, people who will do the work are committed to the estimate.

These two major advantages may be offset by the estimate's major disadvantage. They may be relatively time consuming, requiring that the details of the project design are known.

Variations. These are generally changes in the layout of the tool, most often providing more or less detailed information about elements of the estimate such as labor, overhead, or materials cost.

No Bottom-up Estimate, No Job!

"We develop perfect quality software" was an informal motto of SP Group, a unit of a privately held company. Its clients, divisions of the same company, agreed: SP Group was doing a great job of developing software applications that had almost no bugs. Happy with the quality, the clients didn't care much about the actual costs of the projects. For a project to be approved and paid for by the client, SP Group would simply submit an order-of-magnitude estimate ranging from 1,000 to 10,000 resource hours—and that's it. Then, the company went public and the trend of profit orientation and demonstrated cost efficiency took over. Unable to respond to the trends, all division managers were forced out and new, profit-oriented division executives were brought in. The game of cost estimating also changed. "Sharks," as project managers called the new division managers, flatly refused to even look at the order-of-magnitude estimates. Having profit-and-loss responsibility, sharks wanted to manage their cost and required Bottom-up Estimates to approve a project. Lacking the expertise to develop such estimates, the large majority of project managers were also forced out. Apparently, the time to learn how to develop a Bottom-up Estimate has come to SP Group.

Customization Action	Examples of Customization Actions
Define limits of use.	Use Bottom-up Estimates for the authorization of funds for execution of all internal and external projects.
Add a new feature.	Have a separate Bottom-up Estimate for materials only. It can include item description, proposed vendor, unit cost, quantity, free onboard cost, freight cost, delivered cost, and so on (this is for projects with lots of materials).
Modify a feature	Include new columns for multiple skill categories of labor; for instance, senior analyst, junior analyst, and so on (this is for labor-intensive projects).

Customize the Bottom-Up Estimate. Since we have described a generic Bottom-up Estimate that is tailored to meet the needs of projects across industries, it may not be appropriate for every project. If so, adapt it to gain more value out of it. The examples on page 247 provide ideas on how to adapt it.

Summary

In this section we dealt with the Bottom-up Estimate, where we estimate the cost of individual work items, then add them up to obtain a project total cost. Both small and large projects, whether simple or complex, are good candidates for Bottom-up Estimates. Typically, the application occurs just before project execution, or even in earlier phases if the required information inputs are available. They are valued for their capacity to produce estimates of good accuracy, which is higher than that of any other estimating tools. Customizing the estimate for specific project purposes adds to its value. In summarizing the use of the Bottom-up Estimate, the following list offers key points for its development.

Bottom-up Estimate Check

Check to make sure you developed a proper Bottom-up Estimate. The estimate should show

- Cost code
- Item descriptions
- Item quantities
- Item labor cost
- Item overhead cost
- Item materials cost
- Total cost for the line item
- Total project cost

Cost Baseline (Time-Phased Budget)
What Is the Cost Baseline (Time-Phased Budget)?

The Cost Baseline is a time-phased budget used to measure and monitor cost performance on the project [11]. Developed by summing estimated costs by time period, the baseline reflects estimated costs and when they were supposed to occur, if executed in a specific way (see Figure 7.11). Frequently, the baseline is displayed in the form of an S-curve (see Figure 7.13). Many projects, mostly large ones, may have multiple Cost Baselines expressing different facets of cost performance. For example, the baseline may measure expenditures (cash outflows), received payments (cash inflows), or committed costs. In contrast, other projects may have only one Cost Baseline—a labor S-curve that measures how labor hours are expended.

Developing a Cost Baseline (Time-Phased Budget)

Collect Necessary Inputs. A smooth process of constructing a Cost Baseline depends on the availability of such quality inputs as:

▲ Cost estimate
▲ WBS
▲ Project schedule

Work Packages/Tasks	Item Totals 000 $	JAN	FEB	MAR	APR	MAY	JUN	JUL	AUG	SEP	OCT	NOV	DEC	JAN
1.01 Select Concept	12	8	4											
1.02 Design Beta PC	8		1	3	3	1								
1.03 Produce Beta PC	8		1	3	3	1								
1.04 Develop Test Plans	2		1	1										
1.05 Test Beta PC	6					3	3							
2.01 Design Production PC	18						3	6	6	3				
2.02 Outsource Mold Design	16						1	7	7	1				
2.03 Design Tooling	3						5	10	10	5				
2.04 Purchase Tool Machines	16									20	140			
2.05 Manufacture Molds	80									10	10	60		
2.06 Test Molds	8									8				
2.07 Certify PC	18											18		
3.01 Ramp Up	30												30	
	396	8	7	7	6	5	12	23	23	47	150	78	30	

Figure 7.11   An example of the Cost Baseline (Time-phased Budget).

A simple definition of cost baselining as the sheer spreading of the cost estimate items over time hints that having a documented cost estimate that includes all cost items is a starting point. Hopefully, these items can be arranged in tune with WBS as we have chosen to do. If done so, the knowledge of the project schedule—indicating planned start and expected finish dates for work elements—enables the assignment of the cost to the time period when the cost will be incurred.

Identify the Type of Cost Baseline and Cost Items. Which types of cost are typically included in a Cost Baseline? That, of course, depends on the type of the baseline being developed. As mentioned earlier, several are available, but the size and nature of the project are major determinants of the baseline type you may need. If the target is to prepare a baseline focusing on project expenditures (also called project spending plan or cash outflows or project budget), which is our focus here, consider including a broad menu of cost items, some of which are as follows:

- ▲ Salaries and wages of project personnel (in simplest cases this is the only item to include in in-company projects)
- ▲ Overhead expenses
- ▲ Payments to contractors
- ▲ Payments of vendors' invoices for purchases of equipment, materials and services
- ▲ Interest payable on loans, loan repayments, tax payments, shipping fees, duties, and so on.

In case you are after the baseline measuring cash inflows, examples of some cost items that may be included are as follows:

- ▲ Payments from customers for delivered equipment, materials and services
- ▲ Loans from financial institutions
- ▲ Tax refunds, grants, and so on

COST BASELINE (TIME-PHASED BUDGET) CRITERIA			
Cost or Payment Item	Schedule Trigger Event or Information	Interval Between Trigger Event and Payment	Comments
Management and design team	Per labor schedule	1 month	
Vendor subcontracts	Schedule milestones	1 month	This is company policy.
Vendor's invoices for equipment and materials purchases, lump sum	On-site delivery milestones	2 weeks	This is set by the design team to motivate vendors.

Figure 7.12 Examples of criteria for cost baselining in a product development project.

If the intent is to manage cash flow, you will need items for both cash outflows and cash inflows. Once the cost items to include are identified, it is time to set criteria for cost baselining.

Set Cost Baseline Criteria. The preparation of a Cost Baseline is essentially an act of establishing the relationship between the cost estimate and timing. For this to be possible, there must be clear criteria that determine which project events trigger payments of cost items included into the baseline and how long the intervals between the trigger events and the related payments are (see Figure 7.12). For payments of vendors, for instance, the trigger events are usually milestones defined in the contractual terms that stipulate how and when the payments are to be made. In other times, such as paying salaries of project management team members, labor schedule of their engagement is what triggers their payment at the end of the month. The intervals, whether for payments within or outside the organization, are dictated by the time needed for internal and external communications, approvals and administrative procedures, and company policies bent to keep money as long as they can [19]. Performing an appropriate analysis of the criteria and defining them in a written form is highly advisable for it becomes a crucial foundation for tabulating costs by periods in the process of cost baselining.

Allocate Cost Items to Time Periods. When the type of the baseline is chosen, cost items to be included into the baseline identified, and criteria for baselining defined, the foundations for allocating cost to time periods are laid down. Then you should address coding and arrangement of cost items. Preferably, the project would have its own cost codes (column 1 in Figure 7.11) that are consistent with the company's cost coding system or industry standards. This may not be possible when the project is externally funded, because the customer may mandate the use of its own cost codes. Items from column 2 may be arranged in different ways. If a Cost Baseline is being developed on the basis of the Bottom-up Estimate, the items can be arranged in line with WBS, as we have done in Figure 7.11, using work packages from the WBS for a project. When an analogous or Parametric Estimate is being used to construct a Cost Baseline, other methods to arrange the items can be deployed.

Column 3 provides cost estimates for the items, which will now be allocated to certain time periods in a nine-month project. Since the reporting is on a monthly basis, time periods are months represented in columns 4 to 12. Item 1.01, Develop Specs, will be carried out in months 1 and 2, so part of the estimated $10K will be expended in month 1 and the remaining part in month 2. How much will be allocated to each month hinges on the following factors [19]:

▲ The project schedule, indicating the planned start and end dates of the item, along with resource histograms specifying resource requirements by time period (see the box that follows, "Simple S-Curves in a Complex Microprocessor Project")

▲ The contractual terms

▲ Intervals between trigger events and payments

Working off these information inputs, from the estimated $10K for item 1.01, $8K will be distributed to month 1 and $2K to month 2, and entered over a bar in Figure 7.11. Similarly, estimates for the remaining items are spread over their months of execution and entered over their bars. These bars are actually the schedule for the project in the form of bar chart. The schedule is rarely drawn on the Cost Baseline, but we included it in Figure 7.11 to make the baseline easier to comprehend.

Sum Estimated Cost by Period. Once all item cost estimates are allocated to specific time periods, the next action is to sum estimated cost by periods as is shown in the last row of Figure 7.11. Apparently, this provides information about incremental expenditures by time periods—that is, expenditures for each month—which will be used in the next step to display the Cost Baseline graphically.

Display the Cost Baseline. S-curve is a popular way of displaying Cost Baseline formatted as cumulative expenditures (see Figure 7.13). To figure out cumulative expenditures, add the incremental expenditures for the first period to those of the second period. These are the cumulative expenditures for the first two periods. Add this number to the third period's incremental expenditures to obtain the cumulative expenditures for the first three periods, and continue with this procedure for the remaining periods. When done, graph the cumulative expenditures on y-axis over time (x-axis) to develop Cost Baseline in the form of an S-curve. As in any type of the cost estimate, this is the time to check and review the Cost Baseline. Once the project is over, there is a lot of value in studying how the initial baseline worked over the life of the project, learning the lessons, and using them to improve future Cost Baselines.

Simple S-Curves in a Complex Microprocessor Project

When a microprocessor development project started, management was very clear: they wanted to see a Time-phased Budget for the engineering talent for the project. Starting from the project schedule, the project manager responded by

● Specifying the monthly needs for each type of desired engineers.

● Translating the needs into S-curves for each type of engineers and the aggregate S-curve for the whole design team composed of all of these engineers, who at one time peaked at almost 300.

The S-curves were used for multiple purposes. The aggregate S-curve including person-months over project schedule served as a baseline for a simplified version of the earned value measurement. Also, the curve was expressed in dollars that management had to secure for paying the engineers' salaries. Since all of the engineers were not available in the company, the S-curves for individual engineering types were also used for timely preparation and hiring in the job market.

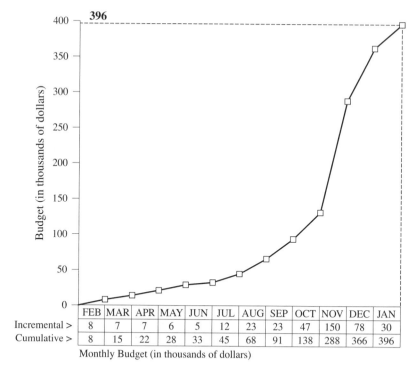

	FEB	MAR	APR	MAY	JUN	JUL	AUG	SEP	OCT	NOV	DEC	JAN
Incremental >	8	7	7	6	5	12	23	23	47	150	78	30
Cumulative >	8	15	22	28	33	45	68	91	138	288	366	396

Monthly Budget (in thousands of dollars)

Figure 7.13 A Cost Baseline displayed as an S-curve.

Utilizing Cost Baseline

When to Use. Most experts believe that cost baselining is definitely a redundancy in small projects, because the cost of its preparation may easily outweigh the benefits [19]. In contrast, other projects do have a need for the Cost Baseline. Typically, the baseline is developed as part of initial project planning to forecast its cash flow. Considering that the Cost Baseline may be based on an analogous, parametric, or bottom-up cost estimate, sometimes as the estimates evolve and become more accurate, so do the Cost Baselines. They are reissued at regular or irregular intervals and may even constitute part of project reports submitted to higher management or external customers [1]. For details about updating and changing the baseline, see the box that follows, "When Should You Update or Change the Budget?"

Time for Development. As a function of the size and complexity of the project and its schedule, resource requirements and cost estimate, the time to develop a Cost Baseline may widely vary. The development of a Cost Baseline based on a low-detail Analogous Estimate and summary schedule may consume an hour or two of the skilled team's time. On the other hand, an experienced project team may spend tens of hours constructing a cost baseline based on a very detailed Bottom-up Estimate, with hundreds of activities in the schedule, much less, though, than what it would take a less skilled team.

When Should You Update or Change the Budget?

Dogmatically sticking with the initial Cost Baseline or Time-phased Budget when there is a need to alter it serves no purpose and is unrealistic. This need for alteration is of a managerial or control nature and is really triggered by several factors, leading to minor (updates) or major revisions (changes) of the baseline. Updates may occur because of factors such as [1]:

- Cost estimate evolvement. As the project progresses, more information becomes available, helping develop more accurate estimates from analogous or parametric to the bottom-up one. Such changes in estimates should lead to the update of the baseline.

- Project changes. Management of these changes may require new expenditures, which should be added to the baseline. Changes may be due to unforeseen conditions or from customer-generated changes.

- Schedule changes. Changes of time-phasing of project activities during the execution stage are frequent and result in inevitable modification of the baseline.

In addition to these updates (minor revisions), there may come times when a major revision of the baseline may occur. During the project implementation, major unplanned schedule, cost, or technical problems may occur. Or, there may be a need to change the project strategy. These, typically, prompt major revision of the project plan, including a major revision of the Cost Baseline. Such changes to the Cost Baseline may happen very rarely, once or twice in the life of a project, if at all. Whether dealing with updates or changes to the baseline, the key is to manage all these modifications and related factors in a proactive rather than a reactive way and maintain the control of the project.

Benefits. The lack of an effective Cost Baseline, even if a cost estimate and labor requirements are available, poses a major threat to a project—organizing measurement of performance and cash flow is difficult, if not impossible. Therefore, constructing the baseline offers benefits of using it as a performance measurement baseline (see the "Earned Value Analysis" section in Chapter 13). In this capacity, the baseline is a basis for comparing actual costs (and when they occurred) with planned costs (and when they were supposed to occur) [1]. This, then, is a way to gauge efficiency and progress, attracting management's attention to any deviations from planned progress and estimated costs.

Cash flow forecasting is another major value that an effective baseline may create. It informs management or the customer in advance of the funds that must be made available in order to procure resources and use them to sustain project progress. When properly performing this role in the course of the project implementation, the Cost Baseline should be modified to reflect performance and progress to date. Some risky consequences of not managing cash flow in a project, and how to avoid them, are described in the box that follows, "The Museum Design Company (MDC)."

Advantages and Disadvantages. The operation of cost baselining is relatively simple, whether performed manually or with the help of a computer. Additionally, the visual power of displaying it in an S-curve format is impressive, further strengthening the case of its simplicity. Contrast these two major advantages with its major disadvantage, Cost Baseline's often time-consuming nature, a possible burden on an already busy project team.

Variations. As explained earlier, Cost Baselines come in different shapes, some measuring expenditures (cash outflows), others received payments (cash inflows) or committed costs, still others indicating expended labor hours. Since the project schedule is a critical input in cost baselining, these may be based on project activities' early start dates, or late start dates, or their averages. Other names for Cost Baseline/Time-phased Budget are cash flow schedule, project budget, and performance measurement baseline.

The Museum Design Company (MDC)

MDC found itself in what appeared to be a paradoxical situation: It had a bunch of contracts but no positive cash flow. How was this possible? Loaded with top design talent and known for a strong track record of superb technical quality, MDC had no difficulty landing project contracts to design the military museum exhibits. But John Riddle Jr., CEO of MDC and an accomplished designer, had to borrow from his bank almost each month to make the payroll. Puzzled by this, Jr. asked for professional help. He was advised to look into cash inflow and outflow S-curves for each project. Since those were not available, project managers developed the curves with the help of the consultant. The majority of them looked like the ones in Figure 7.14a. Apparently, the difference between the funds obtained from the customer (cash in) and payments paid for designers' salaries, overhead, and loan interests (cash out) was negative cash flow all the way through the project except at the end when it was zero. This was the source of the paradox, which eventually could make MDC go out of business and made MDC borrow from the bank, eating up its own profits. Jr. concluded that this negative cash flow situation must be avoided at all costs in future projects. The idea was to make the difference between cash in and cash out positive (Figure 7.14b), enabling MDC to eliminate the need for costly loans and stay in business. After a careful explanation, the customer accepted the idea.

(a) Negative cash flow   (b) Positive cash flow

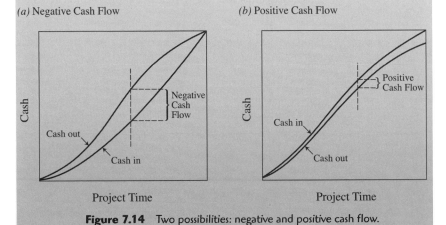

Figure 7.14   Two possibilities: negative and positive cash flow.

Customization Action	Examples of Customization Actions
Define limits of use.	Adjust to the standard used in the industry. For example, use Cost Baseline expressed in labor hours rather than expenditures for internal software projects. Base on averages of early and late start dates. Use cash flow (inflow and outflow) for externally funded projects. Base on early start dates of activities.
Add a new feature.	Enter 100 percent on the top of the right vertical axis of the S-curve. Add the remaining percentages in 10 percent intervals. These percentages indicate percent complete in the project (popular with technically oriented project managers favoring percent complete for progress measurement and reporting).
Modify a feature.	Include a new column in the form in Figure 7.11 between code and items columns to designate the control account plan for each item (this is for companies employing formal earned value measurement).

Customize the Cost Baseline. Given the available variations of Cost Baseline, adapting them to best account for your specific needs is a good proposition. See simple examples of such actions to help you visualize possible customization on page 257.

Summary

This section covered the Cost Baseline, a Time-phased Budget used to measure and monitor cost performance on the project. Typically, the baseline is developed in larger projects as part of initial project planning to forecast its cash flow. Developing the baseline offers benefits of using it as a performance measurement baseline. In that way, the project team can gauge efficiency and progress, attracting management's attention to any deviations from planned progress and estimated costs. In the end, key points in developing the Cost Baseline are listed in the box that follows.

Concluding Remarks

This chapter featured five tools with clearly designed purposes. For two of them, Cost Planning Map and Cost Baseline, the purposes are so distinct that they do not compete with other tools for project manager's attention. While the Cost Planning Map strives to establish a systematiq methodology for cost planning, the Cost Baseline aims at providing a time-phased budget. The remaining three tools may be used in combination or the most appropriate one of them can be chosen. That calls for matching the project with situations in the list that follows and understanding marks that show how each situation favors the tools. If these are not sufficient to characterize the project, add more of them and mark how they favor the tools use. The tool with the highest number of marks is probably the most appropriate to deploy.

Cost Baseline Check

Check to make sure you developed a proper Cost Baseline. The baseline should show

- Cost code
- Item descriptions
- Item cost estimates
- Item budget by period
- Project budget by period
- Cumulative, to date budget for the project
- S-curve (if a choice of display).

A Summary Comparison of Cost Planning Tools

Situation	Favoring Cost Planning Map	Favoring Analogous Estimate	Favoring Parametric Estimate	Favoring Bottom-up Estimate	Favoring Cost Baseline
Provide cost planning methodology.	√				
Show the amount of estimated funds.		√	√	√	√
Show time-phasing of estimated funds.					√
Organizations with stream of small projects	√	√	√	√	
Organizations with large projects	√	√	√	√	√
Based on past experience		√			
Higher accuracy required				√	
Lower accuracy required		√	√		
A few hours to prepare		√			

Medium time to prepare		✓	
Longer time to prepare			✓
Need estimate for project screening, forecasting.	✓	✓	
Need estimate for budget authorization.		✓	✓
Need estimate for cost proposal /change orders.			✓
Make decisions very early in project life cycle	✓		
Estimate in project definition/early design.		✓	
Before execution, design substantially complete			✓

References

1. Harrison, F. L. 1983. *Advanced Project Management*. Hunts, U.K.: Gower Publishing Company.

2. Roetzheim, W. H. and R. A. Beasley. 1995. *Software Project Cost and Schedule Estimating: Best Practices*. Upper Saddle River, N.J.: Prentice Hall.

3. Stewart, R. D., R. M. Wyskida, and J. D. Johannes. 1995. *Cost Estimator's Reference Manual*. 2d ed. New York: John Wiley & Sons.

4. Westney, R. E. 1997. *The Engineer's Cost Handbook*. New York: Marcel Dekker.

5. Boehm, B. W. 1984. "Software Engineering Economics." *IEEE Transactions on Software Engineering* 10(1): 4–21.

6. Pressman, R. S. 1992. *Software Engineering: A Practitioner's Approach*. 3d ed. New York: McGraw-Hill.

7. Gulledge, T. R. and W. P. Hutzler. 1993. *Analytical Methods in Software Engineering Economics*. Berlin: Springer-Verlag.

8. Kile, R. L. and U.S.A.F.C.A. Agency. 1991. *REVIC Software Cost Estimating Model User's Manual version 9.0*. Arlington, Va.: Revic Users Group.

9. American Association of Cost Engineering. 1990. *Standard Cost Engineering Terminology; AACE Recommended Practice and Standard No. 10S-90*. Morgantown, W.V.: American Association of Cost Engineering.

10. Birrell, N.D.A. 1985. *Practical Handbook of Software Development*. Cambridge, Mass.: Cambridge University Press.

11. Project Management Institute. 2000. *A Guide to The Project Management Body of Knowledge*. Drexell Hill, Pa.: Project Management Institute.

12. Humphreys, K. K. and L. M. English. 1992. *Project and Cost Engineers' Handbook*. 3d ed. New York: AACE and Marcel Dekker.

13. Kemerer, C.F . 1997. *Software Project Management*. Boston: McGraw-Hill.

14. Smith, N.J. 1995. *Project Cost Estimating*. London: Thomas Telford.

15. Institution of Chemical Engineering, I. and A.o.C. Engineers. 1977. *A New Guide to Capital Cost Estimating*. Rugby, U.K.: The Institution of Chemical Engineers.

16. Stewart, R. D. 1991. *Cost Estimating*. 2d ed. New York: John Wiley & Sons.

17. Reifer, D. J. 1993. *Software Management.* 4th ed. New York: IEEE Computer Society Press.

18. Marciniak, J. J. and D. J. Reifer. 1990. *Software Acquisition Management.* New York: John Wiley & Sons.

19. Lock, D. 1990. *Project Planner.* Aldershot, U.K.: Gower Publishing.

Quality Planning

The best is the enemy of the good.

Voltaire

This chapter focuses on the following tools:

- ▲ Project Quality Program
- ▲ Flowchart
- ▲ Affinity Diagram

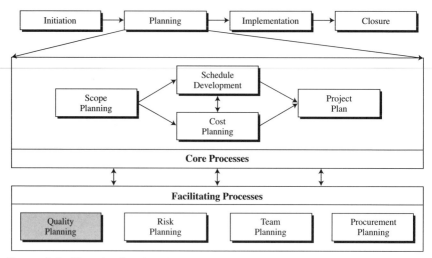

Figure 8.1 The role of quality planning tools in standardized project management process.

The purpose of these tools is to help build predictability into PM processes. Quality planning begins by establishing quality goals, policies, and standards related to the project scope. It continues with the identification and design of PM processes, actions, and responsibilities needed to accomplish the goals and standards. In this effort, the role of the quality planning tools is crucial (see Figure 8.1). They make the planning possible, turning all this information into outcomes such as quality plans and PM processes. As these outcomes are pursued, quality planning needs to be synchronized with the core processes of scope, schedule, and cost planning, as well as the facilitating processes of risk, team, and procurement planning. This chapter aims at helping practicing and prospective project managers acquire the following skills:

- ▲ Learn how to use various quality planning tools.
- ▲ Choose quality planning tools that are appropriate for their project situation.
- ▲ Customize the tools of their choice.

These skills are critical in project planning and building a standardized PM process.

Project Quality Program

What Is a Project Quality Program?

A Project Quality Program is an action plan striving to ensure that the actual quality of a project will meet the planned one (see Figure 8.2). Using WBS as a skeleton for integration, the program sets a quality level based entirely on customers' expectations and requirements. With such a strong customer focus, the program translates the requirements into tangible quality standards, for whose accomplishment a set of tasks is defined. Explicitly defined responsibilities and time lines for the performance of the tasks add necessary elements to use the program as a project quality roadmap. In a nutshell, the program states that this is what this project has to do to ensure that the quality of its deliverables is meeting our customers' requirements.

Developing a Project Quality Program

Although the Project Quality Program justifiably projects an image of simplicity, its development calls for a skillful coordination and integration of multiple concepts and pieces of information. Our limited experience with modern practices for managing project quality is one more reason to have our guard up when developing the program as follows.

Collect Information Inputs. The quality of the Project Quality Program is heavily dependent on the quality of its inputs. In particular, the following inputs are known for their impact on the program:

- ▲ Quality policy and procedures
- ▲ Voice of the customer
- ▲ Scope statement and WBS

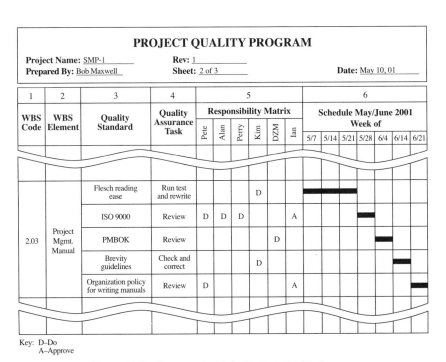

Figure 8.2 An example of the Project Quality Program.

The foundations of how quality is managed in an organization are described in its quality policy. Defined, documented, and supported by management, the policy is a statement of quality principles, beliefs, and key objectives for projects that set a general framework to carry out quality management actions in the organization [5]. This framework is further detailed in quality procedures. Together, the policy and procedures set a direction for the program. For example, if the procedures mandate compliance with ISO 9000 standards, the program will have to comply.

Listening to the voice of the customer will not only help discover customers' needs; it will also help decipher customers' needs and translate them into the recognizable language of the project scope, establish units to measure customers' needs, and express them as quality standards, a crucial piece of the program. For this reason, the Project Quality Program needs to be closely coordinated with the voice of the customer.

Finally, when you are setting project goals, scope statement also sets a quality goal for the project. Along with this input goes the WBS that defines the project work for which a Project Quality Program is developed. Therefore, both the scope statement and WBS are significant inputs to the quality program preparation.

Select a WBS Element. Quality planning should be performed around the skeleton provided by WBS. Taken overall, the basic place for quality programming is the work package level of WBS. Summing up quality programs for a work element above the work package level is summing up quality programs for its constituent work packages. Continuing with this summation up the WBS hierarchy will provide us with a Project Quality Program for the total project. Turning this logic into practice in our example in Figure 8.2, we pick a work package, "project management manual," for which we will develop a quality program. This work package is part of a project for the deployment of an organization-wide project management process.

Establish Quality Standards. What are quality standards for the work package we have selected in our example in Figure 8.2? And why do we need them? We selected five standards for this work package. First, the customer wants the manual and process described in it to comply with their ISO 9000 procedures, an international standard. Also, the process in the manual needs to be in tune with the *PMBOK (Project Management Body of Knowledge)* Guide, a U.S. standard. To be used by project teams, the manual needs to meet three in-company standards: brevity (no longer than five pages), compatibility with the organizational policies for manual writing, and ease of reading (Flesch Reading Ease score larger than 70 points).

These examples help us answer questions such as these: Why quality standards? Who establishes them? What standards? Although there exist various definitions of quality, we define quality as customer-driven: quality is meeting or exceeding customer expectations (see the box that follows, "What Is Your Definition of Quality?"). As we applied voice-of-the-customer tools, we identified the expectations for this specific work package and need to measure and respond to them (see the box "More Effort Should Go into Quality Planning," under "Utilizing a Project Quality Program" in this section). The purpose of using quality standards in the Project Quality Program, then, is to measure the expectations. In other words, to deliver a work package (PM manual) per customers' expectations, we need to accomplish the established quality standards. Since customers are the final judges of quality, they need to be involved and accept the standards.

What Is Your Definition of Quality?

Often projects include members from cross-functional backgrounds. In performing their roles, they speak different functional languages, often unknown to their team peers. For example, while "four P's" (product, price, promotion, place) is a crucial term to a marketing team member, it may mean little to an engineer-designer. These role and language differences create a potential for confusion and use of different definitions of quality, described in the table that follows.

Per the *transcendent definition,* quality is both absolute and universally recognizable [4]. Therefore, it may not be defined precisely; rather, you know it when you see it—for example, in a Rolex watch. When someone believes that differences in quality reflect differences in quantity of some product attribute, that person uses a *product-based definition* of quality. In that case, a 833-MHz logic chip is of higher quality than a 366-MHz one.

Quality Definition	Often Used By
Transcendent	Customers
Product-based	Customers
User-based	Marketing
Manufacturing-based	Manufacturing
Value-based	Design

In the *user-based* definition, quality is defined as fitness for use, or how well the product performs its function. For example, Primavera and MS Project are both fit for use but are used by different project managers with different needs and, thus, different quality standards. Conformance to specifications is a *manufacturing-based* definition, where specifications include targets (e.g., part thickness is 1 inch) and tolerances or allowable variation (1inch+0.01). A quality product—per *value-based* definition—is one that is as useful as competing products and is sold at a lower price. Apparently, in this relationship of usefulness to price, "everyday" low prices offer more long-term quality than buying whatever is on special. All of these definitions are necessary to reflect views of the cross-functional team members, resulting in a product that will satisfy customers' needs. Still, many organizations find them conflicting and prefer using a customer-driven definition: quality is meeting or exceeding customer expectations [6-9]. That is also our definition of choice in developing a Project Quality Program.

In our example of the PM manual, we used international, national, and in-company standards. Using the first two is both what we are accustomed to and what is easier to use because of the wide recognition of such standards. Most of the time, however, projects have to resort to in-company standards. When these are of the quantitative nature—such as the score of 70 (out of 100) points on the Flesch Reading Ease scale—their ease of use is higher. At times, we may need to establish standards of the qualitative or perceptual nature. For example, how satisfied is the customer with our timeliness? For this we may use a perceptual scale (also called a Likert scale) from 1 to 10, where 1 is "Not satisfied" and 10 is "Very satisfied." The meaning of all other levels between 1 and 10 should be clearly defined to make the scale better. Assuming that they are well designed, these scales are very reliable (even statistically) and are used widely. Another important question that project teams often ask is about the number of standards they need to establish for a work package. There are no hard-and-fast rules here. Answers range from an idealistic "As many as one needs" to a pragmatic "Pick one major standard."

Define Quality Assurance Task. Once we have established the quality standards, a natural question is what we should do to meet them so we can deliver the work package per customer expectations. We identify quality assurance tasks that we need to perform and complete successfully to be able to meet the quality standards. Take, for example, our PM manual work package. For our standard of "Flesch Reading Ease score larger than 70 points," the task may be recurring; as we write the manual, we may periodically check the reading ease by running the test in Microsoft Word, for example, and rewrite if we are below 70 points. On the other hand, a task for PMBOK compliance of our PM manual includes assembling a group of experts to judge the degree of the compliance, probably using a qualitative standard.

Assign Responsibilities, Determine Schedule. Identifying quality assurance tasks leads us to ask who is to perform them? And, when will they be performed? (see the box that follows, "But We Have a QA Group!").To build accountability into our quality system, we need to determine who else in addition to the person performing the task will contribute to the task completion and how (see the example in Figure 8.2). It is for this purpose that we have a responsibility matrix built in the Project Quality Program, including responsibilities such as "Do," "Approve", and "Must be consulted." Once responsibilities are clear, the equation of the Project Quality Program will be complete when a time line is established for each quality assurance task.

Obviously, a Project Quality Program contains a responsibility matrix and a schedule (typically a Gantt Chart). Some may wonder whether these are redundant, since we have a responsibility matrix and schedule for the overall project. Put differently, should the matrix and schedule from the quality program be taken and merged with the project's matrix and schedule? The answer to both questions is no. If we take a look at a typical project responsibility matrix and schedule, there is a high probability that no quality assurance tasks will be found in them. What does this tell us? We still do not treat quality as a project priority. Given this, we prefer to have a separate quality program with its responsibility matrix and schedule as long as it takes to build a system that appreciates quality.

Utilizing a Project Quality Program

When to Use. Today customers reign [10]. Satisfied customers represent a real economic asset to a company, which is critical for economic returns. This is apparent from American Customer Satisfaction Index survey from 1997, where the higher-scoring companies have created over 100 percent more shareholder wealth than the lower-scoring companies [11]. Obviously, customers are willing to pay when satisfied with quality of a product or service. Following these facts, there is no doubt that each project can benefit from having a Project Quality Program, whether large or small. Planning for quality pays off (see the box that follows, "More Effort Should Go into Quality Planning"). For their larger budgets and resources for PM, larger projects should devotedly develop a program early in the planning stage and carry it through the end of the project. Smaller projects faced with scarce time budgets may opt to concentrate on time-affordable quality programs with only several major work packages.

But We Have a QA Group!

When blessed with enough resources and professional project managers who understand quality, larger projects tend to do a decent job in using a Project Quality Program. As project size and complexity reduce, the tendency slowly disappears and resistance to the use of the Project Quality Program grows. We have heard a whole spectrum of excuses for this. In one case the excuse of a project team was "we have a QA (quality assurance) group. They should control quality." If we believe in this philosophy, our project quality will suffer from the lack of ownership and prevention. Those who perform the project activities—the project team—know them best and have the highest probability to discover errors in activities and eliminate them. They should be empowered to act as both doers and thinkers, solving project problems and owning project activities. If they do so, they will develop a Project Quality Program, determine owners of all quality tasks, and catch and correct quality defects. In that role, they will also take a preventive approach, continuously improving the work process and taking it to a higher level. That will improve quality, really. Relying on QA groups won't do it. First, they have no knowledge of the project activities that your team has and feel no ownership for what your team does. So their best bet is to rely on the inspection, a corrective but not preventive approach, capturing errors when it is too late—they are already made. Without the ownership and preventive approach on the part of QA groups, the project is doomed to live with substandard quality, a practice of the past. To reach quality of the present and future, we need to own and practice a preventive approach. Then we can use QA groups as a helping resource.

More Effort Should Go into Quality Planning

Joseph Juran, a guru of quality management, developed Quality Trilogy, a prescription for quality planning, control, and improvement (see Figure 8.3). Quality planning begins with identifying the list of project customers [1]. The point here is that project team members should know who uses their project products. Knowing this, the members proceed with discovering customers' needs, usually expressed in their language that has to be translated into project language. Also, the way to measure customers' needs is established. The discovery, translation, and measurement can be done by means of voice-of-the-customer tools covered earlier in the book, in Chapter 4. In response to customers' needs, the next step includes the development of the project product features and goals, certainly part of the project scope definition. Then, there comes an effort to develop an integrated set of project processes for the design and delivery of the project product. Apparently, this combination of knowing what customers need, designing the product per such need, and delivering the product through a seamless process are the essence of quality planning. After finding that quality control (corrective approach) rather than quality planning (preventive approach) received priority attention in organizations, Juran felt more effort should go into quality planning.

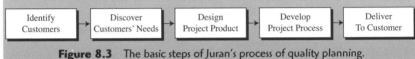

Figure 8.3   The basic steps of Juran's process of quality planning.

Time to Use. Quality planning leading to the development of a Project Quality Program takes time. Once all inputs are prepared, the activity of developing a program is not a time killer. Rather, a program for a 20-work package project, with a quality standard or two (column 3, Figure 8.2) per work package, and a quality assurance task (column 4, Figure 8.2) per quality standard may take from one to two hours to complete. This assumes a skilled team is involved. Less skilled and larger projects will require more time. If they are time-pressed and want to focus on several major work packages, the team is likely to prepare a program in 15 to 20 minutes.

Benefits. The real value of this tool comes from its proactive nature. Instead of taking quality as an implicit concern that we will address as we come across quality problems, the tool anticipates the quality path in the project and charts actions to follow the path. As a result, the actions are geared toward the prevention of the quality problems, rather than their correction once they occur. In addition, the core of a Project Quality Program is built on the premise of customer focus. Simply, the proverbial "Actions speak louder than words" is put to work—all quality assurance actions are taken to deliver per customer expectations.

Advantages and Disadvantages. A Project Quality Program instills a feeling of simplicity. Appearing as a simple table, the program offers advantages in that it is easy to read and follow, benefiting from the visual effect of the responsibility matrix and Gantt-shaped schedule.

Disadvantages are that it is

▲ *Time-consuming.* That it may take time from a scarce time budget of the project team to develop a Project Quality Program may be perceived as a disadvantage.

▲ *Difficult for some project teams.* Projects that are not used to quality standards—usually nonengineering projects—may find difficulty in developing them to a meaningful level.

Customize the Project Quality Program. We have offered a generic Project Quality Program in this section, designed to address the needs of various projects. Its design, however, may not address specific needs of a particular project. In that case, we can expect more value out of this tool when we adapt it to our project needs. The following examples are provided to give you some ideas about adapting the tool.

Customization Action	*Examples of Customization Actions*
Define limits of use.	Use the Project Quality Program with all work packages (large projects).
	Use the Project Quality Program for the three most important work packages; one package-one standard-one task (for small projects).
Add a new feature.	In addition to planning the time line of quality assurance tasks, use the quality schedule for tracking the schedule as well.
Modify a feature	Replace quality standards with specifications (typical for engineering projects).
	Replace the responsibility matrix with a column showing the name of task performer only (typical of small projects).

Project Quality Program Check

Check to make sure you developed a proper Project Quality Program. The program should show

- Quality standards for work packages
- Quality assurance tasks to support the accomplishment of each quality standard
- Responsibilities for quality assurance tasks
- A time line for quality assurance tasks.

Summary

The focus of this section was on the Project Quality Program, an action plan striving to ensure that actual quality of a project will meet the planned one. Each project can benefit from having a program, whether large or small. The real value of this tool comes from its proactive nature. It anticipates the quality path in the project and charts actions to follow the path. As a result, the program is of a preventive rather than a corrective nature. You can derive additional value by customizing the program for your needs. The box on page 271 recaps the key points of structuring the Project Quality Program.

Flowchart

What Is a Flowchart?

A Flowchart is a pictorial specification of the steps or tasks in a process (see Figure 8.4). Using boxes or other symbols to represent steps/tasks, it describes how a process really works, uncovering how the steps relate to one another. Since this step-by-step picture of the process creates a common language and ensures common understanding about sequence, it can serve different purposes. Whether used to document a standard method to perform, to design, or to examine and modify a PM process, the Flowchart's range of application in projects is vast—from the point of project initiation to the point of project termination. In such endeavors the intent of using a Flowchart is to plan, ensure, or improve the quality of PM processes, the basic building blocks of projects.

Flowcharting a Process

Flowcharting a process begins with the process owner. Whether the owner is the project team or an individual, the owner needs to deploy an efficient procedure for designing the process, starting with necessary information inputs.

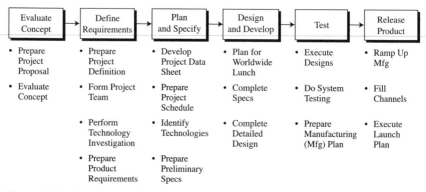

Figure 8.4 An example of the top-down Flowchart for a product development project.

Collect Information Inputs. Making proper decisions about the process design hinges on quality of information about the following:

- ▲ What is the purpose of the process?
- ▲ What does the voice of the customer say?
- ▲ What are the crucial inputs and outputs of the process?

Knowing the purpose of the process means identifying the crucial outputs—that is, products or services—that the process will produce. In the case of the new product development process, for example, aside from outputs such as new products, a major output is the business plan for the new product project. Since the goal of the process is to develop a capable course of action to satisfy customer requirements, there is an apparent need to know who the customers are and what they require. Major customers of the process, for instance, are senior managers who need the business plan in order to decide whether or not to approve the project [12]. This indicates how important is to identify crucial inputs and understand how they get converted into the business plan. One such input is the sales forecast, which will be used to calculate net present value, a vital part of the business plan. In a nutshell, when information about the purpose of the process and customer voice, and crucial inputs, conversion, and outputs is available, boundaries of the process are set. This is a fundamental prerequisite to efficient flowcharting

Select the Type and Level of Detail of the Chart. Flowcharts come in various formats, each one designed for a specific purpose. Therefore, knowing the purpose or intended use of the chart is the major determinant in selecting an appropriate type of chart. Basic purpose, major advantages, and disadvantages of several types of Flowcharts are described in Table 8.1.

Apparently, the Flowchart may be as simple as a basic chart that outlines the major steps in the process or as thorough as a deployment chart that lays out every tiny action and the involved project people (see Figure 8.5). Note, however, that each type of the chart may be more or less detailed, prompting us to decide how detailed we want the process to be to make it understandable. In describing the process of flowcharting, we will focus on the top-down chart, although we make references to other types of Flowcharts when necessary.

Symbols used to draw Flowcharts may vary. For example, a detailed Flowchart may include these symbols.

◯ An ellipse may be used to identify inputs (task, information, or materials) necessary to start the process or indicate outputs (results) at the end of the process.

▢ A step, task, or activity in the process can be shown with a box or rectangle. While several arrows can enter a box, most of the time one arrow comes out of it.

◇ A diamond designates a point where a decision has to be made.

Table 8.1: Basic Purpose, Major Advantages, and Disadvantages of Several Types of Flowcharts

Type of the Flowchart	Basic Purpose	Major Advantage	Major Disadvantage
Basic	Outlines the major steps in the process	Good when an overview of main steps is sufficient	Oversimplifies the process
Top-down	Shows both the few major steps and substeps	Minimizes details to focus on vital steps	Does not include steps that can derail the process (e.g., control steps)
Detailed	Captures details of at least some parts of the process	Captures details necessary when standardizing or improving the process	Capturing details adds complexity, takes time, and has a cost
Deployment	Indicates the detailed steps and people involved in the steps	Useful to clarify roles and dependencies in processes with many handoffs	Capturing details adds complexity, takes time, and has a cost
Opportunity	Separates value-added from cost-added-only process steps	Highlight opportunities for improvement	Capturing and separating steps adds complexity, takes time, and has a cost
Systemigram	Captures the process using semantics of natural language	Can be much more memorable than standard language of project manuals	May appear complex, costly to develop

○ A circle with a letter, number, or both shows a continuation point that will have a corresponding point elsewhere in the Flowchart.

→ An arrow defines the direction or flow of the process.

Unlike the detailed Flowchart, the top-down Flowchart uses boxes and arrows. Other Flowcharts may use still other symbols, as indicated in examples in this section. The key is that once the type of the Flowchart is selected, the appropriate symbols are used to draw the chart.

Draw the Flowchart. There are several actions in drawing the Flowchart:

▲ Begin by brainstorming a list of six or seven major steps carried out in transforming the inputs to outputs.

▲ Sequence the steps in the order they are performed. Then draw them in a plain flow of boxes/rectangles across the top of the page, and connect them with arrows [13].

▲ Break down each major step into substeps, which can either be listed beneath their major step or put in the boxes and connected with arrows. In the example in Figure 8.4, we chose to the do the former. This way, the substeps provide us with a more detailed sequence of tasks to execute the process.

For a method to draw the detailed Flowchart, see the box that follows, "Designing Detailed Flowcharts by Backward Chaining".

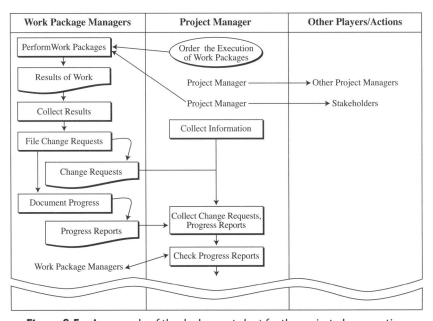

Figure 8.5 An example of the deployment chart for the project plan execution.

Designing Detailed Flowcharts by Backward Chaining

ATT pioneered an interesting method, called *backward chaining*, to design a detailed Flowchart. It starts with the outputs, asking what customer requirements are, moves back through the process to define the crucial steps in producing each of the outputs, and stops when the inputs are reached. The following procedure is followed [3]:

- Begin by asking what is the last essential subprocess that produces the output of the process?

- Continue by asking what input does that subprocess need to produce the process output? Double-check to make sure that the input is really needed.

- Identify the source of each input. Quite often, that input is the output of the previous subprocess. At times, the input will come from external suppliers.

- Keep going backward, one subprocess at a time, until each input originates from the external supplier.

When the process—from the customer outputs to supplier inputs—is complete, the same backward chaining can be applied to define a more detailed Flowchart of each subprocess, further detailing the Flowchart for the overall process. Figure 8.6 illustrates an example of the detailed Flowchart.

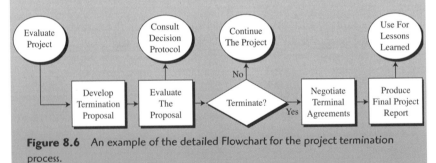

Figure 8.6   An example of the detailed Flowchart for the project termination process.

Mistake-Proof the Process. The purpose of the work on the Flowchart is to make the process efficient and capable of delivering high-quality outputs in a consistent and repeatable manner. For this reason, take a hard look at the developed Flowchart and the process it represents. Consider these questions [6]:

- ▲ Are the process steps identified clearly and arranged in a logical sequence? Is there any need to reorder the steps?

- ▲ Does each step add value? Would leaving out some steps and adding others enhance my process performance? Would combining some steps help the cause?

- ▲ Is there any step that is a bottleneck, slowing down the process? In other words, are capacities of all steps in balance?

▲ What skills and resources do we need to perform the process flawlessly? Are these available, or do we have to replace them with what is available?

▲ Could technology be used to automate any of the steps and increase the performance level?

Being diligent in searching for good answers to these questions is vital for the PM process quality. All too often these questions are not answered because of daily time pressures. Do not make that mistake, or the process quality will suffer.

Determine How to Evaluate the Process. The PM process needs to be controlled to deliver per expected performance level. Also, there may be a need to improve the process, and knowing the quality of the process is the first step to the improvement. This requires establishing measurements and controls, defining points at which quality of the process will be measured. For example, suppose a company uses project definitional index (PDI) as a measure of the scope definition process. Measured at the point when management approves the scope, PDI shows on a scale 1 to 10 how well project team members understand the scope definition.

PM Processes: This Road Leads to Predictability

A PM process is a sequence of steps/tasks meant to create value for project customers. Examples include project planning or project implementation processes. To manage processes means to *design, control,* and *improve* them. In designing, the focus is on the prevention of quality problems by determining the proper sequence of steps/tasks that are capable of achieving high levels of performance. For example, a stable project implementation process for a certain type and size of project should result in a reliable cycle time (project completion time), naturally fluctuating around some average level, say, 12 months. When the abnormal conditions strike—for instance, the project manager is replaced in the middle of the project—a deviation from the average cycle time occurs and the project gets completed in 14 months. Eliminating such abnormal conditions and maintaining the performance level of the process is the job of control. Even so, your project customers may not be happy with the 12-month cycle time. This is where the improvement steps in, where you search for ways to speed up the process through removal of waste and redundancy, defects, and errors. As a result, PM process management leads to better quality and improved project performance through shorter cycle times, improved flexibility, and faster customer responsiveness. With these, you will be able to confidently predict what is to come in your projects, when it will come, and how much it will cost, and still deliver per your predictions. This is what PM process management does for you—create value for your project customers.

Utilizing the Flowcharts

When to Use. Ideally, each PM process should be designed, controlled, and improved. When that happens, we believe that the process is managed, as opposed to evolving on its own. The Flowchart is used as a tool for management of PM processes, a major issue in managing project quality (see the box on page 277, "PM Processes: This Road Leads to Predictability"). More specifically, projects should employ Flowcharts for designing and improving processes [14]. As design tools, they enable management to study and show unexpected complexity, problems, redundancy, and where simplification and standardization are possible, prior to implementation [6]. When a Flowchart for a PM process already exists, it can be conveniently used to compare and contrast the actual ("as-is") versus ideal ("should-be") flow of the process to pinpoint opportunities for improvement (see, for an example, the box "A Process Is Worth 1000 Steps. Or Is It?" coming up in the chapter). A way to search for such opportunities is to ask questions such as the following:

- ▲ How does this step/task impact the project customer?
- ▲ Is it possible to eliminate or improve this step/task?

Time to Use. Designing and improving a PM process is a task demanding the skills of project teams [15]. The diligent study of the process, combined with numerous interactions of team members involved in the flowcharting, is bound to exact a time toll on the participants. How big the toll is depends on the boundaries and complexity of the process. As an example, designing an overall PM process (such as one shown in Figure 8.4) for an organization that manages 10 to 15 projects a year, these projects ranging from $50K to $1M, took a whole day for a five-member team, with extra time spent to document and describe the process. Large and more diverse processes are likely to exact more time.

Benefits. To understand quality problems, project teams need to comprehend what a process is chartered to do and how it really works. In that sense, the Flowchart helps them develop a shared language and understanding of the process, focusing their attention on relevant defects in the process. By pinpointing the defects, bottlenecks, and other problems, project teams can collect and study data about them and find the causes of the problems [16]. Equipped with such facts, they can take actions to error-proof and streamline the process, eliminating non-value-added steps [17].

As a visual representation of how the process works, the Flowchart provides an effective communication vehicle for picturing and understanding the process better and more objectively. Because of this feature, project team members can visualize how they fit in the process, who their suppliers and customers are and how they interact with them. This visualization helps improve communication among all project participants. If, as should be the case, project team members are involved in the process flowcharting, they are destined to develop the sense of ownership in the process, increasing their commitment to further improve the process. Used in this manner, Flowcharts may become a foundation for performance management, project personnel training, and even job descriptions.

Advantages and Disadvantages. Major advantages of Flowcharts are as follows:

- ▲ *The ease of constructing them.* With a little training, project participants can learn to quickly use and build the Flowcharts.
- ▲ *The power of their visual impact.* Graphical presentation of the Flowcharts makes them picturesque, strikingly expressive, and vivid.

Disadvantages of the Flowcharts are as follows:

- ▲ *There exists a constant threat that they will become overly detailed, making the process more complex than it should be.* This tendency is usually driven by the desire of the process developers to design a perfect type of process.
- ▲ *Even without being complex, flowcharting may be time-consuming.*

Variations. Basic purpose and major advantages and disadvantages of five different types of Flowcharts are described in Table 8.1. While other charts from the table have been in use for some time now, *systemigrams* are relatively new to project managers (see the box that follows, "The Flowchart as a Project Management Manual"). Their uniqueness is first reflected in their language. Systemigrams identify agents, who are project participants involved in the process and activities—that is, the tasks that agents will perform. Audiences—the recipients and beneficiaries—will use the resulting products of the activities, called *artifacts*. By linking activities, agents, artifacts, and audiences, systemigrams visually capture the dynamics of complex processes while maintaining the semantics of natural language.

A Process Is Worth 1000 Steps. Or Is It?

One company had over 1000 steps in their product development project management (PDPM) process! This was a stunning finding for a PDPM team when they finished flowcharting their "as-is" process, one that has evolved since their company's founding. Painstakingly, with an amazing level of rigor and discipline, they analyzed all steps and divided them into "value-added" and "cost-added only." The intent of this exercise was clear: separate the wheat from the chaff, or retain the necessary value-added steps while eliminating the unnecessary, cost-added ones. Once the chaff was out, the "should-be" PDPM process became evident. It had slightly over 50 steps. The impact of the PDPM process redesign was also stunning; the lead time for a typical project was reduced from 18 to 12 months. A simple moral to the story is that designing a project management process as opposed to letting it evolve can make a huge difference.

The Flowchart as a Project Management Manual

Many companies use PM manuals to document and direct their policies, goals, strategies, procedures, and processes for managing projects. The length of these narrative manuals may span from as many as 10+ to 100+ pages [2]. Generally, the trend is toward reducing the size of the manuals, making them easier and less costly to use and update. One company took a step further, replacing its narrative PM manual with a set of special Flowcharts called systemigrams (see Figure 8.7) Systemigrams represent PM processes of the company in a visual and simple fashion that is in tune with the organizational culture. To project personnel the systemigrams are their PM language.

Customize the Flowchart. This section addresses a wide variety of general needs for Flowcharts. Without clarifying what charts our projects need and how they will be used, we may not see much value from using the Flowcharts. The following examples offer some ideas on how to customize Flowcharts to account for specific project needs.

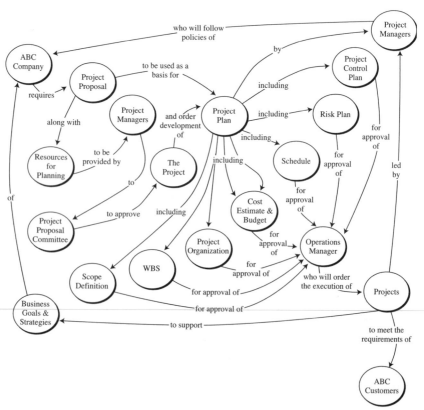

Figure 8.7 An example of a systemigram for the project planning process.

Customization Action	Examples of Customization Actions
Define limits of use.	Use only top–down Flowcharts for PM processes (organizations with simpler projects).
	Use a set of systemigrams as a replacement for PM manual (organizations with complex projects sensitive to "red tape" mentality)..
Add a new feature.	Add a symbol for project documents.
Modify a feature.	Limit each activity box to having one arrow in and one arrow out (to simplify the chart).

Summary

This section explained the Flowchart, a pictorial specification of the steps or tasks in a PM process. To understand quality problems, project teams need to comprehend what the process is chartered to do and how it really works. For that purpose the Flowchart helps them understand the process, focusing their attention on relevant defects, bottlenecks, and other problems. Equipped with such understanding, they can take actions to error-proof and streamline the process, eliminating non-value-added steps. The following box highlights the key points about structuring the Flowchart.

Affinity Diagram
What Is an Affinity Diagram?

An Affinity Diagram is a tool for the efficient organization of information through classification of ideas or facts (see Figure 8.8). It is designed to help you generate or gather a large number of ideas or facts, sort through them efficiently, and identify natural patterns or groupings in the information [6]. This makes it possible for project teams to narrow down the key issues rather than being distracted by large volumes of unstructured information [18].

Flowchart Check
- Check to make sure you developed a proper Flowchart. The Flowchart should include the following:
- Correctly used symbols
- Clearly defined process steps (inputs, outputs, tasks, decisions, etc.)
- Closed loops, meaning that each path takes the process either forward to the next or back to another step
- For each continuation point, a corresponding point somewhere in the Flowchart.
- Verify the Flowchart with people who perform the process tasks as well. Their suggestions should be used to finalize the Flowchart.

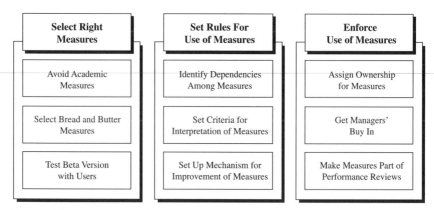

Figure 8.8 An example of the Affinity Diagram customer requirements for the project scorecard.

Developing an Affinity Diagram

Collect Information Inputs. An Affinity Diagram is usually employed in a project designing or improving a project product/service or a PM process. Consequently, the information about the project context, requirements, and process needs to be collected for the proper understanding and development of the Affinity Diagram.

Define the Issue under Consideration. The work on affinity diagramming begins with a clear understanding of the key issue that we want to address. Picture, for example, a project designing a system for performance measurement of product development projects in a company. The project team determined that the most important requirement for project customers—senior management—is the project scorecard, an executive-level report about monthly performance of projects. Accordingly, the team defines the issue under consideration as "What are the key elements of the project scorecard?"

Generate Ideas. Through interviews with executives, the team members collect a lot of ideas, recording each one on a Post-it note. Team members who collected the ideas read their ideas to the team, spurring other ideas that grew the list, part of which is shown in Figure 8.8. The Post-its are placed on the wall (you may use a flip chart or computer screen). Very often the nature of the assignment is such that team brainstorming generates the ideas.

Sort and Group Ideas. Each team member groups Post-its into the logical categories, where they fit best. Creating new groups and rearranging existing ones is welcome. Because the number of ideas is large, the task of making sense of all of them appears overwhelming at first, only to become less challenging with the elimination of redundancies and consolidation of similar ideas.

Name Groupings. With all ideas grouped, the team proceeds to build a quick consensus on a succinct preliminary phrase to express the central meaning or theme of each grouping. After a quick review of the phrases, and a few changes, the phrases become final headers, indicating that the key customer requirements for the project scorecard are selecting the right performance measures, setting rules for the use of the measures, and enforcing the use of measures (see Figure 8.8). When the groupings appear to be large, breaking them down them into subgroups and defining sub-headers is a good option.

Draw and Review the Final Affinity Diagram. Finally, the team draws a final Affinity Diagram, linking the final headers with the groupings. At this time, each team member takes another close look at their notes of interviews with the executives and ideas refined from them. A comparison with the final diagram shows no need to make any adjustments, a frequent occurrence in their previous use of the Affinity Diagrams.

Using the Affinity Diagram
When to Use. An Affinity Diagram can be used for both project quality planning and improvement (see the box that follows, "Tips for Affinity Diagramming"). Its power as a quality planning tool is in its ability to orga- nize information about customers' quality requirements, enabling its easier and more transparent incorporation into the design of project product and processes [19]. As a result, the project product and processes better meet customer requirements. This organizing excellence also comes into play when there is a need to structure any large group of issues that need to be resolved to successfully improve project quality problems. Beyond these two applications, an Affinity Diagram may be used in any situations in need of meaningful organization of large volumes of information. For example, it is used for bottom-up development of the Work Breakdown Structure (see details in the "Work Breakdown Structure (WBS)" section in Chapter 5). Its particularly successful use is in team settings.

Tips for Affinity Diagramming
- Record ideas on Post-it notes, with large print visible from 6 to 10 feet, with several words only — at minimum, a verb and a noun.
- Shoot for a typical diagram with 20 to 30 ideas, but don't be surprised if you see others having over 100 ideas.
- Sort and group in silence in order to achieve better concentration when thinking about how the ideas are related.
- Create duplicate Post-its for an idea that excessively gets moved around different groups. Place duplicate(s) in those groups.
- Work hard to develop headers that depict the essence of all ideas in a grouping.

Time to Develop. A smaller size team with reasonable skill may spend 30 to 45 minutes in creating a three-level, 15-idea Affinity Diagram. Time requirements are bound to grow should the size of the Affinity Diagram and team become bigger.

Benefits. Through the process of generating and grouping ideas, the Affinity Diagram spurs creativity, bringing down communication barriers and encouraging ownership of the emerging results. Such ambiance is a fertile ground for the work by consensus and natural breakthroughs, as well as for overcoming an inability to function, typical of teams faced with an overpowering multitude of ideas.

Advantages and Disadvantages. Its qualities of being simple to use and understand, and graphical in nature, make the Affinity Diagram a very visual communication tool. These advantages may be seriously compromised when the diagramming becomes time-consuming in applications dealing with many levels of grouping.

Variations. The Affinity Diagram is a simple form or main ingredient of the KJ method, developed in 1960s by Kawakita Jiro, a Japanese anthropologist [6]. Unlike the Affinity Diagram, the KJ method may include both ideas and facts. Also, the refinement process leading to the creation of the diagram is more structured.

Customize the Affinity Diagram. Making the best use of the Affinity Diagram takes adapting it to one's needs. Following are some ideas for such customization.

Summary
In this section we reviewed the Affinity Diagram. This is a tool designed to help generate or gather large number of ideas or facts, sort through them efficiently, and identify natural patterns or groupings in the information. When this is done with information about customers' quality requirements, the information is easier to use in the design or improvement of the project product and processes. The following box will help you keep this tool's key points in focus.

Customization Action	*Examples of Customization Actions*
Define limits of use.	Use the Affinity Diagram for the development of WBSs in new, innovative projects in which the organization has little experience.
	Use the Affinity Diagram for organizing ideas in problem solving situations.
	Use the Affinity Diagram for organizing customer requirements.
Add a new feature.	Include both ideas and facts in the diagram in problem-solving situations

Affinity Diagram Check

Check to make sure you developed a proper Affinity Diagram, including the following:

- Ideas
- Groupings of ideas
- Header cards
- Header cards connected with their groupings

Concluding Remarks

Three tools in this chapter—Project Quality Program, Flowchart, and Affinity Diagram—are developed to serve very distinct purposes. Because their purposes are complementary, rather than competitive, all three can be used in combination. Project Quality Program enables the project manager to build quality early in the project planning, starting off customer requirements. One way to ensure that the planned quality gets delivered is to design and, when necessary, improve PM processes. This is what the Flowchart is used for. In the process of preparing customer requirements for a Project Quality Program and using a Flowchart to design/improve the processes, for example, you will generate many ideas. Sorting through and organizing the ideas is the purpose of the Affinity Diagram. Note that the use of the tools is formal and documented in large projects, while informal in small projects.

More details on situations in which to use each of the tools are offered in the summary comparison that follows. In particular, we identified several project situations and marked how each one favors the use of the tools. If they do not describe the project sufficiently, brainstorm to identify more such situations. Mark how they favor the tools. A tool with many marks means you probably need it. Also, carefully review the material covered in this chapter.

A Summary Comparison of Quality Planning Tools

Situation	Favoring Project Quality Program	Favoring Flowchart	Favoring Affinity Diagram
Small and simple projects	√	√	√
Large and complex projects	√	√	√
Build customer focus into project plan	√		
Need proactive approach to quality	√		
Plan and design PM processes		√	
Improve PM processes		√	
Organize ideas on customer requirements			√
Organize ideas for solving quality problems			√
Short time to train how to use the tool	√	√	√

References

1. Juran, J. M., 1988. *Planning for Quality*. New York: Free Press.

2. Kerzner, H. 2000. *Applied Project Management*. New York: John Wiley & Sons.

3. AT&T Quality Steering Committee. 1991. *Reengineering Handbook*. Red Bank, N.J.: AT&T Bell Laboratories.

4. Garvin, D. A. 1984. "What Does Product Quality Really Mean?" *Sloan Management Review*. 26(1): 25–43.

5. Hunt, V. D. 1992. *Quality in America*. Homewood, Ill: Business One Irwin.

6. Evans, J. R. and W. M. Lindsay. 1999. *The Management and Control of Quality*. 4th ed. Cincinnati: South-Western College Publishing.

7. Reeves, C. A. and D. A. Bednar. 1994. "Defining Quality: Alternatives and Implications." *Academy of Management Review*. 19(3): 419–445.

8. Seawright, K. W. and S. T. Young. 1996. "A Quality Definition Continuum." *Interfaces*. 26(3): 107–113.

9. Smith, G. F. 1993. "The Meaning of Quality." *Total Quality Management*. 4(3): 235–244.

10. Hammer, M. and J. Champy. 1993. *Reengineering the Corporation*. New York: Harper Business.

11. University of Michigan Business School and A.S.f. Quality. 1998. *American Customer Satisfaction Index: 1994-1998*. Ann Arbor: University of Michigan Press.

12. Gryna, F. M. *Quality Planning and Analysis*. 4th ed. 2001. Boston: McGraw-Hill.

13. Scholtes, P. R., B. L. Joiner, and B. J. Streibel. 1996. *The Team Handbook*. 2d ed. Madison, Wis.: Joiner Associates Inc.

14. Sellers, G. 1997. "Using Flowcharts for Performance Improvement." *Quality Digest* 17(3): 49–51.

15. Burr, J. 1990. "The Tools of Quality, Part I: Going with the Flow (chart)." *Quality Progress* 23(6): 64–67.

16. Welsh, F. 1997. "Charting New Territory." *Quality Progress* 29(2): 63–66.

17. Heather, H. 1996. "Cycle-Time Reduction: Your Key to a Better Bottom Line." *Quality Digest* 16(4): p. 28-32.

18. Alloway, J.A. 1997. "Be Prepared with An Affinity Diagram." *Quality Progress* 30(7): 75–77.

19. Brassard, M. and D. Ritter. 1994. *The Memory Jogger II*. Salem, N.H.: GOAL/QPC.

Risk Planning

> If you take no risks, you will suffer no defeats. But if you take no risks, you win no victories.
>
> *Richard Nixon*

Major topics in this chapter are risk planning tools:

- ▲ Risk Response Plan
- ▲ Monte Carlo Analysis
- ▲ Decision Tree

These tools are instrumental in identifying risks to the project, assessing their potential impact, and developing actions to mitigate them (see Figure 9.1). The preparation of such a risk baseline requires two-way exchange of information with scope, cost, and schedule baselines. Also important in this process is the coordination with other tools of organizational, quality, and procurement planning. During the implementation phase, the fortune of risk control will start off and be measured against the risk baseline. In a nutshell, risk planning tools create a strategy to fend off undesired events in projects.

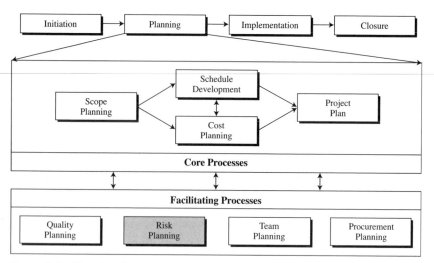

Figure 9.1 The role of risk planning tools in the standardized project management process.

The purpose of this chapter is to support practicing and prospective project managers in their quest to

▲ Learn how to use various risk planning tools

▲ Pick risk planning tools that are most appropriate for their project situation

▲ Customize the tools of their choice

Internalizing these skills plays a central role in project planning and building the standardized PM process.

Risk Response Plan

What Is the Risk Response Plan?

The Risk Response Plan assesses risks and identifies actions to increase opportunities and reduce threats to project goals [3]. To be effective, the plan must be realistic (as to the severity of the risk), timely, cost-conscious, bought into by all involved players, and owned by the appropriate person. Perhaps more than anything else the plan must be proactive, developing actions ahead of the risk occurrence (see Figure 9.2). Instead of being viewed as having a complete control of events, it should be seen as an advanced preparation for possible adverse future events [1].

Developing the Response Plan

Making decisions is perhaps the toughest of jobs that project managers have to take on. That wouldn't be very challenging in a situation of total certainty, when all information that they need for decision making is already available and the outcomes of their decisions are also known. Project managers' life, however, is much more complex, and most of their decisions are made with incomplete information and uncertain outcomes. This is the realm of project risk management. Beyond it lies the region of total uncertainty, with complete absence of information, where nothing is known about outcomes. This total certainty (knowns)—risk (known unknowns)—total uncertainty (unknown unknowns) continuum is illustrated in Figure 9.3.

RISK RESPONSE PLAN

Project Name: Genesis   **Rev #:** 2   **Page #:** 1 of 1
    Estimate Date: Apr. 25, 02

Work Package/ Task	Risk Event # & Description	Probability (%) (A)	Risk Impact (days, $, etc.) (B)	Risk Event Status (A) x (B)	Criticality*	Impacted Risk Events Number	Actions — Preventive	Actions — Trigger Point	Actions — Contingent	Owner
General assess. document	2. No internal expert for business analysis	90%	60 days	54 days	C	4	Hire an expert	Expert not hired by 5/20/02	Borrow the expert from Unit B	Peter P.
Code and Unit tests	9. Internal S/W testing slow	90%	30 days	27 days	C	11 & 12	Outsource	Vendor not chosen by 6/1/02	Use old vendor	Marsha M.
All six walk-throughs	14. Executives not available for milestone review	95%	70 days	66.5 days	C	6-16	Reduce # of executive reviews	No approval by 5/1/02	Request sponsor's intervention	Jamie V.

*   C–Critical
    NC–Near-Critical
    NNC–Noncritical

Figure 9.2   An example of the Risk Response Plan.

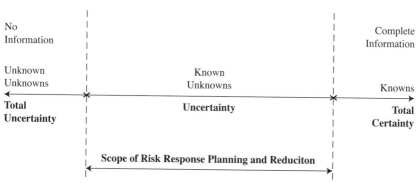

Figure 9.3   The uncertainty continuum and risk response planning.

From *Project and Program Risk Management: A Guide to Managing Project Risks and Opportunities* by R. Max Wideman. Copyright © 2000 by Project Management Institute. Reprinted with permission of Project Management Institute.

Our focus is on the risk area by developing a Risk Response Plan through a simple cycle of identify, assess, respond, and document. Basic definitions are included in the box that follows, "Basic Risk-Related Definitions."

Prepare Information Inputs. An ideal in risk planning is to take it as rigorously as we take cost or schedule planning. In pursuit of such an ideal, you should start with rigorous inputs, including the following [5]:

- Risk management plan
- Project planning outputs
- Risk categories
- Historical information

The risk management plan is a document developed in the beginning of the project that provides a roadmap for dealing with risk throughout the project's life. Included into the plan may be the following [5]:

- Methodology. Identifies and describes approaches, tools, and data sources that may be used to handle risks.
- Roles and responsibilities. Define who does what in risk management in the project, from project team members to members of the company's risk management teams.
- Budgeting and timing. Specify the budget for risk management for the project, as well as the frequency of the risk management processes.
- Tools. Describe which specific methods for qualitative and quantitative risk analyses to use and when to use them
- Reporting and tracking. Defines the format of the Risk Response Plan and report, how the results of risk activities will be documented, communicated to stakeholders, and preserved for purposes of lessons learned.

Obviously, the purpose of the risk management plan is not to address individual risks in a project; rather, it is to guide a project team in developing a Risk Response Plan and then monitoring its performance.

Project planning outputs—performance baseline including scope, cost, time, and quality baselines—are what is at risk. Having full knowledge of them is crucial in developing response plans to counter risks to which the outputs will be exposed. These risks can be organized into different categories. For example, risks can be classified according to their effect on the project—scope, quality, schedule, and cost risks (in other words, failure to complete the project tasks within planned scope, quality, schedule, and cost performance). Another way is to categorize risks per their primary source into external (but unpredictable), external predictable (but uncertain), internal non-technical, technical, and legal [1]. This perspective looks for the balance between the internal and environmental impacts. The point is that the firm and its projects need a consistent risk categorization, suiting its business and culture, which can serve as a framework for a systematic identification and treatment of risks. Since risk management tends to be data-intensive, reliable historical information such as past project records, postmortems, or published sources (e.g., product benchmarking studies) is vital. Although it furnishes precious knowledge of what might go wrong, it is important to read past experience correctly when applying it in foreseeing risks to the current project.

Identify Risks. This step's purpose is to identify all the potential risks that may significantly influence the success of the project. Multiple ways for accomplishing this step are available, ranging from engaging the project team in a brainstorming session to consulting experienced team members to requesting opinions of experts not associated with the project. In any of them, several things have to be taken into account. First, risks vary across the project life cycle. Typically, risks tend to be relatively high early in the project because so many resources remain to be invested. Similarly, later in the project when most of resources are invested, most of unknowns are turned into knowns and risks are relatively lower. Also, some risks occur only in certain project stages; for example, risks related to project acceptance tests are typically encountered at the project end. Sometimes, even assumptions may become a source of risk (see the box that follows, "Is an Assumption a Risk?"). This dynamic nature of risks makes the identification process iterative, requiring that once risks are identified early in the project, they need to be continuously reviewed, with appropriate adjustments [6].

Basic Risk-Related Definitions

- Project risk. The cumulative effect of the chances of uncertain occurrences adversely affecting project objectives [1].
- Risk event. The description of what might happen to the detriment of the project.
- Risk probability. The likelihood that a risk will occur.
- Risk impact. Severity of its effect on the project objective. Also called risk consequences or amount at stake.
- Risk event status. A measure of importance of a risk event. Also called criterion value or its ranking.
- Contingency reserve. The amount of money or time normally included into the project cost or schedule baseline to reduce the risk of overruns of project objectives to a level acceptable to the organization [6].
- Management reserve. The amount of money or time that is not included into the project cost or schedule baseline, which is used by management to allow for future situations that are not possible to predict.

Is an Assumption a Risk?

"Is an assumption a risk?" This was a question that a project manager asked in a risk response meeting. Assumptions are factors that are not entirely known or are uncertain but for planning purposes are considered to be true or certain. For example, a firm launched a project to develop and market a product in a Pacific Rim country. A major assumption was that the country's annual market growth rate would continue to be around 10 percent. Per its assumption management practice, the firm first documented the assumption by defining it, nominating its owner, and identifying a monitoring metric [3]. Next, the project manager instructed the owner to periodically test the metric in order to ensure that no change of assumption occurred. Seeking to be proactive, the owner defined at which time the assumption becomes a risk (trigger point) and potential risk response actions.

A few months later the country was hit by a recession and the growth rate turned negative. The project team revisited the assumption and, since the recession was expected to last for some time, decided that the assumption changed into a risk, immediately invoking the Risk Response Plan. So, "Is an assumption a risk?" It is not. It is rather a source of the potential risk.

Second, risk events rarely strike independently. Rather, they tend to interact with other risk events, combining into larger risks. Looking for such interactive possibilities is important in risk identification. Finally, since risks come in all types of packages, planners should conduct risk identification in a systematic way so that no stone is left unturned internally in the project and externally in the environment, including management of stakeholders. A huge help in this respect may be received from risk categorization. For this purpose, our example in Figure 9.2 uses WBS as a systematic framework for risk identification (column 1). Beginning with the first work package, we may ask our project team, "What may go wrong?" meaning what risk events can hit our work package (column 2). Or, we can use any of the risk categorizations—formed according to the effect or primary source—as a checklist to identify possible risks for our first work package, and then continue similarly for all other work packages. Relying on WBS for this purpose also provides benefits such as cross-project consistency and comparison and creation of historical risk database.

Assess Qualitatively. A usual problem at this time of risk assessment is that a large number of risks might have been identified. Which ones deserve attention? Apparently, those that have both the highest impact on the project and are most likely to occur. While a preliminary clue is to look at the WBS to spot these critical risks—they are probably in most important portions of WBS—we still need to analyze impact, probability, and severity (criticality) of each risk.

In the qualitative assessment, we tend to use a nonnumeric probability scale—for example, a five-level scale, where 1=very unlikely, 2=low likelihood, 3=likely, 4=highly likely, and 5=near certain [7]. If you don't have much experience or data to reliably assess quantitative probabilities—

addressed later in the quantitative assessment — nonnumeric scales are sufficiently good. Consequently, you will qualitatively assess each risk's probability on this nonnumeric scale. When this is done, the next step is to assess the impact of each risk, again on a discrete scale. One example is a scale such as 1=very low impact, 2=low impact, 3=medium impact, 4=high impact, and 5=very high impact. To illustrate its use, let's assume that a risk to be assessed has three impacts: project costs can increase, schedule can slip, and quality can be reduced. For each one of them, the scale can define the levels of impact, as shown for the schedule risks in Table 9.1. After each of the three impacts is rated, the overall impact rating of the risk is the largest of the three impacts [8].

When all risks are assessed like this, it is time to use a formula to combine their risk probability and impact to establish a measure of severity. Although nonlinear formulas can be employed, linear formulas such as Severity = Probability + N x Impact are easier to apply. For example, N can be equal to 2, meaning that impact is twice as important as probability in establishing risk severity. In this case, the assessed probability and impact for each risk would be entered in the formula Severity = Probability + 2 x Impact and the obtained value input into the Probability-Impact (P-I) matrix consisting of 5 x 5 squares (see Figure 9.4). The matrix is usually divided into red, yellow, and green zones, representing high severity (critical—first priority), medium severity (near-critical—second priority), and low severity (noncritical—lower-priority) risks, respectively, based on the organization's thresholds for risk severity. If you have a large number of risks, the position of a risk in the matrix determines its ranking and, thus, severity. Squares with the highest values have the highest ranks, and squares with equal value may be ranked in order of their impact level [8]. Still, a question remains: How many of the highest-ranking risks should we deal with?

Some larger projects commonly focus on the top ten highest-ranked risks. In contrast, some smaller projects decide to manage the top three risks, arguing the lack of resources to take on a larger number of risks. Both may be dangerous. If these projects have more than ten and more than three risks in their red zone, respectively, they are bound to disregard some critical risks. On the other hand, if both of them are facing only one risk in their red zone, and others are in the green zone, they are wasting their resources looking at the top ten and three risks.

Table 9.1: Example of Rating a Risk Impact on Schedule on a Five-Level Scale

Scale	1 Very Low	2 Low	3 Medium	4 High	5 Very High
Risk impact on schedule	Slight schedule delay	Overall project delay < 5%	Overall project delay 5-14%	Overall project delay 15-25%	Overall Project delay > 25%

Probability	Risk Score = P + 2 x I					Key:
NC = 5	7	9	11	13	15	▨ High Severity
HL = 4	6	8	10	12	14	▨ Medium Serverity
L = 3	5	7	9	11	13	☐ Low Severity
LL = 2	4	6	8	10	12	
VU = 1	3	5	7	9	11	
	VL = 1	L = 2	M = 3	H = 4	VH = 5	
	IMPACT					

Figure 9.4 *Segregation of risks into low, medium, and high severity by Probability-Impact matrix.*

So, what is a reasonable way out? The answer is in the P-I matrix. Respond to the highest-ranked risks in the matrix, down to an agreed level [8]. For example, focus on handling risks down to risk score 11 (see Figure 9.4), and treat other risks as noncritical. With this approach, you neither squander resources nor disregard significant risks.

As mentioned earlier, if you don't have much experience or data to reliably assess quantitative probabilities, qualitative assessment of risks based on nonnumeric scales is good enough. They should proceed to the respond step. For those with experience and reliable data, the next action is quantitative risk assessment.

Assess Quantitatively. This step analyzes numerically the probability of each risk, its consequences on project objectives, and the extent of overall project risk [5]. It can be used separately or together with qualitative assessment. For example, if the available time and budget permit, and both qualitative and quantitative analysis are desired, the "together" approach may be the clear choice. This is our choice here as well.

The process begins from the results of the earlier risk identification step. For each of these identified risks, you need to quantify the probability of occurrence by asking, "What is the probability that this risk will happen?" "Ninety percent," the team decides. This means that there is a 10 percent probability that the risk will not occur. Clearly, the probability that the risk will occur plus the probability that it will not occur equals 1. Assessing the probability is no more than an estimate based on solid historical information from similar experiences in past projects or considerate opinion of experts. In our example from Figure 9.2 (risk event 2, column 3), the team reviewed past records and solicited inputs from several experienced project mangers. On that basis, each team member developed an estimate of probability, and after a team discussion, a consensual number of 90 percent was adopted.

The next step is to determine the risk impact. "What is the severity of consequences if this risk occurs?" is the question you should ask. While the impact may be expressed in almost any units, from percent of lost market share to customer fallout percent, the real emphasis here is to estimate schedule or cost severity of the risk. In our example, the project's goal is all about delivering on schedule; therefore, the impact focuses on schedule (Figure 9.2, column 4). With this data, the risk event status (also called criterion value or ranking) can be determined by the following relationship [1]:

$$\text{Risk Event Status} = \text{Risk Probability} \times \text{Impact}$$

In our example (Figure 9.2, risk event 2, column 5), this relationship gives the risk event a status of 54 days. When the status is calculated for all risk events, the natural question is which of them are really vital and deserve attention and which are trivial. To answer this question, we will use principles similar to those on the issue of severity in qualitative assessment. First, establish numerical intervals of severity that determine whether a risk event status is critical (potential showstoppers), near-critical (soon to be potential showstoppers), or noncritical (minor risks). For example, in a smaller project the risk event status exceeding 15 days was critical, between 7 and 14 days near-critical, and below 7 days noncritical. Second, respond to the highest-ranked risks, down to an agreed level. In an instance this meant focusing on the top ten risks only.

Right after you have determined the risk impact, a question emerges: "Does this risk event impact any other risk event?" If such is the case, identify the impacted risk events. The rationale here is that many smaller risks may interact, snowballing into a risk impact that significantly exceeds the sum of the individual risk impacts. To preempt such a possibility, the information about impacted risks will be used in the next step to define response actions.

Respond. The whole Risk Response Plan culminates into its most creative part—determining the proactive response actions from a range of possible responses to a risk event in order to reduce threats to the project. Such an action should be rooted in risk policies and procedures established in the risk management plan. In particular, a proactive response action includes three steps of implementation: preventive action, trigger point, and contingent action (see Figure 9.2, columns 8, 9, and 10). The preventive action is the primary strategy, or plan A, of responding to a risk. When executed, however, it may or may not work. The point at which we establish that the primary strategy doesn't work is the trigger point. At that time, the backup strategy or contingent action (plan B) is taken to counter the risk. For example, the preventive action for risk event 9 in Figure 9.2 is outsourcing software quality testing from a larger and well-staffed firm. If by June 1 the vendor to provide the testing is not selected and the purchase order issued, the preventive action is not successful and should be suspended. That's the trigger point at which we introduce the contingent action by going to our old vendor, which although not large, still has time to get the testing done.

Any suitable proactive response action essentially falls into one of the four broad categories of response strategies: avoidance, transference, mitigation, and acceptance of risk [5]. Changing the project plan or condition to eliminate the selected risk event is risk avoidance. When faced with the risk of not having an available expert to perform quality business analysis, the risk was avoided by hiring such an expert (see row 1 in Figure 9.2). Risk transfer simply involves shifting consequences of a risk event to a third party, along with the ownership of the response [5]. This means that a project exposed to a risk of slow software quality testing internally can transfer the risk by hiring a professional firm to do the testing (see row 2 in Figure 9.2). Mitigation's intent is to lower the probability or impact (or both) of an unfavorable risk event to an acceptable threshold. In our example in Figure 9.2 (row 3), the risk of busy executives slowing down the project is reduced by reducing the number of major milestone reviews they have to attend and approve/disapprove the continuation of the project. The three response strategies—avoidance, transference, and mitigation—are deployed when risks they are responding to are among the highest-ranked risks. Obviously, these responses will be incorporated in the project plan.

How Much Reserves and Allowances to Plan For?

Let's think back about total certainty (knowns)—risk (known unknowns)—total uncertainty (unknown unknowns) continuum. What kind of reserves do we need in order to respond when any of these categories hit? First, because of their totally certain nature, the knowns do not require any reserves.

How do you allow for risk consequences of the known unknowns? Many firms add them to the baseline estimate as a separate fund for schedule and cost *contingency allowances*. Others incorporate them into the individual activities. While we favor the former, the latter approach — which, by the way, is too risky to use because of activity owners' tendency to use up allowances liberally— appears to have wider presence. How is the fund formed? Popular methods include applying standard allowances and percentages based on past experience [1]. We argue that the use of the Risk Response Plan (Figure 9.2) may be a very appropriate way to compute the fund. Take a risk from the plan that is *not* among the highest-ranked risks—let's call it a lower-ranked risk. Multiplying its risk probability by risk impact provides the risk event status, which may be expressed in cost or schedule terms. These numbers, given in column 5 of the plan in Figure 9.2, are essentially cost and schedule reserves or allowances for the risk event. Adding up allowances for all of the lower-ranked risk events in the plan creates a project contingency allowances fund. A great advantage of the approach is in integrating a proactive Risk Response Plan with cost estimating and scheduling. A firm, for example, calls this fund "AFC" (allowance for change). When any of the risk events occurs, the owner applies to AFC for cost or schedule allowance.

Finally, what about reserves for the unknown unknowns? Although they are absolutely not possible to foresee, such things will happen [1]. Therefore, some firms develop *management reserves* involving cost or schedule to allow for such future situations when cost or schedule objectives may be missed. Once the reserves are used, the cost baseline gets changed. Managing management reserves is in the domain of higher management, typically the project sponsor.

For those risks that are not among the highest-ranked risks, a risk acceptance strategy is used. It implies that project managers have decided to not change the project plan or are unable to articulate a feasible response action to deal with a risk [5]. A typical example of the risk acceptance is the establishment of contingency allowances. For an explanation of how the allowances are formed, see the box on page 296, "How Much Reserves and Allowances to Plan For?"

An integral part of the response development is the identification and assignment of risk owners—individuals or parties responsible for each preventive action, trigger point, and contingent action. In so doing, one should recognize that while some risks are independent, leaving their owners fully responsible for their management, some risks might be interdependent. If so, their preventive actions, trigger points, and contingent actions should be developed and owned interdependently.

Document. Summarizing the results of the risk response planning into a document with conclusions and recommendations allows managers to take several important actions [1]. To begin with, they can make project decisions fully recognizing the involved risks. Also, they can continue to evaluate risks on the current project. Finally, this document will serve as a baseline for risk management analysis in performing a postmortem review, a great source of information for historical risk databases. For example, one manufacturing company requires that risk assessments are archived with other project documentation in the Web site, shared drive, and project blue book (documents retained after the project completion, post-project project file).

Utilizing a Risk Response Plan

When to Apply. There is no project that cannot benefit from developing the plan. Small projects typically rely on the qualitative assessment and P-I matrix, often deciding to handle only a few highest-ranked risk events. Not surprisingly, the dominant mode of their planning is informal, as is the periodic reevaluation of the plan throughout the project. Although at times it may be overly simplistic for large and complex projects, the Risk Response Plan is nevertheless widely used, with more formality and stronger orientation on quantitative assessment. Focused on the larger number of highest-ranked risk events, larger projects also tend to do more formal, periodic reassessments of the plan.

Time to Complete. Lots of teams running small and simple projects are able and can afford to expend one to a few hours of their time to conduct a session and develop the plan. This time proportionately rises as projects get bigger and more complex. Tens of hours may be necessary for a team in charge of a large and complex project to devise a Risk Response Plan of this type.

Benefits. The Risk Response Plan helps sift through the myriad of uncertainties, pinpoint and highlight the project areas of highest risk, both before work has begun and throughout the project [9]. This offers you an opportunity to identify effective ways of reducing those risks in a proactive manner, rather than being confronted by them. Resulting from such opportunity is a benefit of incorporating risks directly in the process of planning and executing a project. This further enables better understanding of the project goals, scope, and feasibility [1]. In addition, the plan generates information for a more realistic project plan and implementation, a more reasonable contingency planning, and an early warning of risk.

Advantages and Disadvantages. The plan's major advantage is its

> ▲ *Simplicity*. This is especially true of the qualitative portion with its easy-to-discern color messages of severity levels. Rid of advanced statistics, its quantitative part is still simple and adequate for many projects.

But this simplicity also results in a major disadvantage:

> ▲ *Individual focus*. The plan's major focus is on individual risk events, and although it recognizes the risk of interacting risk events, it does not provide a reliable mechanism to deal with them. When its reliance on single-point estimates for probability and risk impact are added, it becomes apparent why critics suggest that if the plan is used in larger projects, it should not be used alone but in conjunction with other, more complex tools, Monte Carlo simulation, for example.

Variations. Variations of the P-I matrix and Risk Response Plan abound. The matrix goes by such other names as *risk matrix* [8] and *P-I table*, for example, while *risk register* is often used as a synonym for the plan [4]. These variations typically have different scales for the assessment of the probability and risk impact, formulas for the establishment of the measure of severity, and methods for the division of the matrix into severity zones. One such difference that is used in the P-I matrix enables ranking risks that have multiple impacts—schedule, cost, and quality, for example. When a single risk is identified, each of its multiple impacts is rated on the probability and impact similarly to the method from Table 9.1. The rating for schedule impact, for example, is used to determine a weighting for its probability (W_{sp}) and impact (W_{si}). When this is repeated for cost and quality impact, the overall risk rating is $(W_{sp} \times W_{si}) + (W_{cp} \times W_{ci}) + (W_{qp} \times W_{qi})$. These scores are then used for risk ranking. Vose refers to this approach as semiquantitative [4]. Similar approaches can be used for the calculation of risk event status and risk ranking in the risk response plan.

Customization Action	Examples of Customization Actions
Define limits of use	Make the plan's use informal in small projects, focusing on the top five risks.
	Make the plan's use formal in large projects, focusing on the highest-ranked risks, down to an agreed level.
Adapt a feature.	Adapt P-I matrix zoning to the company's thresholds of risk.
Add a feature.	Add semiquantitative approach to the P-I matrix.
	Add the time line for preventive/contingent actions.

Customize the Plan. As a generic tool, the Risk Response Plan may help a company to a certain extent. To derive more value from it will require adapting it to the company's specifics and projects. In the following we offer several ideas on how to go about the customization.

Summary

In this section we presented the Risk Response Plan, a tool that assesses risks and identifies actions to increase opportunities and reduce threats to project goals. Each project can benefit from developing the plan. Small projects typically rely on the informal, qualitative assessment, often deciding to handle only a few highest-ranked risk events. Focused on the larger number of highest-ranked risk events, larger projects also tend to do more formal, periodic reassessments of the plan. In this manner, the plan helps pinpoint and highlight the project areas of highest risk and identify effective ways of reducing those risks in a proactive manner. It also enables more reasonable contingency planning and an early warning of risk. In short, here are the key points in structuring the Risk Response Plan.

Risk Response Plan Check

Check to make sure that the plan is appropriately structured. It should include the following:

- Identified risks
- P-I matrix
- Risk probability and impact
- Risk event status
- Impacted risk events
- Preventive action, trigger point, and contingent action
- Name of the risk owner

Monte Carlo Analysis

Monte Carlo Analysis (MCA) uses a model of a project—project network diagram, for example—to analyze its behavior. For this purpose, MCA randomly samples probability distribution of each project activity to "perform" the project hundreds or even thousands of times [5]. This provides statistical distribution of the calculated project durations and approximates the expected value of the project duration, as illustrated in Figure 9.5. With these distributions, you can quantify the risk of various schedule scenarios, alternative implementation strategies, activity paths, or even individual activities. For example, as Figure 9.5 indicates, there is a 40 percent probability that the project will be finished before or on May 20.

Performing Monte Carlo Analysis

Typically, MCA deals with schedule, cost, and cash flow risks, although other facets such as the quality of the final project product can at times be analyzed. Taken overall, performing a schedule risk analysis is more complex than a cost analysis, simply because you need to establish dependencies between project activities in order to identify the critical path. For this reason, our focus is on looking at the MCA process in schedule risk analysis (see Figure 9.6).

Prepare Information Inputs. Four inputs play a key role in MCA:

- ▲ Risk management plan
- ▲ Risk Response Plan
- ▲ Project logic diagram or schedule
- ▲ Probability distributions

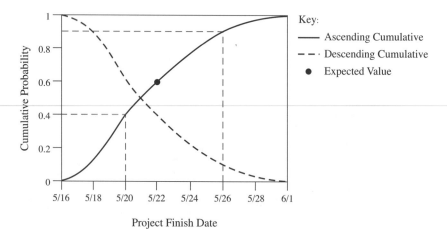

Figure 9.5 Cumulative distribution of project duration produced in Monte Carlo Analysis.

From *Risk Analysis: A Quantitative Guide* by David Vose. Copyright © 2000 by John Wiley & Sons Limited. Reprinted with permission of John Wiley & Sons.

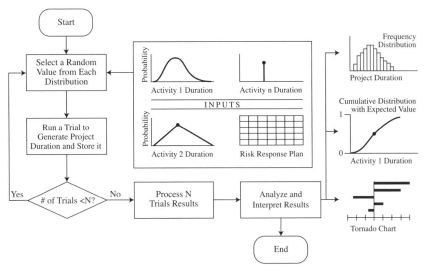

Figure 9.6 *Process of Monte Carlo Analysis for schedule risk.*

Since a risk management plan provides a roadmap for dealing with risk throughout the project's life, it is logical to expect it to offer directions for how to use MCA in the project. Another type of directions is contained in the Risk Response Plan. There, in particular, individual risks are identified, described, and analyzed to assess, rank (P-I matrix), and quantify them, along with preventive and contingent response strategies. All of this information will be fed into MCA to generate a range of possible project durations (see the box that follows, "Basic Language of Monte Carlo Analysis"). For that purpose, we also need the project network diagram that sequences project activities while indicating dependencies between the activities. Some prefer to start with the deterministic project schedule instead of the diagram, for example, formatting the schedule as the time-scaled arrow diagram, the cascade type (see Figure 9.7). Using the CPM schedule is also a valid option.

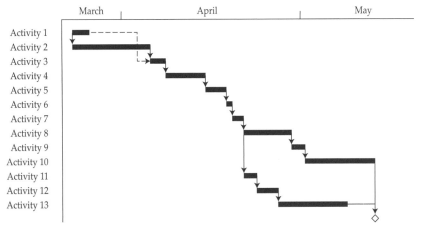

Figure 9.7 An example of the Time-scaled Arrow Diagram for risk analysis with Monte Carlo.

Basic Language of Monte Carlo Analysis

- *Chance event* is a process or measurement for which we do not know the outcome in advance.
- *Continuous distribution* is used to represent any value within a defined range of values (domain).
- *Discrete distribution* may take one of a set of identifiable values, each of which has a calculable probability of occurrence.
- *Deterministic model* is where all parameters are fixed, having single-valued estimates.
- *Expected value (EV)* is the probability weighted average of all possible outcomes. Synonyms: mean, average.
- *Mode* is the particular outcome that is most likely; the highest point on a probability distribution curve.
- *Model* is a simplified representation of a system of interest such as project CPM chart. It projects project outcome (e.g., project duration) and outcome value (e.g., 18 months).
- *Probability* is the likelihood of an event occurring, expressed as a number from 0 to 1 (or equivalent percentages). Synonyms: likelihood, chance, odds.
- *Probability distribution* represents mathematically or graphically the range of values (e.g., from 2 to 14 days) the variable (e.g., activity duration) can take, together with the likelihood that the variable will take any specific value. Synonyms: probability density function, probability function.
- *Project scenario* is a future state of the project. Synonyms: iteration, trial.
- *Random sampling* is a process generating a random number between 0 and 1, which determines the value of the input variable from the probability distribution.
- *Random variable* is a measure of a chance event. Synonyms: chance variable, stochastic variable.
- *Single-valued estimate* has one value only. Synonym: point estimate.
- *Standard deviation* is the square root of the variance.
- *Stochastic model* is a model that includes random variables. Synonym: probabilistic model.
- *Variance* is the expected value of the sum of squared deviations from the mean.

Generating a range of possible project durations and their probabilities is not possible without preparing probability distributions for project activity durations. This preparation process may begin with the question "How long does it take to complete a project activity?" Let's assume that you performed this activity many times and each time it took ten days to get it done. If asked to estimate the duration of that same activity in a future project, you would likely put it at ten days. If each project activity would have such a single point estimate (also called single value estimate) as an input to calculate project schedule duration, the duration would also have only one value. There is not much uncertainty in project activity durations in this single-valued deterministic model—they are all fixed. In the majority of today's projects, such a scenario is not realistic. More realistic is the following probabilistic (stochastic) model.

Imagine that you repeated activity 1 an extremely large number of times (trials, iterations, scenarios) and its duration ran from 5 to 39 days (range of outcomes). You recorded the fraction of times that each duration value (outcome) occurred. The fraction for a particular outcome is approximately equal to its probability (p) of occurrence for Activity 1. When we have these approximate probabilities (the more trials you do, the closer the fraction becomes to the true probability) for all possible outcomes, we can chart them as probability distributions (see the activity 1 duration curve in Figure 9.6). Assume that experience-based probability distributions are also available for some other activities in the project as well (see activity 2 duration in Figure 9.6). If we really had such probability distributions (see the box that follows, "Frequently Used Probability Distributions"), they would be close to objective probabilities, which are defined as being determined from complete knowledge of the system and are not affected by personal beliefs.

Some activity durations, however, may be single-valued (e.g., activity n duration in Figure 9.6). Hence, we can have a combination of activity durations that are distributed and those that are single-valued. How does this impact project duration? As long as one or more activity durations (inputs to the model) are probability distributions, project durations (the outputs) will be probability distributions also [10].

Although some companies do have experience-based databases with the approximate distributions of their project activities, that is the exception rather than the rule. What, then, do we do to prepare probability distributions for activity durations? We will do what is a dominant practice in real-world projects—prepare and rely on subjective probabilities—someone's belief whether an outcome (activity duration) will occur.

The most adequate way for this is to enlist the help of experts, or experienced project participants. Brainstorming with activity owners, studying durations of similar activities in past projects, and consulting other specialists in the company who are not involved in the project all help determine the probability distributions or single values for the activity durations [4].

These distributions from past projects can be modified to reflect the information from Risk Response Plan mentioned earlier as a crucial input to MCA. For example, risk event status is determined for each individual risk event treated in the plan. If deemed appropriate by the project team, it is possible to expand the past-based distribution for the related activity to include the impact of the risk event status. In the case of the general distribution for a particular activity duration, that could mean increasing the maximum to reflect the risk event status. Updating the probability distributions with the information from the Risk Response Plan is the real way to determine the probability distributions most suited to each project activity based on the actual risks in the particular project.

Frequently Used Probability Distributions

Three values are used to describe very simple and popular triangular distribution (see Figure 9.8.a): Triang (5, 10, 20), minimum L (5), most likely M (10), and maximum H (20). Numbers in parentheses are project activity durations in days. The mean equals (L+M+H)/3. Beta distribution (see Fig. 9.8.b) that has been used for a long time to estimate activity durations in PERT requires the same three parameters as triangular distribution, minimum (5), most likely (10), and maximum (20). The mean is (L+4M+H)/6. Two parameters describe lognormal distribution (see Fig. 9.8.c)—mean (10) and standard deviation (2). Known for its flexibility, the general distribution (see Fig. 9.8.d) allows shaping the distribution to reflect the opinion of experts [4]. It is described by an array of values (7, 10, 15) with probabilities (2, 3, 1) that fall between the minimum (5) and maximum (20).

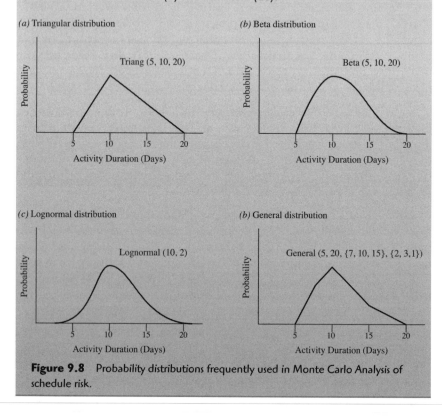

Figure 9.8 Probability distributions frequently used in Monte Carlo Analysis of schedule risk.

Select Randomly a Value from Each Distribution. When the probability distributions are available for all project activities (variables), the stage is set for the next step. For each activity within its specific range of duration values, select one duration value randomly. The key word here is *randomly*. Using random sampling technique, MCA generates a random number between 0 and 1, which is fed into a mathematical equation that determines the activity duration value to be generated for the distribution [4]. All of these selected values constitute a random sample of values that will be used to generate project durations. Sampling can also be done with other efficient methods such as Latin hypercube sampling [4]. Whatever the method used, random sampling from probability distribution is performed in a manner that reproduces the distribution's shape [4].

Run a Trial to Generate Project Duration and Store It. Having a random sample of activity duration values means that for each activity in the project schedule there is one value only. Plugging this combination of activity duration values in the project network diagram will produce a scenario for project duration. In essence, this is a deterministic schedule with a single value for project duration, built on single-value durations for each activity. At this time, we will store this project duration until the time comes to use it again.

Repeating this sequence of random sampling many times and running a trial will produce as many scenarios for project duration, each one plausible. This prompts the question, "How many trials do we need?" Typically, trials (iterations) go until the predetermined number is reached (number N in the decision box in Figure 9.6). That number depends on the number of variables (activities) and the degree of confidence required but typically lies between 100 and 1000 [4]. The idea here is that sufficient number of trials preserves the characteristics of the original probability distributions for activities and approximates the solution distributions for project duration [10].

Process Results. When the trials are complete, our "storage" will contain N project durations. Each one is a possible case for the behavior of the project schedule. Processing them by means of a software program can produce many forms of results, whereas our focus is on the following (see the right-hand side of Figure 9.6):

- ▲ Expected value (EV) of the project duration. Averaging trial values for project durations approximates the expected value, the probability weighted average of all possible outcomes. However, the higher the number of trials, the higher the precision of the EV and the approximations of probability distribution shape for project durations.

- ▲ Frequency distribution. This is a histogram plot showing relative frequency obtained by grouping the data generated for project durations into a number of bars or classes. Frequency is the number of values in any class. Dividing the frequency by the total number of values will produce an approximate probability that the project duration (output variable) will lie in that class's range (see Figure 9.9).

- ▲ Cumulative frequency. It has two formats: ascending and descending (refer again to Figure 9.5). The former indicates the probability of the project duration being less or equal to the value on the x-axis. Conversely, the latter shows the probability of the project duration being greater than or equal to the value on the x-axis. Since the ascending format is more frequently used, we will stick with it. EV is marked on the plot with a black dot.

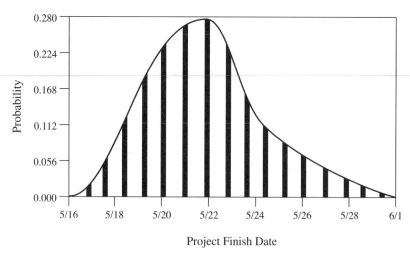

Figure 9.9 Frequency distribution histogram of project duration.

▲ Tornado chart. This chart shows the extent to which the uncertainty of the individual activities' duration impacts the uncertainty of project schedule duration (see Figure 9.10). Specifically, the bar represents the degree of impact the activity (input variable) has on the project schedule (model's output). Therefore, the longer the bar, the greater the impact that the project activity has on the project duration. Per standard practice, bars are plotted from top down in decreasing degree of impact. When there are both the positive and negative impact, the chart is a bit reminiscent of a tornado, hence the name. To avoid the chart looking overly busy, Vose suggests limiting the plot to those activities (variables) that have an impact of at least a quarter of the maximum observed impact [4].

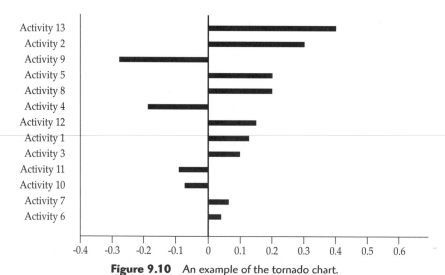

Figure 9.10 An example of the tornado chart.

From *Risk Analysis: A Quantitative Guide* by David Vose. Copyright © 2000 by John Wiley & Sons Limited. Reprinted with permission of John Wiley & Sons.

Analyze and Interpret Results. The results of the schedule risk analysis must be interpreted in a way that clearly provides answers to the questions the analysis was initiated to answer. For that reason, it is beneficial to follow four principles for schedule risk analysis [4]:

▲ Focus on the problem

▲ Keep statistics to a minimum

▲ Use graphs whenever appropriate

▲ Understand the model's (e.g., time-scaled arrow diagram chart) assumptions.

To demonstrate these principles, let's assume that the project team set out to perform the schedule risk analysis in order to answer these questions:

▲ How likely is it that the team will satisfy the project deadline (May 20) imposed by management?

▲ If the probability is lower than 90 percent (team's preferred probability), what do we do to negotiate the deadline with management?

▲ If we successfully negotiate the deadline issue, which are the top three activities in terms of their impact on project duration?

First, the team goes to the cumulative frequency graph (Figure 9.5). To obtain the answer to the first question, they will

▲ Enter the x-axis at the deadline date imposed by management (May 20).

▲ Move upward to the cumulative curve.

▲ Move left to the corresponding value on the y-axis. This y-axis value is the probability of completing their project before on the imposed deadline date (40 percent).

Clearly, the probability is very low, way lower than the preferred 90 percent. To answer the second question, the team decides to ask management for a new option of adding a contingency of six days to the imposed deadline, in which case the project would be finished by May 26 and would be 90 percent probable. To build a better case for negotiations with management, the team develops another new option. They combine the schedule crashing (throw in more resources in the project activities) and fast-tracking (overlap activities as much as possible) approaches. With this approach, they obtain the new project logic diagram and probability distributions for activities, and run MCA again. Going again through the cumulative frequency plot, they find out that the probability to meet the imposed deadline date of May 20 is 90 percent. Armed with the plot less redundant statistics, they approach management with both new options and successfully obtain new resource commitments from management for the second new option. Then, a look at the tornado chart (see Figure 9.10) reveals the three key activities with the highest impact on the deadline—these are top priorities to keep an eye on during the implementation, an answer to their question three. The team understands that their MCA is founded on (a) the changed logic of the time-scaled arrow diagram chart and (b) the changed probability distributions for activities that are based on new resource commitments. In a nutshell, the four principles for schedule risk analysis directed the team in its approach.

Utilizing Monte Carlo Analysis

When to Use. Traditionally, it has been the large and complex projects that have most often enjoyed the benefits of MCA. The belief was that, unlike smaller projects, larger projects had more important goals and could also afford necessary resources for performing MCA. It took some significant events for this view to start changing. First, the trend toward "management by projects" led to a proliferation of smaller but important projects. Also, very powerful desktop computer programs for MCA have become affordable. Finally, project offices capable of supporting many projects with MCA have become a frequent organizational unit. All of these events helped put MCA within the reach of smaller projects, changing the application pattern of MCA in corporations. Nowadays, MCA is used in both larger and smaller projects to respond to certain situations. For example, if the projects are sensitive to a completion deadline, MCA is a preferred option. Similarly, if there are many project scenarios and what-if analysis to explore, MCA is favored over decision trees [10]. When facing a situation where very small nuances may dictate which project wins the contest, for example, in the project selection process within the company, using MCA is the right move.

Time to Use. Roles in MCA crucially influence how much time it takes. A typical approach is to task the project office specialists or administrative assistants to perform MCA using the data provided by the project team. For a smaller project of 50 activities, for example, data entry and running MCA may take 10 to 30 minutes. Assuming that a project logic diagram exists, preparing activity probability distributions through team brainstorming and formatting them into a table to be fed to the project office may take an hour or two. The growing size and complexity of the project is bound to increase the time for MCA.

Benefits. Original project schedules and budgets are often unrealistic or, more precisely, inadequate. A major reason for this inadequacy is the uncertainty surrounding the project activities. In response to this uncertainty, many project teams assign an arbitrary duration or cost to the activities and hope for the best [8]. Contrast this approach with MCA, which allows richer, more detailed representation of the risk problem—important in some situations (see the box that follows, "Confident or Probable?"). Take, for example, a situation where a company's fortunes are at stake. If their new project does not develop a product to hit the market before their competitors, they may lose the market leadership. MCA would significantly improve quality of their decisions by offering a clear analysis of their risks, different scenarios, and probabilities of reaching the goal. Still in another situation, a project manager may be asked to finish a project by an unrealistic date or budget. To her, MCA provides the roadmap to lowering the risk by arguing against the unrealistic approach and obtaining resources necessary to complete the project as desired. In summary, MCA's value is in its ability to examine each project scenario, including the extreme scenarios, to see what conditions give rise to their results. That helps not only to validate the project realism but also to differentiate between what is possible and what is not possible, and most importantly, how to change what is not possible into what is possible [10].

Advantages and Disadvantages. MCA offers several advantages [4]:

- *Basic mathematics.* Mathematics needed to use MCA is quite basic. Even including complex mathematics or modeling correlation and other interdependencies (e.g., power functions, logs, IF statements, etc.) poses no extra difficulty.

- *Easy to use.* There are very good and commercially available software packages that automate the tasks involved in MCA. That means the computer does the work necessary to produce the outcome distribution.

- *Easy to change.* Making changes to the MCA model is a quick process that provides a great opportunity of comparing with previous models. Examining the behavior of each of the models is also straightforward.

- *Legitimacy.* MCA is seen as a well-reputed tool, increasing the likelihood that project teams and management will accept its results.

Often MCA is criticized for its disadvantages:

- *Complex.* Although based on basic mathematics, MCA is built on concepts of probability, a concept many project managers have yet to master.

- *Approximate technique.* Its outcome distributions are approximate, critics claim. While this is true when the number of iterations is small, you can easily overcome this problem by increasing the number of iterations until reaching the required level of precision.

- *Takes time.* There are complaints about the time a computer needs to generate iterations.

Customize the MCA. MCA is a risk analysis tool of great value that deserves to be used in the majority of projects facing an uncertain implementation environment. For that to be possible, the generic type of MCA that we have described has to be customized to fit the situation of an organization's projects. It is for that purpose that we offer several ideas for the customization.

Confident or Probable?

John Glenn, a senior program manager in a high-tech company, told this story of risk. "My company has had for a long time this practice of asking how confident project managers were they would finish the project by the deadline. The point was in having them assess the risk of not completing on time, something like, 'I am 70 percent confident I will have the project finished by May 1.' The word *confident* was really used to mean *likely* or *probable*. Their answer was pretty much based on the gut feeling, since we never equipped them with any consistent tools to make the confidence assessment. Still, we felt we were very successful at creating the awareness of risk and the need to have a Risk Response Plan.

"In my graduate classes I learned more about the two words. The confidence factor is a number on a 0-to-1 scale that indicates the confidence in an assertion or inference. '0' means 'no information or knowledge,' while '1' means 'with complete confidence.' On the other hand, probability means the likelihood of an event occurring. Although the two terms have very different meanings, I didn't try to change the way we use them in our company. Why? First, that would create a lot of resistance among people who have come to perceive it as a great methodology. Second, everyone used 'confident' clearly and consistently to mean 'probable.'

"So I left the terminology alone. I thought, however, that relying on the gut feeling as the way to assess confidence was rather primitive and inconsistent. Therefore, I worked hard to introduce and marry Monte Carlo Analysis (MCA) with our 'confidence' system. Now, we do MCA to assess how probable our project deadline is, but we do not use the word probable. Rather, we still say it the old way—that we are confident."

Customization Action	Examples of Customization Actions
Define limits of use.	Use MCA in all projects taking over 1000 person-hours (major projects).
	Use MCA for risk analysis focusing on project schedule (in companies competing on time to market).
	Use MCA for risk analysis focusing on project cost (in companies competing on cost leadership).
	Use MCA focusing on Net Present Value when selecting new projects.
	Have project teams own the MCA, which will be performed by the project office (for organizations with a project office).
Adapt a feature.	Use the descending format of the cumulative frequency chart instead of the ascending format.
Add a feature.	MCA in larger projects must be accompanied by a risk analysis report including a model's assumptions, graphical representation of results, and conclusions, if any [4].
	Add spider and trend plots when the situation mandates so.

Monte Carlo Analysis Check

Check to make sure you performed a good Monte Carlo Analysis. It should be based on inputs:

- Risk Response Plan
- Project model
- Probability distributions for variables

and produce results that at least include

- Expected value
- Frequency distribution histogram plot
- Cumulative frequency curve

in order to interpret the results of the risk analysis in a way that clearly provides answers to the questions the analysis was initiated to answer.

Summary

This section centered on Monte Carlo Analysis (MCA), a tool that uses a model of a project—a project network diagram, for example—to analyze its behavior and risks. The purpose is to lower the risk by arguing against the unrealistic approach and obtaining resources necessary to complete the project as desired. To accomplish this, MCA examines each project scenario, including the extreme scenarios, to see what conditions give rise to their results. That helps not only to validate the project realism but also to differentiate between what is possible and what is not possible. Still, it is the large and complex projects that enjoy these benefits of utilizing MCA. To recap the essence of this section, the box above offers the key points in performing the Monte Carlo Analysis.

Decision Tree
What Is the Decision Tree?

A Decision Tree is a graphical device for analyzing project situations under risk. Reflecting the decision process (see Figure 9.11), the tree displays sequential decisions in the form of the branches of a tree, from left to right, originating from an initial decision point and extending all the way to the end outcomes [11]. A path through the branches is composed of a sequence of separate decisions and chance events. The way to evaluate decisions is to calculate the expected value of each path, folding the tree back from the end points to the initial decision point [12]. For the description of a typical Decision Tree's components, see the box that follows, "Five Components of a Typical Decision Tree."

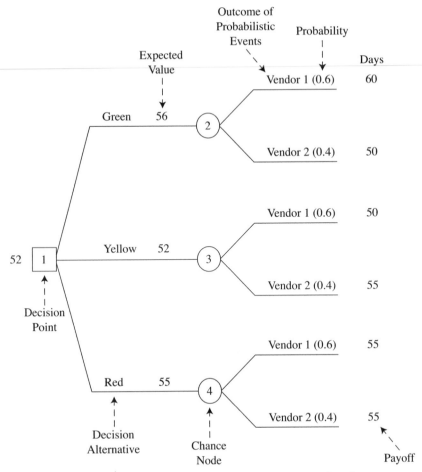

Figure 9.11 Decision Tree for a project situation under risk.

Analyzing the Decision Tree

Literature on Decision Trees often looks at minimizing the risk by selecting decision alternatives that offer maximum Net Present Value or minimum cost. The abundance of such examples prompted us to take a different approach. In particular, we look at the general process of analyzing Decision Trees on an example related to minimizing schedule duration. This is in response to risks faced by a large group of today's projects that must deliver the fastest possible schedule as an unconditional priority.

Prepare Information Inputs. Three major inputs have the central role in building and analyzing the Decision Tree:

- ▲ Risk management plan
- ▲ Risk Response Plan
- ▲ Other project information

Five Components of a Typical Decision Tree

- Decision nodes. Points in time when decisions are made or alternatives chosen. Shown as square boxes here, they are controlled by the decision maker. Also called decision points. The starting node is called *root*.
- Chance nodes. Times when the result of a probabilistic events occurs. Decision makers have no control over them. Also called probabilistic nodes or points (a circle).
- Branches. Lines connecting the decision and chance nodes in a sequential manner. The branches leading out of a decision node represent possible decisions, while those stemming from chance nodes represent possible outcomes of probabilistic events.
- Probabilities. The probabilities of the probabilistic events shown on the branches representing those events. Mostly they are conditional and for any particular chance node must sum to 1.
- Payoff (outcome values). The outcome of each alternative placed at the end of the branch. They may represent present values discounted to the date of the root decision or cost.

As a roadmap for dealing with risk throughout the project's life, the risk management plan specifies how to use Decision Trees in project decisions under risks. Information contained in the Risk Response Plan about individual risks and their response strategies is also crucial. In our example from Figure 9.11, this information will be funneled into calculating durations for each of the outcomes in the Decision Tree. For these calculations, we also need other project information such as network diagrams of the decision alternatives for the central module design.

Describe Decision under Risk. Common sense dictates that in order to make the best decision, we first need to understand the decision context and related risks a project is facing. A convenient way for this is to describe the decision. Here is an example.

Fast Corporation competes on time-to-market capabilities. As they are entering the design phase for its new enterprise server development project, their goal is to finish it as soon as possible and hit the market before the competition. While literally each project day is considered to be of extreme priority, the development cost is given low importance. In such a situation, the project team is attempting to decide on the appropriate design approach to use for this product. A major uncertainty regards how many days it will take to design the central module in the product. Three major alternatives are being considered, each one identified with a single word:

- *Green (G)*. Incorporate the routing rules early in the central module design.
- *Yellow (Y)*. Predict the routing rules early and modify them at the end of the central module design.
- *Red (R)*. Incorporate the routing rules at the end of the central module design.

The second uncertainty is related to an off-the-shelf part that goes into the central module, whichever the alternative. Two different vendors that cannot be influenced produce the part. It is well known that both companies are racing each other to put out the newest upgrade of the part in the market, and they have announced the same release date. To represent this decision description and enable its analysis, the team must first structure the model.

Structure the Model. The model is drawn from left to right (for a better understanding refer again to the "Five Components of a Typical Decision Tree" box). Therefore, draw a decision node (the square marked 1), then add to its right-hand side three branches, for three available design alternatives—the green, yellow, and red (Figure 9.11). Put a chance node (the circles marked 2, 3, and 4) at the end of each branch, followed by two branches, each one for outcomes of probabilistic events—vendor 1 hitting the marker first and vendor 2 hitting the market first. As monotonous as structuring the model appears, it is precious in sharpening our understanding. As a decision guru put it, these models help formalize common sense for decision models that are too complex for informal use of common sense [12].

Assess the Probability of the Possible Outcomes. Fast Corporation's server development project team cannot wait for the actual release of the first-to-market part by vendor 1 or vendor 2. That would significantly extend the module design schedule, jeopardizing the whole server project. Therefore, the project team decides to assess the probability of who—vendor 1 or vendor 2—will release the product first. Their research and past performance of the vendors led them to assess that there is 60 percent probability that vendor 1's part will first reach the market. The probability for the part from vendor 2 is 40 percent. Everybody on the team is clear that these are subjective probabilities influenced by their perceptions and beliefs. Adding the probabilities to the tree model is useful (see Figure 9.11).

Two Simple Steps for Solving a Tree

The procedure for solving a tree is called "rolling back" or "folding back" the tree. Simply, it starts at the end of branches at the right, and working back to the left, we solve for the value of each node, annotating it with its expected value (EV), which can be the expected monetary value (EMV) or EV cost, if the value is measured in currency, or EV schedule, if the value is measured in time units (e.g., days). Two simple steps for solving a tree are [2] as follows:

- At each chance node, calculate the EV as the sum of each branch's outcome value (payoff) multiplied by probability. This is the value of the node and the branch leading to it.

- At each decision node, we find the best EV alternative. This is the alternative with the highest EMV (when dealing with present values) and the lowest alternative for EV (schedule or cost).

When the folding process is completed, the alternative with the best EV for the leftmost decision nodes becomes the best alternative.

Determine Payoffs of the Possible Outcomes. The project team has developed initial network schedules for each of the design options—green, yellow, and red—as if the vendor part is already available. Essentially, the sequence of design activities involved in each option is different, as well as some of the activities. Also, although both vendors' parts can be used for the central module, the process of their incorporation into the design options is different, causing the duration of each outcome to differ. Because the team's expectation is that the first-to-market vendor part will be released sometime midway through the module design, they need to evaluate how such a release is going to change the initial network schedules. The product of their evaluation is a set of possible outcomes values, also called payoffs. These schedule durations of the outcomes expressed in days are added at the end of each branch (see Figure 9.11).

There are two conceptually different parts to a Decision Tree analysis. Included in the first part are structuring the model, assessing the probabilities of possible outcomes, and their payoffs. This is a particularly unstructured task, requiring a significantly greater proportion of effort. The second part—evaluate alternatives and select the strategy—is the easy part of the model and the heart of the decision analysis under risk. We address it next.

Evaluate Alternatives and Select the Strategy. Our objective is to evaluate possible outcomes and design alternatives and select one with the shortest-possible schedule. To accomplish this, we need to solve the tree (see the box on page 314, "Two Simple Steps for Solving a Tree"). When these steps are applied:

> *Step 1.* Chance Node 2 EV (Expected Value): (0.60 x 60 days) + (0.40 x 50 days) = 56 days
>
> Chance Node 3 EV (Expected Value): (0.60 x 50 days) + (0.40 x 55 days) = 52 days
>
> Chance Node 4 EV (Expected Value): (0.60 x 55 days) + (0.40 x 55 days) = 55 days
>
> *Step 2.* The best alternative is the alternative with lowest EV, the shortest schedule – 52 days. That means the team goes with the yellow option.

Decision Tree analysis enables more than just identifying the best alternatives. Sensitivity analysis, as well as tornado and spider charts, can also be developed to better understand the decision under risk [2]. They, however, are beyond our scope.

Utilizing Decision Trees
When to Use. Theoretically, we can use a Decision Tree to evaluate *any* decision under risk, regardless of its complexity, as long as the decision and probabilities of the possible outcomes are specified [12]. Practically, this is not the case. Rather, practicing project managers see the tree as a method to address daily problems, requiring a straightforward and quick selection

of the best alternative. Why is this? The issue here is that complex decision situations lead to a "combinatorial explosion" and the time associated with it. As we add decision and chance nodes, the trees tend to grow exponentially [10]. For example, multiple alternatives with multiple uncertainties can explode Decision Trees into hundreds of paths, which is where Monte Carlo has a significant advantage. Constructing and solving for such trees may take hundreds of hours. Who can really afford this when our projects are in the fast lane? Perhaps only very large projects with generous budgets can do so. Since such are exceedingly rare, practicing project managers go to decision trees primarily when they need to swiftly evaluate simple alternatives, pick up the best, and go on with their daily routine [10]. Still, the majority of such situations occur in larger projects, although we have seen small project managers using two-alternative with four to six paths very informally with minimum time consumption.

Time to Use. Two extremes can help fathom the time requirements of Decision Trees. Spending 10 to 15 minutes to construct and evaluate a two-alternative Decision Tree with four paths seems realistic to many project managers. On the other hand, tens of hours may go into the construction of a Decision Tree with hundreds of paths. The assumption here is that information necessary to estimate probabilities and outcome values is already available. The analysis of large trees may be a matter of minutes, given the power of professional software necessary to use large trees. In case of smaller and medium projects Excel spreadsheet does a fine job.

Benefits. Two major benefits seem to motivate project managers to use Decision Trees. First, Decision Trees reduce an evaluation and a comparison of all decision alternatives under risk to a single value metric. Simply, this metric indicates the degree of support of project goals [10]. In our example, that metric was schedule calendar day, gauging progress toward Fast Corporation's quest for time-to-market speed. In other cases, an additional convenience stems from the fact that most of the time this single value metric is expressed in monetary terms, a universal language of business and projects. Then, this single monetary value combines cost, schedule, and performance criteria.

The second benefit lies in the belief of many project managers that the real value of Decision Trees is not in the numerical results but rather in their ability to help us gain insights into decision problems. With or without the numerical results, the users should understand that Decision Trees do not provide an entirely objective analysis. In the absence of sufficient empirical data necessary for a complete analysis, many facets of the analysis are rooted in personal judgment – structuring the model, assessing probabilities or payoffs, for example. Still, experience has shown that the Decision Tree tool is very beneficial [12].

Advantages and Disadvantages. Decision Trees are characterized by some advantages that are difficult to match:

▲ *Convenience.* Decision Trees are very convenient to visualize and analyze project decisions with one or a few paths. When more paths are added, the power of the trees is in their ability to augment our intuition.

▲ *Visual impact.* When seemingly intangible decision alternatives under risk are displayed in a Decision Tree format, they appear tangible for the tree's clarity and visual impact.

These should be weighed in against disadvantages such as:

▲ *Low rate of use.* Surprisingly, Decision Trees have not seen much use by practicing project managers. This is unfortunate because they offer value in risk deliberations that other tools cannot match, making us believe that in promoting this tool to project managers we should position it as a quick and informal tool for simpler decisions under risk.

▲ *Complexity.* Given how larger Decision Trees may become cumbersome and complex, it is understandable that they turn regular practicing project managers off. Application of such trees in projects should be the specialists' province of work.

Variations. A payoff table can easily represent very simple Decision Tree problems. Table 9.2, for example, is a tabular variation of the Decision Tree from Figure 9.11. It presents decision alternatives, probabilities of possible outcomes, payoffs, and expected values in a less visual format than the tree. Also, as the number of nodes and paths grow, the tabular format becomes increasingly difficult to use.

Customize the Decision Tree. The generic use of Decision Trees as described here will certainly provide value to the user. Customizing trees to fit one's projects' needs may further enhance the value. The following ideas may help you get a better feel for the customization.

Table 9.2: Tabular Representation of Decision Tree Problem in Figure 9.11

Probabilities	0.60	0.40	
	Scenarios		
DECISION			
Alternative	*Vendor 1 first*	*Vendor 2 first*	Expected Value
Green	60	50	56
Yellow	50	55	**52**
Red	55	55	55

Customization Action	Examples of Customization Actions
Define limits of use.	Use Decision Trees with the help of professional decision analysts and professional software (large projects).
	Use small and simple Decision Trees either informally (e.g. on the back of an envelope) or with the help of Excel to quickly make decisions (small projects).
	Use the expected value (EV) schedule metric (time-to-market companies), EV cost metric (cost-conscious companies), or expected monetary value (EMV) metric as your company strategy mandates.
Adapt a feature.	Use a tabular format (if that's easier for project team members).

Summary

This section focused on the Decision Tree, a tool for analyzing project situations under risk. Decisions Trees can be used in both large and small projects. Practicing project managers see the tree as a method to address daily problems, requiring a straightforward and quick selection of the best alternative. A major benefit of the trees is in reducing an evaluation and a comparison of all decision alternatives under risk to a single value metric. Also, the trees help us gain insights into decision problems. In summary, these are the key points in performing the Decision Tree analysis.

Decision Tree Check
Check to make sure you performed a good Decision Tree analysis. It should

- Describe the decision under risk.
- Structure the model, including decision nodes, chance nodes, branches.
- Assess probabilities of outcomes.
- Determine payoffs and solve a tree by rolling it back to produce results that at least.
- Show expected values (EV) at each chance node.
- Indicate the best EV alternative at each decision node.
- Identify the best alternative, one with the best EV for the leftmost decision nodes.

Concluding Remarks

Two aspects are crucial in selecting when to use the Risk Response Plan, Monte Carlo Analysis, and Decision Trees. First, a Risk Response Plan is a convenient choice for treatment of risk events that are assumed to be independent of each other (which may not be the case), whether in small or large projects, formally or informally. Its highest value is when applied in conjunction with Monte Carlo Analysis or Decision Trees, as complementary tools, because the latter two are able to account for interacting risks. Second, Monte Carlo and Decision Trees are often seen as alternatives. So, which one is more appropriate to use? Of course, it depends on the project situation. Review project situations in the table that follows to understand how each one favors the use of the two tools, and select those that match yours. If necessary, identify more situations in addition to those listed and mark how each favors the tools. The tool that has the higher number of marks is probably a better choice for your project.

A Summary Comparison of Risk Planning Tools

Situation	Favoring Risk Response Plan	Favoring Monte Carlo Analysis	Favoring Decision Trees
Small and simple projects	√ (informal)		√ (informal)
Large and complex projects	√ (formal)	√	√
Projects sensitive to completion deadline		√	
Explore many project scenarios and what–if analysis		√	
Short time to train how to use the tool	√		
Take little time to apply	√		
Treat risks as being independent from each other	√		
Treat risks as interacting with each other		√	√
Need focus on most critical risks	√	√	√
Require a single value criterion (e.g., expected monetary value)	√		√
Require criteria other than value (e.g., schedule)	√	√	
Optimize continuous decision variables (e.g., schedule duration)		√	
Need output distributions		√	
Need to show subsequent decisions in models			√
Work with few decision + chance nodes, hand solution possible			√
Work with more than 5 chance nodes		√	

References

1. Wideman, M. 1992. *Project and Program Risk Management*. Newton Square, Pa.: Project Management Institute.

2. Winston, W. L. and W. S. Albright. 2001. *Practical Management Science*. 2d ed. Pacific Grove, Ca.: Duxbury.

3. Royer, P. S. 2000. "Rock Management: The Undiscovered Dimension of Project Management." *Project Management Journal* 31(1): 6–13.

4. Vose, D. 2000. *Risk Analysis: A Quantitative Guide*. 2d ed. New York: John Wiley & Sons.

5. Project Management Institute. 2000. *A Guide to The Project Management Body of Knowledge*. Drexell Hill, Pa.: Project Management Institute.

6. Couillard, J. 1995. "The Role of Project Risk in Determining Project Management Approach." *Project Management Journal* 26(4): 3–15.

7. Graves, R. 2001. "Open and Closed: The Monte Carlo Model." *PM Network* 15(2): 48–52.

8. Graves, R. 2000. "Qualitative Risk Assessment." PM Network 14(10): 61–66.

9. Hulett, D. T. 1995. "Project Schedule Risk Assessment." *Project Management Journal* 26(1): 21–31.

10. Schuyler, J. 2001. *Risk and Decision Analysis in Projects*. 2d ed. Newton Square, Pa.: Project Management Institute.

11. Cleland, D. I. and D. F. Kocaoglu. 1983. *Engineering Management*. New York: McGraw-Hill.

12. Eppen, G.D., et al. 1998. *Introductory Management Science*. 5th ed. Upper Saddle, N.J.: Prentice Hall.

Team
Building

This chapter is contributed by Hans J. Thamhain, Bentley College.

A castle is only as strong as the people who defend it.

Sun Tzu—The Art of War

This chapter focuses on the tools for identifying, organizing, and developing the project team:

- ▲ Four-Stage Model of Project Team Building
- ▲ Stakeholder Matrix
- ▲ Skill Inventory
- ▲ Commitment Scorecard

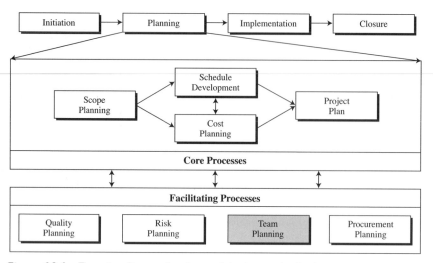

Figure 10.1 The role of team-planning tools in the standardized project management process.

These tools focus specifically on the human side of project planning, helping to analyze the resource requirements, and to identify, recruit, and build the people needed into a unified, high-performing project team. In addition, many of today's projects require team building as a continuing process throughout their life cycles. Therefore, consistent with the broad concept of stakeholder management and the need for developing learning organizations, these tools overlap other phases of the standardized PM process shown in Figure 10.1. The purpose of this chapter is to help practicing and prospective PM professionals to

- ▲ Understand and build the skills needed to function effectively on their project teams.
- ▲ Select project leaders and team members.
- ▲ Build cross-functional communication networks and alliances.
- ▲ Deal with conflict and power sharing.
- ▲ Build project ownership and commitment.
- ▲ Develop a learning organization.

Mastering these tools is part of the core competencies of PM and is necessary for effectively leveraging company resources toward established business objectives.

Four-Stage Model of Project Team Building
What Is the Four-Stage Model of Project Team Building?

The Four-Stage Model is a tool for organizing and systematically developing project teams. The concept was originally suggested by B.W. Tuchman in 1965 [2], who identified four stages that all teams have to pass through in their transformation to a unified, effective work team: *Forming, Storming, Norming and Performing.* Each stage is associated with specific characteristics, team activities, and managerial guidance. The concept has been refined and developed by many, most noticeably by P. Hershey and K. Blanchard [3]. Today, the Four-Stage Model enjoys wide acceptance as a powerful tool for team planning and development. It often becomes the starting point for identifying critical success factors, skill profiles, and potential team members. It also serves as a roadmap for project team development. The model shares a "phase-based" philosophy as part of contemporary PM approaches, such as the rolling wave concept, phased developments, phase reviews, voice–of–the customer, and stage-gate developments. The basic structure of the Four-Stage Model is shown in Figure 10.2. Inputs to each stage come from both the organizational and project environment. These inputs affect the team formation process and its subsequent development. They also affect the team dynamics, measured in critical success factors, decisions, action items, and resource requirements.

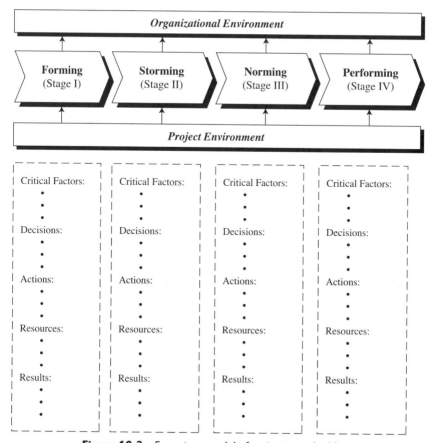

Figure 10.2 Four-stage model of project team building.

Constructing the Four-Stage Model

As shown in Figure 10.2, the team development model is typically presented as a series of four serial steps: forming, storming, norming, and performing. However, its application is for both serial and parallel work processes, including concurrent engineering, design-builds, fast tracking, rolling wave developments and multiorganizational joint developments. The labels for each step can be chosen differently and the number of steps can be varied. For each stage, a set of *management parameters* should be defined. At the minimum, these parameters should include (1) *critical success factors*, important for the team to perform effectively, (2) *decisions* to be made by management, the team and joint decisions, with responsible individuals and timelines clearly defined, (3) *action items* to be taken by management, the team and joint actions, with responsible individuals and timelines clearly defined, (4) *resources required* and their approval process, and (5) *specific results* to be delivered during, or toward the end of the stage.

Preparing Information Inputs. Involvement of all stakeholders is important! To be meaningful and realistic, inputs for these stage/management parameters should come from a broad spectrum of stakeholders, including team members, functional support groups, company management, contractors, partners, and customers. However, at the formation of the project and its team, the known stakeholder population is small, but it increases with the progression of the project through its life cycle. Efforts should be made at any stage to reach into the various groups of stakeholders to cast the widest possible net for input gathering. Specific group communication tools, such as brainstorming, focus teams and nominal group technology, can support the input data gathering process and help in organizing the information into a meaningful and reliable set of management parameters as a basis for organizing and developing the project team.

Communicating the Team Organization. The evolving project team organization should be documented as part the project plan, and updated as the project moves through its life cycle. Conventional tools such as (1) task roster, (2) task matrix or responsibility matrix, (3) N-square interface chart, and (4) Project Charter can be useful to document and communicate task responsibilities and their organizational interfaces.

Team Characteristics and Leadership during Team Development. The characteristics of the workgroup changes as it goes through the stages of development. Therefore, team leadership style and effectiveness depend on the project situation and stage. At the beginning of the team formation, a more directive management style is needed than during the later stages when the team is approaching maturity. To provide additional guidelines for developing the inputs for the Four-Stage Model, following are descriptions of each stage, together with some leadership perspective shown in Figure 10.3 that summarizes Hershey and Blanchard's [3] popular model of team leadership:

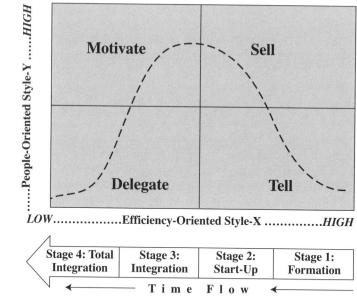

Figure 10.3 Leadership style versus team integration level.

▲ Forming stage. At this stage, core team members are being defined and introduced to the project and its mission. Communication flows by and large one way, from the designated project leader, senior management or the project sponsor, to the emerging team members. The emerging workgroup is not yet a team, but just a collection of people from different organizations and functional backgrounds. Anxieties, confusions, and role ambiguities are predictably high in this stage, while mutual trust, respect, task involvement, and commitment are very low. This stage requires a leadership style that relies on clear directions, guidance, strong image building, vision sharing, close supervision, and considerable top-down decision making.

▲ Storming stage. At this stage, also called the startup stage, many of the project team members have been defined and signed on to the project. Members of the workgroup begin to get involved with the project assignment, try to understand the project scope and requirements, and sort out specific roles and responsibilities. As projects success often depends on innovation, cross-functional teamwork, member-generated performance norms, and decision making, the team must work out highly complex processes of work integration and deal with the issues of self-direction and control. At this stage, the team usually experiences very high degrees of anxieties and conflict, while mutual trust, respect, task involvement, and commitment are lowly evolving. Typical questions that should be expected to be asked by team members and addressed by team leaders are listed in the box that follows, "Typical Questions to

Be Addressed During Storming Stage"). However, in addition, project leaders must pay much attention to the human side, dealing effectively with the high levels of conflict, facilitating interactions among team members, building cross-functional interfaces, providing feedback, and fostering an environment conducive to mutual trust and respect, buy-in, and purpose. Hershey and Blanchard [3] define this mixture of "directive and relationship-oriented leadership" as "selling style."

▲ Norming stage. At this stage, which is also called the "partial performing stage," most team members have been signed on and are working as an integrated work team toward the project objectives. Team members begin to feel comfortable with their roles and assignments, and rely on other members' expertise. The team as a whole starts to unify and enjoys the team environment. Mutual trust, respect, task involvement, and commitment to established goals strengthen during this stage together with communication effectiveness throughout the project organization and its interfaces. Expertise-based decision making and leadership begin to emerge, allowing the project manager to share power with the team members and to encourage a broad range of self-direction and control. However, project leaders must still pay strong attention to the human side, building confidence in the team capability, dealing effectively with issues of workload, performance measures, interfaces, and integration. The project manager must continue the team building process toward the performing stage that characterizes the fully integrated, high performing team.

▲ Performing stage. At this stage, the team is highly unified and committed to achieve the established project objectives. By definition, a team that reached the performing stage becomes "self-directed" as described in the box that follows. Leadership evolves from within the team based on trust, respect, and credibility. The need for external supervision and administrative support is minimum. Yet maintaining this very delicate balance of power and control requires highly sophisticated PM skills and leadership, and active support by senior management.

Utilizing the Four-Stage Model of Project Team Building

When to Use. The model is useful for project team development at any point during the project life cycle. However, its benefits are especially leveraged during the early phases of project and team formation. This is the time period of "forming" and "storming," which is by-and-large ineffective with regard to producing any useful deliverables, a time period that should be minimized to benefit cost, time to market, and strategic focus. From yet another perspective, the model is especially recommended for organizing and developing teams of large, more complex, and technology-intensive projects. However, regardless of project size and complexity, the use of the model will expedite the team formation and ultimately improve overall project performance. The final output from the Four-Stage Model is a *team development plan*.

Typical Questions to Be Addressed During Storming Stage

- What are the specific objectives to be achieved with this project?
- Who is in charge?
- Whom do I report to? What kind of dual accountabilities do I have?
- What is the priority of this project and the management commitment to it?
- What communication channels exist?
- What type of project procedure, plan, or protocol do we follow?
- What type of project reviews are being expected and when in the project cycle should they occur?
- What factors are critical for project success?
- What are our roles and responsibilities regarding tasks and interfaces with other functions?
- How is our performance being evaluated and who does the evaluation?
- Is this project work part of my career path or a diversion?
- What kind of conflict should be expected and who will resolve it?
- How can problems be escalated?
- What kind of training and help is available?
- How do we deal with risks and uncertainties?

Time to Develop. The Four-Stage Model can be used at different points in the project life cycle, each one with a different purpose and different time requirement.

Self-Directed Teams

Definition. A group of people chartered with specific responsibilities for managing themselves and their work, with minimal reliance on group-external supervision, bureaucracy and control. Team structure, task responsibilities, work plans, and team leadership often evolve based on needs and situational dynamics.

Benefits. Ability to handle complex assignments, requiring evolving and innovative solutions that cannot be easily directed via top-down supervision. Widely shared goals, values, information, and risks. Flexibility toward needed changes. Capacity for conflict resolution, team building, and self-development. Effective cross-functional communications and work integration. High degree of self-control, accountability, ownership, and commitment toward established objectives.

Challenges. A unified, mature team does not just happen, but must be carefully organized and developed by management. A high degree of self-motivation, and sufficient job, administrative, and people skills must exist among the team members. Empowerment and self-control might lead to unintended results and consequences. *Self-directed* teams are *not* necessarily *self-managed,* they often require more sophisticated external guidance and leadership than conventionally structured teams.

▲ Initial modeling of team formation and startup. This could be a single-handed effort by the project manager, thinking through the issues, and writing down critical issues and requirements. *Time required:* 30 minutes.

▲ Detailed modeling of team formation and startup. This can be accomplished by a small group of initially available team members or core team members, including the project manager and functional support managers. Brainstorming, focus grouping, and benchmarking of previous team developments will lead to a *team development plan. Time required:* 2 to 4 hours.

▲ Team-generated development plan. This is the most effective, most beneficial, but also most time-consuming use of the tool. All team members participate in brainstorming, focus grouping, and benchmarking of previous team developments, which will lead to a *team development plan. Time required:* 3 to 6 hours.

Benefits. With today's demands on high efficiency, speed, and quality, project teams have become increasingly more important for dealing with the technical complexities, cross-functional dependencies, and need for innovative performance. The Four-Stage Team Development Model is often being used by managers as a framework for analyzing the team development process. It has become an important tool for supporting the organization and development of multifunctional teams that can effectively deal with these challenges.

Advantages and Disadvantages. Advantages include the following:

▲ *Provides a template* for visual summary of the team organization and development process.

▲ *Involves team members and management* in critical thinking about needs, processes, and requirements for effective teamwork and project execution.

▲ *The team involvement itself provides a vehicle* for building mutual trust, respect and commitment.

▲ *Helps in understanding of project objectives,* requirements, challenges, and benefits.

▲ *Provides a tool for continuing team development* and organizational learning.

Disadvantages include the following:

▲ *Considerable time and leadership skills required* for developing team-generated inputs.

▲ *Potential for conflict and power struggle* during team sessions.

▲ *Differences in organizational culture and value* may emerge and need to be dealt with.

▲ *Implementation of team development plan may be impossible,* leading to frustration and disillusion.

Variations. Many variations of the Four-Stage Team Development process exist. For the template, a different number of stages may be used, and input-output parameters can vary to accommodate the specific needs of the project organization. A common variation is the "merger" of stages, combining "forming and storming," "norming and performing," or both. Also, the process of information gathering, team involvement, and plan implementation varies a great deal among organizations to accommodate special organizational needs.

Customize the Four-Stage Model of Project Team Building. An effort should be made to make the model consistent with both the business process of the company and any established PM process. The model, its team development process, and resulting plan will be more readily accepted by the project team and company management if it is relevant to the business environment. Especially sensitive to company culture are the number of stages and their labels. For example, the change of Stage II from "storming" to "clarifying" or "starting" can make a big difference regarding participants' perception of purpose, expected behavior, and results. Similar to any other organizational development, the basic framework and process, the four-stage model should be carefully defined by senior project personnel and management, and tested in pilot runs, before it is endorsed as a formal team-building tool. However, project leaders can take the initiative of using the principle components of the four-stage process in support of their team formation and development efforts, although the process has not yet been formally established.

Four-Stage Model of Project Team Building Check

Make sure that the Four-Stage Model is properly structured and effectively applied. With it, you should be able to

- Develop the basic model framework and process with senior management and project professional involvement.

- Initially, run model in pilot mode, fine-tune, and issue a simple procedure.

- Ensure consistency of the model and its process with the business environment and established project system.

- For each stage input, involve as many team members and their interfaces as possible.

- Make the team part of the resulting team development plan; obtain buy-in to process and results.

- Focus increasingly on cross-boundary relations, delegation, and commitment.

- Facilitate continuous organizational learning and improvement, and fine-tune the model and its process.

Summary
The Four-Stage Model supports project team startup and development efforts. It helps to deal with the enormous managerial challenges of organizing, directing, and controlling teamwork in our increasingly complex project environments. The model helps in identifying the resources needed and creating the involvement necessary for unifying the workgroup toward an effective, objective-focused team. In addition, effective project leaders use the model to take preventive actions early in the project life cycle, and the model fosters a work environment that is conducive to team building as an ongoing process. Customizing the model for a specific organization requires leadership and inputs from both the PM system and top management.

Stakeholder Matrix
What Is the Stakeholder Matrix?

The Stakeholder Matrix is a tool for systematically identifying and developing the total project team. The matrix recognizes the complexity of project communities that include many stakeholders, ranging from host organizations to support groups, experts, vendors, customers, special interest groups, and regulators [4]. It provides a framework for profiling the broad spectrum of stakeholders and the degree to which they influence project success. Project success and performance often depends on factors that exceed the boundaries of traditional contracts and PM [5]. These factors include the goals, needs, and ambitions of the many parties and individuals that hold an interest in the project and its outcome. To be successful, project leaders must understand the needs and influences of these stakeholders and be able to leverage their resources most favorable toward the goals and ultimate success of the project. The matrix provides a framework for assessing the needs and expectations of all major project stakeholders, and thus helps in managing potential contributions and customer relations of the project community. The Stakeholder Matrix has its conceptional roots in the task matrix with contemporary extensions to Affinity Diagrams such as the *KJ methodology* (for discussion and applications of the KJ method, see Zien and Buckler [6]). Its topology is shared by many other interface management tools, such as project interface definitions, project management readiness plan, voice of the customer, quality function deployment, and customer relations management. The basic structure of the Stakeholder Matrix is shown in Figure 10.4.

Constructing the Stakeholder Matrix

As shown in Figure 10.4, the Stakeholder Matrix is a two-dimensional grid. One side of the matrix lists the project stakeholders, while the other side lists the various influence factors to project success. Figure 10.4 shows typical examples of stakeholders and success factors and provides a good template for data collection. However, project leaders must determine the *actual* and *specific* stakeholders, influence factors, and their degree of associations.

Stakeholders	Influence on Project Parameters																					
	Resources						Project Requirements						Project Process						Performance Evaluation & Reward				
	People	Money	Facilities	Material	Information	Knowledge	Priority	Objectives	Specifications	Schedule	Budget	Quality	Logistics	Deliverables	Teamwork	Project Process	Organizational Interfaces	Infrastructure	Technology	Problem Resolution	Project Performance	Project Success	Team Performance	Individual Performance	Team Rewards	Individual Rewards	Job Security
Project Manager																											
Team Leader																											
Team Member																											
Functional Support Mgrs.																											
Senior Management																											
Customer/Sponsor																											
Contractor/Vendor																											
Partner																											
Regulators																											
Special Interest Group																											
Media																											
Other																											

Key: Stakeholder has… ○ *Little or No* influence on project parameter   ● *Some* influence on project parameter
■ *Considerable* influence on project parameter   ❑ *Mission-critical* influence on project parameter

Figure 10.4   Stakeholder Matrix.

Preparing Information Inputs. To be meaningful and realistic, inputs for the Stakeholder Matrix should come from a broad spectrum of senior personnel who understand specific segments of the stakeholder community. After the set of stakeholders and influence parameters has been identified, the degree of stakeholders' influence on project success should be determined and recorded for each interface point. A simple three or four-point scale, such as shown in the legend of Figure 10.4, is suggested to record the degree of influence on each project parameter. Group communication tools, such as brainstorming, focus groups, and Delphi techniques, can help in supporting the input data gathering process.

Communicating Stakeholder Influences. The Stakeholder Matrix should be documented, either separately or as part the project plan and updated as the project moves through its life cycle. It is important to ensure that the Stakeholder Matrix is consistent with other interface documents, such as the task roster, task matrix, N-square interface chart and customer relations management tools, used within the project organization. In addition to simply showing stakeholder influences, a stakeholder influence grid can be generated to support the management of stakeholder relations, as shown in the box that follows, "Stakeholder Influence Grid."

Stakeholder Influence Grid

The influences of various stakeholders on project parameters can be mapped from the Stakeholder Matrix into the stakeholder influence grid (see Figure 10.5). The grid shows for each stakeholder (a) the importance of support and (b) level of commitment. In the example on the left, the A, B, and C bubbles indicate different stakeholders, and the bubble size indicates the degree of influence the stakeholder is perceived with. The grid divides stakeholders into four principle categories:

- Fully on board. Stakeholders who are very important to project success and are fully committed.
- Conscientious objector. Stakeholders who are very important to project success but are not committed.
- Strong believer. Stakeholders who are very committed but are not too important for project success.
- Cheerleader. Stakeholders who are neither important nor committed.

As a refinement, the grid map can show the "current" and "desired" position." The grid can be used to manage stakeholder relations.

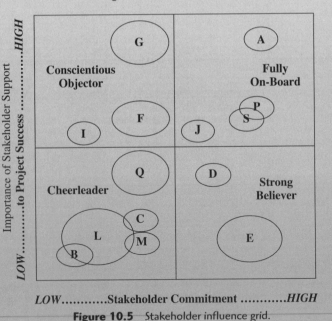

Figure 10.5 Stakeholder influence grid.

Utilizing the Stakeholder Matrix

When to Use. The Stakeholder Matrix is especially useful during the early stages of project team formation and team planning. It provides a tool for identifying the diverse set of key players of the project organization, their influences on the various project parameters and success criteria. The Stakeholder Matrix becomes the framework for developing project interface plans and stakeholder relation management plans as part of the overall *team development plan,* the guiding document for optimizing team integration and effectiveness throughout the project life cycle.

Time to Develop. The Stakeholder Matrix can be developed in two stages:

- ▲ Initial matrix setup. This could be accomplished single-handedly by the project manager, defining the key stakeholders and their impact on principle project parameters. *Time required:* 30 minutes.

- ▲ Detailed matrix setup. This includes the fine-tuning and validation of the diverse spectrum of stakeholders and an in-depth assessment of the influences on project parameters. This can be accomplished by a small group of senior people familiar with the project interfaces. The involvement of stakeholders from the outside of the core team can be especially beneficial in getting a realistic perspective. These stakeholders include functional support groups, customers, contractors, and community interest groups. Brainstorming, focus grouping, Delphi, and benchmarking of similar previous projects will lead to a detailed Stakeholder Matrix. *Time required:* 2 to 4 hours.

Benefits. Successful project teams know how to build cooperative networks and mutually beneficial alliances among all project stakeholders. The Stakeholder Matrix provides a simple tool for the project team to identify and assess the stakeholder influences on specific project parameters. The matrix also provides a framework for the team to identify the conditions necessary for obtaining stakeholder cooperation and commitment in exchange for fulfilling certain needs and wants, a contemporary management concept described as "currency exchange" [7]. The Stakeholder Matrix has become an important contemporary tool for dealing effectively with the intricate web of relations involved in today's project situations, and to build a unified, committed project team that includes all stakeholders.

Advantages and Disadvantages. Advantages include the following:

- ▲ *Provides a template for summarizing* the stakeholders and their influences on project parameters.
- ▲ *Provides a framework for mapping* project interfaces and organizational currencies.
- ▲ *Provides a framework for building* the total project team and for managing stakeholder relations.
- ▲ *Provides a framework for managing* projects in the context of company strategy.
- ▲ *Helps in dealing with project stakeholders* proactively to resolve potential problems, conflicts and contingencies.
- ▲ *The Stakeholder Matrix development* and its application creates an environment conducive to team building and organizational learning.

Mapping Organizational Dependencies

The Stakeholder Matrix can be helpful in mapping dependencies among project team members, support organizations, customers, and other project stakeholders. This is an important step in building the social network necessary for effective interaction of all players toward project success. The Stakeholder Matrix helps to address the important questions such as:

- Whose support and cooperation do we need?
- What organizational conflicts exist between stakeholders?
- Do we have the necessary commitment from a critical support group?
- Who will resolve certain problems?
- What lines of communications exist among interfaces?
- What currencies do we have to influence our stakeholders?
- Do we involve the right people from the ABC special interest group?
- How can we help to build the influence of the KLM team?
- How can we build the influence over resource Q?
- How can we strengthen communications across interface ITF?
- Whose agreement do we need to proceed?
- Who are the key decision makers for this release?
- Whose opposition should we expect?
- What relationship do we have with PQR?
- What alternatives do we have to resource X?
- How do we share these responsibilities with XYZ?
- How do these people evaluate our performance?

Disadvantages include the following:

- ▲ *Some time and leadership skills are required* for developing matrix inputs.
- ▲ *Considerable time and leadership are required* for mapping Stakeholder Matrix data into stakeholder influence grid, team development plan, or project success factor analysis [8].
- ▲ *Potential for conflict* over different perception of stakeholder profile.
- ▲ *Overreliance on the Stakeholder Matrix for PM.*
- ▲ *The matrix may lose relevancy* in changing project environment unless continuously updated.

Variations. Many variations exist in the structure and application of the Stakeholder Matrix. For example, matrix entries can be shown in a more or less quantitative way, such as using alpha or numerical weights, or graphical symbols. Stakeholders can be described "generically," as shown in Figure 10.4, or listed individually. The task matrix format is often being used to summarize stakeholder information. The largest variations are being seen in the graphical presentation of stakeholder relations, such as the stakeholder influence grid (shown previously in the chapter in the box) that derive from the matrix. These charts and graphs map stakeholder influences, management actions, project success criteria, currencies, and other parameters of the project environment in an attempt to build a project success model and to build a winning team. Yet another common variation is the rank ordering of stakeholder parameters [8, 9].

Customize the Stakeholder Matrix. The matrix should be structured to be conducive for capturing the most relevant information for developing the total team, its interfaces, and its stakeholder relations. A thorough understanding of the project environment and business strategy is necessary to define the most relevant matrix parameters that impact project performance. The value of the Stakeholder Matrix is in its application as a team-building tool. To be effective as a team development tool, members of the project community must see themselves as critical players in the project execution. The matrix and its derivative documents must highlight the visibility of these team members and help them to interact and solve problems.

Summary

The Stakeholder Matrix supports team planning and development effort with an emphasis on cross-functional interfaces and relationship building. The matrix helps in identifying the project stakeholders and their influence on the various project parameters, and as such, it becomes a catalyst for developing more comprehensive interface maps and team development plans. In addition, project leaders use the matrix to benchmark team interface performance and organizational development needs.

Stakeholder Matrix Check

Make sure that the Stakeholder Matrix is properly structured and effectively applied. With it, you should be able to

- Identify the elements for each matrix axis: (1) stakeholders and (2) project parameters; then group the elements into categories.
- Mark the degree of stakeholder influence on various project parameters.
- Indicate the degree of influence symbolically or numerically, without "micro-detailing."
- Develop the matrix in multiple steps: first a single-handed, broad overview of stakeholders and project parameters, then involvement of core team members and project interface personnel.
- The axes of the Stakeholder Matrix should be consistent with the actual project organization. Customization of the template shown in Figure 10.5 is absolutely necessary.
- Use the matrix as an information source and catalyst for developing more comprehensive interface maps and team development plans.
- Involve the total project team in the Stakeholder Matrix development and share the final document with all team members.
- Focus the matrix application on cross-boundary relations management, organizational learning, and improvement.

Skill Inventory

What Is the Skill Inventory?

The Skill Inventory is a tool for systematically identifying the skill sets needed by project team members and their leaders. The concept can be used for identifying or benchmarking work skills of any job category or organizational level. It can also be used for personnel assessment and development, team building and broad-scale organizational development. Skill Inventories are an effective starting point for project team planning, staffing, and startup. They span across the strategic domain of PM, connecting the operational area of teamwork with the resource infrastructure of the organization. The basic structure of the Skill Inventory is shown in Figure 10.6.

Skill Inventory for Project Personnel, Class: Project Manager				
Skill Categories and Components	**Definition, Application, and Impact**	**Criticality Rating** (1...4=Most)	**Capability Rating** (1...4=Most)	**Action Plan**
Technical Skills (Category I) Ability to manage the project and its technology Aiding problem solving Communicating with technical personnel Facilitating trade-offs Fostering innovative environment Integrating technical, business, and human objectives System perspective Technical credibility Understanding engineering tools and support methods Understanding technology and trends Understanding market and product applications Unifying the technical team				
Administrative Skills (Category II) Attracting and holding quality people Communicating effectively (orally and written) Delegating effectively Estimating and negotiating resources Measuring work status, progress, and performance Minimizing changes Planning and organizing multifunctional programs Scheduling multidisciplinary activities Understanding policies and operating procedures Working with other organizations				
Interpersonal & Leadership Skills (Category III) Ability to manage in unstructured work environment Action-orientation, self-starter Aiding group decision making Assisting in problem solving Building multifunctional teams Building priority image Clarity of management direction Communicating (written and oral) Creating personnel involvement at all levels Creating visibility Credibility Defining clear objectives Eliciting commitment Gaining upper management support and commitment Managing conflict Motivating people Understanding professional needs Understanding the organization				
Strategic/Business Skills (Category IV) Building alliances, coalitions, and cooperation Dealing with risks and uncertainties Leading in multifunctional leadership Motivating and inspiring others Negotiating and leveraging resources Strategic thinking, planning, and decision making Thinking as an entrepreneur Understanding the business environment Vision				

Figure 10.6 Sample Skill Inventory for a project manager.

Constructing a Skill Inventory

As shown in Figure 10.6, the Skill Inventory is a listing of skill set components for one particular class of personnel. The skill components are grouped into categories, breaking down the complexity of the listing and providing a better overview and analysis of the skill sets. For each skill component, the table provides space for recording a *criticality rating* and a *capability rating*. These are judgment factors of (1) how important each skill component is to team performance and project success and (2) the level of current capability with the particular class of personnel or person under consideration. While under special circumstances elaborate rating schemes can be developed, it is suggested to *keep the rating scale simple*. A four-point Likert-type scale, as shown in the following, may be sufficient for rating the skill criticality and capability level respectively, and is often more effective than an "over-designed" scoring system:

Rating/ Score	Criticality	Capability
1	Little or no importance	Little or no capability
2	Somewhat important	Some basic capability
3	Very important	Effective and proficient
4	Crucial to success	Highly effective and proficient

In addition to the criticality/capability scoring, the Skill Inventory table provides two columns for recording references to other documents, such as skill definitions, impact assessments and action plans for skill development.

First-Time Construction. When the Skill Inventory template is constructed for the first time, the specific skill categories and components must be defined for each class of personnel, such as project manager, senior designer, marketing specialist, and so on.

Subsequent Applications. Once established, the template can be used with a minimum of fine-tuning to fit a new project situation. As a guideline, it is recommended to use the four skill categories listed in the sample inventory of Figure 10.6 as a framework for defining the specific skill sets required for a given personnel class and project situation:

Skill Category I: Technical Skills

Skill Category II: Administrative Skills

Skill Category III: Interpersonal & Leadership Skills

Skill Category IV: Strategic/Business Skills

A general description of the four skill categories is given in the box that follows, "Grouping Skill Sets into Categories." In addition, Figure 10.7

illustrates the relative distribution of skills with increasing managerial responsibilities. The focus, dominance, and criticality of skills shifts from "technical" to "administrative" and eventually to "interpersonal/leadership" and "strategic/business" skills, as the person assumes increasing managerial responsibilities and moves to higher organizational levels. Taken together, the Skill Inventory shown in Figure 10.6 for a project manager, and also fitting for a typical project team, provides a good overall format and convenient starting point for developing or customizing the Skill Inventory, and developing project managers toward the high-performance profile as shown in the text sidebar "Leadership Profile of Project Managers."

Preparing Information Inputs. Both the skill set definition and scoring can be done individually or in groups. To be most meaningful and realistic, however, inputs for the Skill Inventory should come from a broad spectrum of the project community, from people who understand the specific skill requirements and can judge the current level capabilities. After the skill sets have been defined (or fine-tuned) under each of the four categories, the rating of each skill component and action planning can begin.

Inventory for Individual and Group Resources. Skill Inventories can be generated for and applied to

- ▲ Individual resources such as project managers
- ▲ Groups, such as R&D, product development group, or the whole project team.

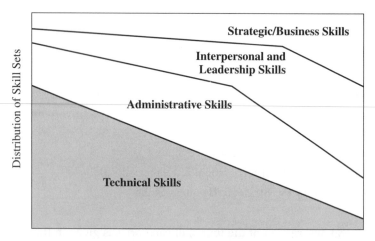

Figure 10.7 Relative distribution of skill sets.

Grouping Skill Sets into Categories

Today's complex, multifaceted, and technology-oriented projects require sophisticated skills in project planning, organizing, and integrating multidisciplinary activities. Organizations are changing into flatter, more flexible structures with higher degrees of power-sharing, distributed decision making, intricate communication channels, and a self-directed workforce. All of this requires more sophisticated skill sets at all organizational levels. Further, project leaders must be competent both technically as well as socially, understanding the culture and value system of the organization in which they work. The days of managers who get by with only technical expertise or pure administrative skills are gone.

There is no magic formula that guarantees success in managing technology-based organizations. But research shows consistently [10, 11] that high-performing PM professionals have specific skills that can be grouped in four principle categories:

- Technical skills
- Administrative skills
- Interpersonal skills and leadership
- Strategic/business skills

Technical Skills. Most projects involve specialized multifunctional work. Project leaders rarely have all the technical expertise involved in the project execution. Nor is it necessary or desirable that they do so. It is essential, however, that project managers and team leaders understand the associated technologies, their trends, the markets, and the business environment so they can participate effectively in the search for integrated solutions. Without this understanding, the consequences of local decisions on the total program, the potential growth ramifications, and relationships to other business opportunities cannot be foreseen by the manager. Furthermore, technical expertise is necessary to communicate effectively with the work team, to assess risks, and to trade off on cost, schedule, and technical issues.

Administrative Skills. Administrative skills are essential. Project leaders must be experienced in planning, staffing, budgeting, scheduling, performance evaluations, and control techniques. While it is important that project leaders understand the company's operating procedures and the available tools, it is often necessary for these senior people to free themselves from the administrative details.

Interpersonal Skills and Leadership. Effective leadership involves a whole spectrum of skills and abilities: clear direction and guidance; ability to plan and elicit commitments; communication skills; assistance in problem solving; dealing effectively with managers and support personnel across functional lines often with little or no formal authority; information-processing skills, the ability to collect and filter relevant data valid for decision making in a dynamic environment; and ability to integrate individual demands, requirements, and limitations into decisions that benefit the overall project. It further involves the manager's ability to resolve intergroup conflicts and to build multifunctional teams.

Strategic/Business Skills. These skills involve the ability to see the business as a whole and to direct and monitor its overall performance. It includes components such as understanding the nature of the industry and having a general manager's perspective and business sense. These skills relate to policy decisions, strategic thinking, and long-term objectives. These skills are critical for shaping organizational directions and leveraging organizational resources. Knowledge in the area of systems theory, strategic planning, and general business management provides the foundation and supporting tools of this skill area.

Leadership Profile of Project Managers

- Bringing together the right mix of competent people which will develop into a team
- Building lines of communication among task teams, support organizations, upper management, and customer communities
- Building the specific skills and organizational support systems needed for the project team
- Coordinating and integrating multifunctional work teams and their activities into a complete system
- Coping with changing technologies requirements and priorities while maintaining project focus and team unity
- Dealing with anxieties, power struggle, and conflict
- Dealing with support departments; negotiating, coordinating, integrating
- Dealing with technical complexities
- Defining and negotiating the appropriate human resources for the project team
- Encouraging innovative risk taking without jeopardizing fundamental project goals
- Facilitating team decision making
- Fostering a professionally stimulating work environment where people are motivated to work effectively toward established project objectives
- Integrating individuals with diverse skills and attitudes into a unified workgroup with unified focus
- Keeping upper management involved, interested, and supportive
- Leading multifunctional task groups toward integrated results in spite of often intricate organizational structures and control systems
- Maintaining project direction and control without stifling innovation and creativity
- Providing an organizational framework for unifying the team
- Providing or influencing equitable and fair rewards to individual team members
- Sustaining high individual efforts and commitment to established objectives

How Many Skill Inventories Do We Need? To minimize proliferation of forms and plans, usually only a small number of Skill Inventories are being used, covering broad segments of the project community. Often only one Skill Inventory is being generated and maintained for the entire project team. Alternatively, Skill Inventories are being generated for the project leader and each of the major team segments, such as R&D, development, and operations. For first-time users, a *single* Skill Inventory, covering the entire project team, is recommended to simplify the process, optimize user benefits, and minimize frustrations over administrative overhead and personal conflict. A simple four-point scale, such as shown in Figure 10.6, is suggested to record the *skill criticality* and *current capability* level as a basis for team development planning. Group communication tools, such as brainstorming, focus groups, and Delphi techniques can help in supporting the data gathering process.

Communicating the Skill Inventory. The Skill Inventory should be documented, either separately or as part the project plan, and updated as the project moves through its life cycle. It is important to ensure that the Skill Inventory is consistent with other documents, such as the project charter, task roster, task matrix, and customer relations management tools, used within the project organization.

Utilizing the Skill Inventory
When to Use. Skill Inventories are useful tools for identifying, assessing, and developing the project team and leveraging its resources. The tool can be especially beneficial during the early stages of project team formation and team planning. It provides a framework for identifying the skill sets and capabilities needed for effective teamwork with perspective on the criteria for project success. The Skill Inventory helps in the process of identifying and selecting team members, and in developing their skill sets as needed. The Skill Inventory also connects the team's capabilities with the business requirements, and as such, it provides a strategic link between the operational and strategic components of the company. In this context, the Skill Inventory provides additional perspective for developing project interface plans and stakeholder relations, and becomes an important part of the overall team development plan for optimizing team performance throughout the project life cycle.

Time to Develop. The Skill Inventory can be developed in two modes:

Mode I: First-Time Construction. When the Skill Inventory template is constructed for the first time, the specific skill categories and components must be defined for each class of personnel. This includes determining the skill sets for each skill category, evaluating the skill components for importance and capability, plus skill development planning. *Time required for developing one Skill Inventory:* 3 to 4 hours.

Mode II: Subsequent Applications. This includes fine-tuning of skill sets in the established template and evaluation of the skill components for importance and capability, plus skill development planning. *Time required for each Skill Inventory:* 2 to 3 hours.

Benefits. The Skill Inventory is an important tool for supporting the organization and development of multifunctional teams. Specifically, the Skill

Inventory can be used as a tool for assessing actual skill requirements and proficiencies. For example, a list can be developed, for individuals or teams, to assess the following for each skill component or set:

- ▲ Criticality of this skill to effective job performance
- ▲ Existing level of proficiency
- ▲ Potential for improvement
- ▲ Required support systems and managerial help
- ▲ Suggested training and development activities
- ▲ Periodic reevaluation of proficiency

Therefore, the Skill Inventory is useful in developing professional training programs and training methods. It also provides a framework for the team to identify the conditions necessary and critical for project success and has become an important contemporary tool for project team-building.

Advantages and Disadvantages. Advantages include the following:

- ▲ *Provides a template* for summarizing the skill sets needed for effective teamwork and project success.
- ▲ *Provides a scorecard* for assessing the criticality and capability of skill sets.
- ▲ *Provides a framework for building* the total project team and for managing stakeholder relations.
- ▲ *Provides a framework for connecting* operational resources with business strategy.
- ▲ *Helps in identifying project stakeholders.*
- ▲ *Creates an environment* conducive to team building.
- ▲ *Leads* to organizational learning.

Disadvantages include the following:

- ▲ *Considerable time and leadership required for constructing* initial Skill Inventory template.
- ▲ *Some time and leadership are required for developing* an inventory update.
- ▲ *Difficulty of standardizing skill sets,* criticality, and capability measures.
- ▲ *Potential for conflict* over different perception of skill sets and scoring.
- ▲ *Misuse of the Skill Inventory* for resource and priority negotiations.

Variations. Many variations exist in the structure and application of the Skill Inventory, ranging from the selection of different skill categories to scoring schemes that are more quantitative or more descriptive. Six major variations stand out regarding their applications and will be referenced more specifically.

- ▲ Single Skill Inventory. One single Skill Inventory is often used for the entire project organization. This is primarily done for keeping the process simple and for saving tool development time. However, a single inventory also minimizes the risk of "micro-analyzing" skill requirements, instead providing a more integrated, unified picture of what "we as the project team stand for and need to succeed."
- ▲ Aptitude testing. The Skill Inventory format can be applied for team and professional development. The example in Figure 10.8 shows a modified skill inventory, formatted as an Aptitude Test for Project Managers.
- ▲ Team readiness. The Skill Inventory format, such as the template shown in Figure 10.6, can be modified for conducting a team readiness test (for specific discussions and formats, see [12, 13]).
- ▲ Team performance measures. The Skill Inventory template, such as the template shown in Figure 10.6, can be modified for conducting a team readiness test.
- ▲ Post project review. The Skill Inventory format (shown in Figure 10.6) can be modified for conducting postmortem reviews, comparing startup against closeout criticality of skill sets and their developments.
- ▲ Project Management Maturity Model. The Skill Inventory can be integrated as part of a broader organization development effort. Templates such as shown in Figure 10.7 can be modified to become part of the Project Management Maturity Model (PM3) and its testing schemes toward PM success (for specific discussions and self-assessment formats see Kerzner [14]).

Customize the Skill Inventory. The template shown in Figure 10.6 provides a good general format and convenient starting point for customizing the Skill Inventory. When the Skill Inventory template is constructed for the first time, the specific skill categories and skill components must be defined for each class of personnel, as previously discussed in the "Constructing a Skilled Inventory" section. Additional customizing can be undertaken by integrating skill sets suggested by stakeholders peripheral to the operational core team, such as customers, contractors, and support groups. However, caution should be exercised not to expand the skill sets beyond 50 items. In the same context, an effort should be made to keep the measurements simple, following the guidelines suggested in the "Constructing" section.

Score 1...10	**Aptitude Test for Project Managers** Use a 10-point scale to indicate your agreement or disagreement with each of the statements (1=Strong disagreement; 5=Neutral; 10=Strong agreement)
	Personal Desire To Be A Manager 1. Managing people is professionally more interesting and stimulating to me than solving technical problems. 2. I am interested and willing to assume new and greater responsibilities. 3. I am willing to invest considerable time and effort into developing managerial skills. 4. I have an MBA (or am working intensely on it). 5. I am prepared to update my management knowledge and skills via continuing education. 6. I have discussed the specific responsibilities, challenges, and skills required with manager. 7. I have defined my specific career goals and mapped out a plan for achieving them. 8. I would be willing to change my professional area of specialty for an advanced managerial opportunity. 9. Managerial and business challenges are more interesting and stimulating to me than technical challenges. 10. Achieving a managerial promotion within the next few years is a top priority, very important to the satisfaction of my professional needs.
	Technical Knowledge 1. I understand the technological trends in both my area of responsibility and the business environment of my company. 2. I understand the product applications, markets, and economic conditions of my business area. 3. I can effectively communicate with my technical colleagues from other disciplines. 4. I can unify a technical team toward project objectives and can facilitate group decision making. 5. I have a systems perspective in my area of technical work. 6. I have technical credibility with my colleagues. 7. I can use the latest design techniques and engineering tools. 8. I recognize work with potential for technological breakthrough early in its development. 9. I can measure work/project status and technical performance of other people on my team. 10. I can integrate the technical work of my team members.
	Administrative Skills 1. I don't mind administrative duties. 2. I am familiar with techniques for planning, scheduling, budgeting, organizing, and personnel administration, and can perform them well. 3. I can estimate and negotiate resources effectively. 4. I can measure and report work status and performance. 5. I find policies and procedures useful as guidelines for my activities. 6. I have no problem delegating work even through I could do it myself even quicker. 7. I don't mind writing reports and preparing for meetings, and I do it well. 8. I can hand change requirements and work interruptions effectively. 9. I am good in organizing social events. 10. I can work effectively with administrative support groups throughout the company.
	People Skills 1. I feel at ease communicating with people from other technical and administrative departments. 2. I can effectively solve conflict over technical and personal issues, and don't mind getting involved. 3. I can work with all levels of the organization. 4. I am a good liaison person to other departments and outside organizations. 5. I enjoy socializing with people. 6. I can persuade people to do things which initially they don't want to do. 7. I can get commitment from people, even though they don't report directly to me. 8. People enjoy working with me and follow my suggestions. 9. I am being frequently asked by my colleagues for my opinion and for presenting ideas to upper management. 10. I think the majority of people in my department would select me as team leader.
	Business Acumen 1. I would be good at directing the activities of my department toward the overall company objectives. 2. I am productive. 3. I enjoy long-range planning and find the time to do it. 4. I am willing to take risks to explore opportunities. 5. I feel comfortable working in dynamic environments associated with uncertainty and change. 6. I would enjoy running my own company. 7. I consider myself more of an entrepreneur than an innovator. 8. In social functions, I tend to get involved more in business discussions rather than technical discussions. 9. I enjoy being evaluated in part, on my contributions to my company's business environments. 10. I have been more right than wrong in predicting the business environment.
_____	**Total Test Score (Divided by 5) = Normalized Score [ ]**

Figure 10.8   Aptitude test for project managers.

Summary

The Skill Inventory is a powerful organization development tool. It helps to identify, recruit, and develop the skills sets needed for bringing together high-performing project teams that can execute projects effectively according to established plans. The Skill Inventory can be customized to provide templates for extended PM developments, such as managerial aptitude testing, team readiness reviews, team performance measures, post project reviews, and PM maturity model assessments. The Skill Inventory offers business leaders a tool for going beyond the obvious and simple methods of developing PM talent. It helps them to recognize the highly multidisciplinary nature of skill requirements and the need for integrating professional skill developments into the business process.

Skill Inventory Check

Make sure that the Skill Inventory is properly structured and effectively applied. It should include the following components:

- The structure should list skill components grouped into skill categories.
- The skill components should be customized for the given project situation.
- Skill components should be judged/scored regarding their criticality and current capability.
- Senior project personnel should participate in the development of the inventory and its scoring.
- Skill requirements should not be micro-analyzed.
- The *inventory* should be used as an information source and catalyst for creating the team development plan.
- *Skill Inventory* development should involved the total project team, and the final document should be shared with all team members.
- The Skill Inventory application should focus on organizational learning and improvement.

Commitment Scorecard

What Is the Commitment Scorecard?

The Commitment Scorecard is a tool for systematically identifying the level of team commitment and buy-in to the project and its established objectives. The Commitment Scorecard is a powerful tool for diagnosing commitment deficiencies and developing stakeholder commitment throughout the project organization. It is especially effective for building commitment during the early stages of team formation and for maintaining commitment over the project life cycle. While virtually all managers recognize the critical importance of commitment for effective teamwork and project performance [15], only one in eight managers feels capable of effectively building and sustaining commitment from the various stakeholders of the project environment [16]. The concept of the scorecard is similar to other performance assessment tools, such as the business scorecard, team scorecard, or project performance scorecard. The specific measurements and their application to project team building were developed by the author, based on extensive field studies over the past ten years [16, 17]. The basic structure of the Commitment Scorecard is shown in Figure 10.9.

Constructing a Commitment Scorecard

As shown in Figure 10.9, the Commitment Scorecard is structured for ready use. It utilizes pretested statements for measuring the level of commitment to a project and its objectives. The statements are divided into *10 primary* and *25 secondary,* or *derivative, drivers,* identifying the most common components that support and drive individual, as well as team commitment. For each statement, the scorecard provides space for recording a *judgment score of agreement* from project stakeholders. The sample scorecard of Figure 10.9 provides two score sections: (1) project leader perception and (2) team perception. The number of columns can be expanded to include additional stakeholders and their judgments. A five-point Likert scale is suggested for scoring the statements, as shown in the following:

-2 = Strongly disagree with the statement

-1 = Disagree with the statement

0 = Neutral or can't tell

+1 = Agree with the statement

+2 = Strongly agree with the statement

Drivers Toward Commitment		Strength of Commitment		Analysis and Actions	
		Leader Perception	Team Perception	Diagnostics	Action Plan
Primary Drivers	Team enjoys mutual trust and respect	-2 -1 0 +1 +2	-2 -1 0 +1 +2		
	Team leader earned respect and credibility	-2 -1 0 +1 +2	-2 -1 0 +1 +2		
	Team leader can inspire, motivate and lead	-2 -1 0 +1 +2	-2 -1 0 +1 +2		
	Project plans and objectives are clear and agreed on	-2 -1 0 +1 +2	-2 -1 0 +1 +2		
	We enjoy working as a team on the project	-2 -1 0 +1 +2	-2 -1 0 +1 +2		
	We are proud of our project accomplishments	-2 -1 0 +1 +2	-2 -1 0 +1 +2		
	Accomplishments are visible and recognized	-2 -1 0 +1 +2	-2 -1 0 +1 +2		
	Our project is heading for success	-2 -1 0 +1 +2	-2 -1 0 +1 +2		
	Management is involved and supportive	-2 -1 0 +1 +2	-2 -1 0 +1 +2		
	Risks, anxieties and job uncertainties are low	-2 -1 0 +1 +2	-2 -1 0 +1 +2		
Summary Score: ∑(Prim) = add all column scores x 2.5					
Secondary Drivers	Project work provides autonomy, flexibility, and freedom	-2 -1 0 +1 +2	-2 -1 0 +1 +2		
	Work directions are clear	-2 -1 0 +1 +2	-2 -1 0 +1 +2		
	Work-related conflict is low and problems are being resolved	-2 -1 0 +1 +2	-2 -1 0 +1 +2		
	Personal conflict, politics, and power play is minimal	-2 -1 0 +1 +2	-2 -1 0 +1 +2		
	Project plans and requirements are achievable	-2 -1 0 +1 +2	-2 -1 0 +1 +2		
	Surprises and contingencies are minimal and handled effectively	-2 -1 0 +1 +2	-2 -1 0 +1 +2		
	Work and team processes are continuously improved	-2 -1 0 +1 +2	-2 -1 0 +1 +2		
	Communication channels work effectively	-2 -1 0 +1 +2	-2 -1 0 +1 +2		
	Project has measurable milestones and regular reviews	-2 -1 0 +1 +2	-2 -1 0 +1 +2		
	Performance measures and evaluations are fair and equitable	-2 -1 0 +1 +2	-2 -1 0 +1 +2		
	We enjoy good team spirit and morale	-2 -1 0 +1 +2	-2 -1 0 +1 +2		
	Work environment is pleasant and supportive	-2 -1 0 +1 +2	-2 -1 0 +1 +2		
	We have the necessary resources and support	-2 -1 0 +1 +2	-2 -1 0 +1 +2		
	Minimum project interruptions and changes	-2 -1 0 +1 +2	-2 -1 0 +1 +2		
	The projects provides opportunities for career advancement	-2 -1 0 +1 +2	-2 -1 0 +1 +2		
	Team has necessary skill sets and opportunity for training	-2 -1 0 +1 +2	-2 -1 0 +1 +2		
	We have pride in our project work	-2 -1 0 +1 +2	-2 -1 0 +1 +2		
	Project work is professionally stimulating and challenging	-2 -1 0 +1 +2	-2 -1 0 +1 +2		
	Team shares risks, provides mutual support and cooperation	-2 -1 0 +1 +2	-2 -1 0 +1 +2		
	Project administration does not interfere with technical work	-2 -1 0 +1 +2	-2 -1 0 +1 +2		
	Top management supports the project and its leadership	-2 -1 0 +1 +2	-2 -1 0 +1 +2		
	Team has winning image and attitude	-2 -1 0 +1 +2	-2 -1 0 +1 +2		
	No excessive time or performance pressures	-2 -1 0 +1 +2	-2 -1 0 +1 +2		
	Organizational environment is stable and predictable	-2 -1 0 +1 +2	-2 -1 0 +1 +2		
	Project is well connected to upper management and sponsor	-2 -1 0 +1 +2	-2 -1 0 +1 +2		
Summary Score: ∑(Secon) = Add all scores, each column					
Total Composite Score: ∑∑(Comp) = ∑(Prim) ∑(Secon)					

Scoring: -2 = Strongly disagree with the statement
-1 = Disagree with the statement
0 = Neutral, or can't tell
+1 = Agree with the statement
+2 = Strongly Agree with the statement

Summary Score for all *Primary* or *Secondary* Drivers:
Maximum = 50;   *Minimum* = -50

Total Composite Score for all Drivers:
Maximum = 100;   *Minimum* = -100

Figure 10.9   Commitment scorecard.

Commitment Is a Natural Characteristic of Project Teams and Can Be Developed!

The components that drive commitment do readily exist in many project environments. They are embedded in the project management process, such as planning, tracking, and reporting. The desire for commitment is stimulated by the work, its challenges, visibility, and accomplishments. Yet, this desire must be carefully cultivated, highlighted, and connected to the team's motivation [1]. Understanding the drivers and barriers to commitment, their sources and secondary influences, is an important prerequisite for gaining and maintaining commitment from the team and, hence, for leading project teams effectively toward desired results.

In addition to the judgment of agreement, the Commitment Scorecard provides two columns for (1) analyzing commitment problems and (2) action plans for strengthening commitment and team performance. Since space in the scorecard is limited, these two columns are reserved just for summary statements and references to other documents that articulate the diagnostics and action plans in further detail. The methods of calculating and interpreting commitment score points are discussed in the section that follows.

Preparing Information Inputs. Regardless of its application for assessing individual or team commitment, the first step is completing the questionnaire portion of the scorecard. Next, the summary scores need to be calculated and interpreted as a basis for further discussion, analysis, and development.

Completing the questionnaire. An individual or group judgment should be obtained on each scorecard statement. Especially for the assessment of *team commitment*, the involvement of task groups or the entire project team in interactive group sessions will produce the most meaningful and realistic results. It also helps to clarify and to build confidence, trust, and respect—factors that ultimately help to strengthen commitment. A simple five-point scale, such as shown in Figure 10.9, is suggested for recording the judgment directly on the scorecard. These scores, together with the composite calculations, become the basis for analysis and action planning. Flip charts, white boards, or Post-it notes are recommended for recording the diagnostics and action plans.

Calculating Commitment Scores. After all statements have been judged, a *composite score* can be calculated by adding up all individual statement scores. *Primary drivers weigh 2.5 times the value of secondary drivers.* Specifically, calculate for each column:

- ▲ Summary score for all primary drivers. Σ (Prim) $= \Sigma$ (all absolute primary scores) x 2.5
- ▲ Summary score for all secondary drivers. Σ (Secon) $= \Sigma$ (all absolute secondary scores) x 1.0
- ▲ Total composite score for all drivers. Σ (Comp) $= \Sigma$ (Prim) $+ \Sigma$ (Secon).

Interpreting Commitment Scores. Given the weighting, the judgment scores range between –50 and +50 for both primary and secondary drivers, and between –100 and +100 for the composite score. The following interpretation of the numerical composite score provides a guideline for the established level of commitment:

- ▲ Composite scores are negative. Virtually no commitment has been obtained from the people surveyed. Unclear project plans, fear of failure, and lack of incentives and interest are often the reasons for lack of commitment.

- ▲ Composite score of 0-50 points. Some commitment has been established, with much room for improvement. The situation is typical for a team that is in the storming stages of development or that has just entered the norming stage.

- ▲ Composite score of 51-75 points. This indicates a reasonably strong level of commitment. Yet further improvements via team development should be targeted.

- ▲ Composite score above 75 points. This indicates very strong commitment to the project and the team. Only 10 percent of all project teams (15 percent of all project leaders) tested reach this level of commitment. Effective leadership is required to refuel and maintain this high level of commitment.

Communicating the Commitment Scorecard. The Commitment Scorecard should be documented separately from the project plan and shared with the project team and stakeholders in management. The template of Figure 10.9 provides a convenient format for "quantitatively" summarizing the commitment status. A memo-report, together with the tabulated results of Figure 10.7, is suggested for communicating the survey findings, including analysis and recommendations for team development toward unified, committed high project performance. This report, together with its action plan, becomes a roadmap for developing commitment during the early stages of team formation. It also serves as a tool for continuing team development, helping to maintain, refuel, and sustain high levels of commitment.

Utilizing the Commitment Scorecard

When to Use. The Commitment Scorecard is a useful tool for assessing and developing the commitment of a project team and other members of the organization throughout the project life cycle. The tool is especially beneficial during the early stages of project team formation and team planning. It provides a framework for clarifying the issues associated with gaining and maintaining buy-in to established project plans and objectives, and for identifying and removing the barriers that impede individual or team commitment.

Time to Develop. The Commitment Scorecard development is a four-step process that can be completed by individuals or in groups:

Process Step	Individual Scorecard (Time Requirement)	Group Scorecard (Time Requirement)
Step 1: Judgment/Scoring of Statements	30 minutes	2 hours
Step 2: Calculation of Summary Scores	5 minutes	10 minutes
Step 3: Diagnostics/ Root-Cause Analysis of Barriers to Commitment	30-60 minutes	1-3 hours
Step 4: Recommendations/ Action Plan for removing Barriers to Commitment	30-60 minutes	1-3 hours

Benefits. The Commitment Scorecard facilitates the systematic identification of commitment by individuals and teams to the project and its established objectives. It provides a framework for diagnosing commitment deficiencies and developing commitment throughout the project organization. The Commitment Scorecard also helps in unifying the team behind the project goals and objectives, and helps to focus the organization on the conditions critical for project success, hence providing a bridging mechanism between the operational focus of the project environment and the strategic focus of the business enterprise. In addition, the Commitment Scorecard provides perspective for developing project interfaces and stakeholder relations, and becomes an important part of the overall team development plan for optimizing team performance throughout the project life cycle. In summary, the Commitment Scorecard provides

- Insight into to drivers and barriers toward commitments
- A framework for project stakeholders identification
- Quantitative measures of commitment
- A framework for systematically diagnosing deficiencies and developing commitment
- An opportunity for benchmarking and organizational learning
- Measures for the Project Management Maturity Model
- A focus on strategic goals and organizational objectives
- Perspective for developing project interfaces and stakeholder relations.

Advantages and Disadvantages. Advantages include the following:

- *Provides a scorecard* for assessing strength of commitment, including all of the above benefits.
- *Provides a template for summarizing* commitment analysis and development plan.

  ▲ *Offers a low-cost and low-risk organization* development process.

  ▲ *Helps in identifying project interfaces* and stakeholders.

  ▲ *Creates an environment* conducive to team building.

  ▲ *Leads* to organizational learning.

Disadvantages include the following:

  ▲ *Time and considerable leadership is required* for applying scoreboard to project team development.

  ▲ *Standardized statements don't fit all project situations;* a need for customization often exists.

  ▲ *Potential for conflict over different perception* of statements and scoring.

  ▲ *Inability to solve all commitment problems* with a scorecard.

Leadership Implications

Project leaders must understand the criteria and organizational dynamics that drive people toward commitment. Building commitment requires a leadership style that relies to a large degree on shared power and earned authority. An effective team leader is a social architect who understands the interaction of organizational and behavioral variables and can foster a can-do climate that is high on active participation and professionally stimulating interactions but low on anxieties and dysfunctional conflict. This requires skills in leadership, administration, organization, and technical expertise. It further requires the project leader's ability to involve top management and to build alliances with support organizations, ensuring organizational visibility, resource availability, and overall project support throughout its life cycle. These are some of the important criteria for gaining and sustaining commitment with project teams and other stakeholders of the project environment.

Recommendations for Obtaining and Sustaining Commitment

Despite the complexities of the commitment process and the differences among companies, certain characteristics of leadership and work environment appear to be most favorably associated with obtaining and maintaining commitment in team-based organizations. These conditions serve as bridging mechanisms for overcoming fears and anxieties that are normal and predictable when commitment is requested from an individual or project team. Some recommendations are discussed with focus on three categories: people-oriented influences; organizational process, tools, and techniques; and work- and task-related influences.

People-Oriented Influences. Factors that satisfy professional interests and needs seem to have the strongest effect on the ability to obtain commitment and eventually sustain commitment over the project life cycle. The most significant drivers are derived from the work itself, including personal interest, pride and satisfaction with the work, professional work challenge, accomplishments and recognition, as well as the trust, respect, and credibility placed in the team leader. Other important influences include effective communications among team members and support units across organizational lines, good team spirit, mutual trust, low interpersonal conflict, and personal pride, plus opportunities for career development, advancement, and to some degree, job security. All of these factors help in building a unified project team that can leverage the organizational strengths and core competencies effectively. These are the factors that seem to create an environment conducive to team commitment and ultimately high project performance.

Organizational Process, Tools, and Techniques. These influences include the organizational structure and technology transfer process that relies on modern project management techniques. Field research specifically points at effective project planning and support systems, clear communication, organizational goals and project objectives, and overall managerial leadership as important conditions for effectively gaining and sustaining commitment. An effective project management system also includes effective cross-functional support, joint reviews and performance appraisals, and the availability of the necessary resources, skills and facilities. Other crucial components that affect the organizational process, and ultimately commitment, are team structure, managerial power and its sharing among the team members and organizational units, autonomy and freedom, and most importantly, technical direction and leadership. Many of the variables related to organizational process are primarily under the control of senior management. They are often a derivative of the company's business strategy developed by top management. It is important for management to recognize that these variables affect the team's perception of the work environment, such as organizational stability, availability of resources, management involvement and support, personal rewards, organizational goals, objectives, and priorities, and they therefore have a direct effect on commitment.

Work- and Task-Related Influences. Commitment also has its locus in the work itself. Variables associated with the personal aspects of work, such as interest in the project, ability to solve problems, minimize risks and uncertainties, job skills, and experience, are significant in driving commitment. Hence, it is important for management to understand the personal and professional needs and wants of their team members and to foster an organizational environment conducive to these needs. Proper communication of organizational vision and perspective is especially important. The relationship of managers to staff and people in their organizations, mutual trust, respect, and credibility all are critical factors toward building an effective partnership between the project team and its management and sponsor organization.

Taken together, project leaders must be able to attract and hold people with the right skill sets, appropriate for the work to be performed. Managers must invest in maintaining and upgrading job skill and support systems, and promote a project climate of high interest and involvement, in order to influence the commitment process favorably.

Commitment Scorecard Check

Make sure that Commitment Scorecard is properly structured and effectively applied:

- The set of questions and their weights, as shown in Figure 10.7, should be considered as a "standard," and modified only as an exception, not a rule.
- The questionnaire should contain statements in two groups: primary and secondary drivers to commitment, each with different weights.
- Each survey instrument should contain only *one score column* to avoid confusion and bias.
- Each question should be carefully judged/scored after thoughtfully considering the question in the context of the given project situation.
- The survey findings, including analysis and recommendations, should be summarized in a standardized format.
- The scorecard survey results should be shared with all project team members.
- The scorecard survey results should be used as a roadmap and catalyst for continuing team development and organizational learning throughout the project life cycle.

Customize the Commitment Scorecard. The template shown in Figure 10.9 provides a good general-purpose format of the Commitment Scorecard. Customization of the questionnaire statements is recommended only after careful consideration, because the questionnaire and score weights suggested in Figure 10.9, are based on extensive field research (e.g., the statements explain over 70 percent of the variance of commitment in project teams; [16]). The probability of covering a larger spectrum of commitment variables with *substitute* statements is low. However, if the scoreboard user is willing to invest more time in the generation of the scoreboard, a *larger set* of statements will likely cover more commitment variables and provide further insight into the drivers and barriers of commitment for a given situation. In a different context, two suggestions for customization are being made. First, the scoreboard that is given to an individual or team as survey instrument should contain only *one* score column (that is the column with the numbers) to avoid confusion and bias. The summary board can contain all columns of all parties surveyed or just their composite scores. Second, the format of the memo-report, which summarizes the survey findings and includes analysis and recommendations for team development, should be specified regarding its structure, contents, and length. Such report standardization will benefit benchmarking and management actions, along with the overhead required for team development effort.

Taken together, the effective use of the Commitment Scorecard involves many organizational and managerial issues with considerable leadership implications, as summarized in the Text Sidebar. Part of these challenges involve the proper understanding of the criteria and organizational dynamics that drive commitment, and the managerial skills of negotiating commitments that align project requirements with the professional and personal needs of the team members. Specific recommendations for obtaining and sustaining commitment are summarized in the final Text Sidebar of this section.

Summary

The Commitment Scorecard is a powerful tool for assessing and developing commitment of a project team and other members of the organization. While most frequently applied during the early stages of project team formation and team planning, the scorecard is a useful and effective organization development (OD) tool for developing and sustaining commitment throughout the project life cycle.

Concluding Remarks

Four tools were covered in this chapter: Four-Stage Model of Team Building, Stakeholder Matrix, Skill Inventory, and Commitment Scorecard. All four tools support team planning, the process of identifying, organizing, and developing project teams.

Effective teamwork is a critical determinant of project success and the organization's ability to learn from its experiences and position itself for future growth. To be effective in organizing and directing a project team, the leader must not only recognize the potential barriers to high team performance work but also know when in the life cycle of the project they are most likely to occur. Effective project leaders take preventive actions early in the project life cycle and foster a work environment that is conducive to team building as an ongoing process.

The Four-Stage Model of Team Building provides an effective framework for team formation and development. It also provides an umbrella for integrating many of the tools and concepts that support team planning via customer relations management, skill building, interface management, and commitment. Building effective project teams involves the whole spectrum of management skills and company resources. Most importantly, managers must pay attention to human side. They must understand the complex interaction of organizational and behavioral variables. Through the interaction with all stakeholders, managers can gain insight into the critical functions and cultures that drive project performance and can identify those components that could be further optimized. Effective project leaders are social architects who understand the interaction of organizational and behavioral variables and can foster a climate of active participation and minimal dysfunctional conflict. They also build alliances with support organizations and upper management to ensure organizational visibility, priority, resource availability, and overall support for the multifunctional activities of the project throughout its life cycle. The tools presented in this chapter should help in the process of team planning and proactively facilitating control of the project startup process.

A Summary Comparison of Team Planning Tools

Situation	Four-Stage Model	Stakeholder Matrix	Skill Inventory	Commitment Scorecard
Provide template for team planning, organizing, and startup.	✓	✓	✓	✓
Identify and select project leader and prospective team members.	✓		✓	
Identify and develop required skill sets for project team and its leadership.			✓	
Deal with team formation and startup challenges.	✓	✓	✓	✓
Map cross-functional interfaces and build communication networks and alliances.		✓		
Identify project stakeholder and assess their influences and needs.		✓		✓
Build project ownership, cooperation, and commitment.	✓	✓		✓
Deal with conflict, power sharing, and organizational politics.		✓	✓	✓
Create environment conducive to high-performance team building.	✓	✓	✓	✓
Facilitate learning organization.	✓	✓		✓
Understand relationship between project operational objectives and strategic goals of enterprise.		✓	✓	✓

References

1. Burgess, R. and S. Turner. 2000. "Seven Key Features for Creating and Sustaining Commitment." *International Journal of Project Management* 18(4): 225–233.

2. Tuchman, B. W. and M. C. Jensen. 1977. "Stages of Small Group Development Revisited." *Groups and Organizational Studies* 2: 419–427.

3. Hershey, P. and K. Blanchard. *Organization and Behavior.* 1995, Englewood Cliffs, N.J.: Prentice Hall.

4. Olson, E.M., et al. 2001. "Patterns of Cooperation during New Product Development among Marketing, Operations and R&D." *Journal of New Product Development* 18(4): 258–271.

5. Keller, R. 2001. Cross-Functional Project Groups in Research and New Product Development. *Academy of Management Journal* 44(3): 547–556.

6. Zien, K. A. and S. A. Buckler. 1997. "From Experience: How Canon and Sony Drive Product Innovation." *Journal of Product Innovation Management* 14(4): 274–287.

7. Cohen, A. R. and D. L. Bradford. 1990. *Influence without Authority.* New York: John Wiley & Sons.

8. Tuman, J. 1993. "Models for Achieving Project Success through Team Building and Stakeholder Management" *The AMA Handbook of Project Management.* Edited by P. Dinsmore. New York: AMACOM.

9. Gray, C. and E. Larson. 2000. *Project Management.* New York: Irwin/McGraw-Hill.

10. Shenhar, A. and H. J. Thamhain. 1994. "A New Mixture of Project Management Skills." *Human Resource Management Journal.* 13(1): 27–40.

11. Shuman, J. and H. J. Thamhain. 1996. "Developing Technology Managers" Chapter 21 in *Handbook of Technology Management.* Edited by G.H. Gaynor. New York: McGraw-Hill.

12. Wellins, R. S., W. C. Byham, and J. M. Wilson. 1991. *Empowered Teams.* San Francisco: Jossey-Bass.

13. Robins, H. and M. Finley. 2000. *Why Teams Don't Work.* San Francisco: Berrett-Kohler.

14. Kerzner, H. 2001. *The Project Management Maturity Model.* New York: John Wiley & Sons.

15. Berman, S., et al. 1999. "Does Stakeholder Orientation Matter? The Relationship between Stakeholder Management Models and Firm Financial Performance." *Academy of Management Journal* 42(5): 488–506.

16. Thamhain, H. J. 2002. "Building Project Team Commitment" at PMI-2002 Conference. San Antonio, Tx.

17. Thamhain, H. J. and D. L. Wilemon. 1999. "Building Effective Teams for Complex Project Environments." *Technology Management* 5(2): 203–212.

Project
Implementation
Tools

Scope Control

Major topics in this chapter are scope control tools:

- ▲ Change Coordination Matrix
- ▲ Project Change Request
- ▲ Project Change Log

The purpose of these tools is to zero in on controlling the scope of a project in the course of project plan execution (see Figure 11.1). Closely coordinated with the schedule and cost control tools, the scope control tools help the project team get their arms around the sound scope changes and updated scope baseline. This information about the updates is used to report project progress and close down the project at the end of the implementation. Helping in this effort are team development, quality, risk, and other controls.

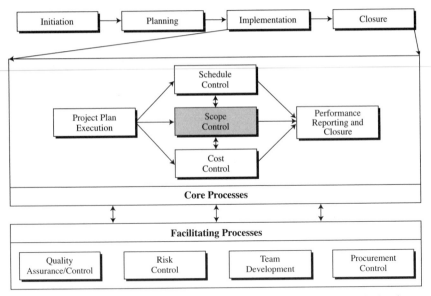

Figure 11.1 This is when and how scope control tools are used in the standardized project management process.

This chapter focuses on preparing practicing and prospective project managers to acquire the following skills:

- ▲ Become familiar with various scope control tools in the proactive cycle of project control.
- ▲ Select scope control tools that match specifics of their situation.
- ▲ Customize the tools of their choice.

The acquisition of these skills is critically important in successfully implementing the project and building the SPM process.

Change Coordination Matrix

What Is the Change Control Matrix?

Think of the Change Coordination Matrix (CCM) as a convenient roadmap to a destination called "project changes properly controlled." To serve this purpose, a CCM helps spell out steps in the change control process, identify actions to be taken, assign responsibilities for the actions, and coordinate those responsible [4]. By weaving other tools such as the Project Change Request (PCR) and Project Change Log (PCL) into the process, a CCM fully translates change control policy and rules of an organization into a practical change control procedure of a how-to-do nature (see Figure 11.2).

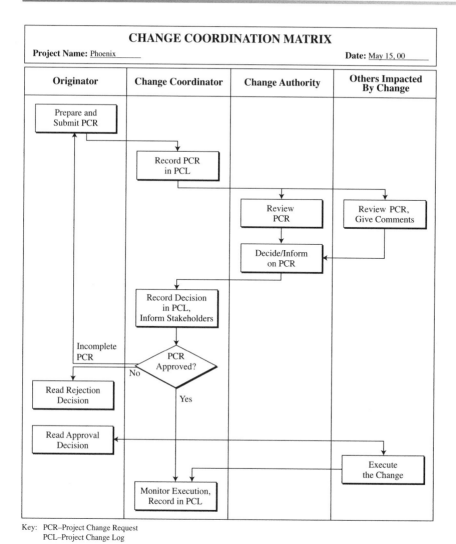

CHANGE COORDINATION MATRIX

Project Name: Phoenix

Date: May 15, 00

Originator	Change Coordinator	Change Authority	Others Impacted By Change

Prepare and Submit PCR

Record PCR in PCL

Review PCR

Review PCR, Give Comments

Decide/Inform on PCR

Record Decision in PCL, Inform Stakeholders

Incomplete PCR

PCR Approved?

No

Read Rejection Decision

Yes

Read Approval Decision

Execute the Change

Monitor Execution, Record in PCL

Key: PCR–Project Change Request
    PCL–Project Change Log

Figure 11.2 An example of the Change Coordination Matrix.

Developing a Change Coordination Matrix

Constructing a CCM that harmonizes steps, actions, owners, and their interactions in the change control business calls for a carefully crafted process that would minimize uncontrolled changes (see the box, "The Fear of Uncontrolled Changes: If You Have to Do It, Do it Early"). Following we describe the process pictured in Figure 11.2.

Prepare Information Inputs. Three inputs play a major role in furnishing information necessary for building a CCM:

▲ The change control plan (described in the "Scope Statement" section of Chapter 5) provides details of the project change rules that have to be incorporated in CCM steps.

▲ An appropriate coordinator must be appointed to conduct project change business (see the box, "Who Is Older: The Coordinator Or the Change Control Plan?").

▲ Scope baseline with Work Breakdown Structure, to which requested changes may be made. It is also presented in the "Scope Statement" section of Chapter 5.

Submit a Project Change Request. Naturally, anyone involved in a project, inside or outside the team, is welcome to submit a PCR, assuming the project scope is not frozen. Details about PCR are provided later in this chapter.

Record in the Project Change Log. After receiving a PCR, the coordinator records it in PCL and distributes copies (see Figure 11.2). Typically, copies go to the change authority (for example, to the members of the change review board) and the originator. Distributing copies to other people who can help evaluate the change is an even more efficient method. In that case, managers and specialists who may be impacted by the requested change may notice implications that the authority have missed.

Review Project Change Requests. The coordinator submits the change requests to the change authority. This is usually done in batches or in any other way that ensures the prompt review of the requests.

Decide on a Project Change Request. The change authority's responsibility is to make an approval or rejection decision, which is recorded in the PCL (Figure 11.2). When A PCR is rejected, a copy is stored in a master file. A copy is returned to the originator, explaining the decision and the grounds for rejection. If the PCR is approved, the coordinator makes the PCR official and sends it to those impacted by the change to carry it out. Also, the coordinator informs the originator. Finally, it is not unusual that the change authority finds a PCR incomplete, prompting the coordinator to request more information from the originator.

Who Is Older: The Coordinator or the Change Control Plan?

Project-driven organizations with well-established processes of change control typically have change control policies in place, often accompanied by a template change control plan. Such a plan, then, gets adapted for a specific project by its team, which already may have a change coordinator aboard, if it is large enough. In such cases, change coordinators tend to be well versed in the change business, spending all or lots of their work time on coordinating changes. Apparently, the change control plan here is older than the coordinator.

Other organizations that are on a steep change control learning curve offer an example of the opposite experience. Not having change control policies or template plans in place, they may expect a certain project team member to take on a part-time role of the coordinator, developing a tentative CCM that also serves as a project change control plan. The coordinator is thus older than the plan.

The Fear of Uncontrolled Changes: If You Have to Do It, Do it Early
Uncontrolled changes are known as project killers, because they

- Cause delay.
- Increase the cost.
- Damage morale and productivity.
- Spoil relationships among project participants.

Why all of this? First, more often than not, changes cause the work to be redone in the impacted activity. Second, any activity related to the impacted one needs to be changed as well. This means that the earlier you make the change, the less impact and less damage you inflict on the project. Early in the project, very few activities are done, and there is not much to redo, when a change hits. In contrast, with the bulk of activities completed in a product development project, a change may require a significant redesign work, repurchasing tooling, fixtures, and materials, remaking prototypes, and so on. Even such a bizarre change as replacing a team member during project implementation may set a team back by a month. The lesson to learn here is to think hard and make changes during the early stage. Once in implementation, sprint hard and do not make changes; it is too costly in every sense of the word.

Monitor the Change Implementation. The life of a change is not over with its approval, although challenges of its implementation may take time and diminish its visibility. This is why a coordinator needs to exercise disciplined monitoring of the change implementation status. A practical tool for this is a PCL.

Update Project Cost and Schedule. Approved changes may change basic project parameters of the total cost and completion date, necessitating their update. Again, a good tool for the update is a PCL, because it provides the necessary visibility. This, of course, does not preclude updating project cost, schedule, and scope documents. Since updates of this type are typically a reactive type of behavior, you might consider a proactive tone to CCM (see the box that follows, "Consider a Proactive Use of CCM").

Utilizing Change Coordination Matrix
When to Use. CCM should be developed in each project that is subject to change, possibly in the early planning stage before the scope is defined [5]. In that case CMM offers the change roadmap when the first Project Change Request emerges, setting the tone of change control throughout the project. This is especially vital in larger projects, in which the rule is often one of: the larger and more complex the project, the stronger need for CCM. Because of their lesser scope and complexity, smaller projects will face a very simplified change control process – the project manager often may be the only person available to handle the change. In circumstances like this, an informal, not documented but well-understood CMM would be more logical than having a formal one.

Consider a Proactive Use of CCM

Liberally accepting changes as they come our way is the essence of a reactive approach to change control. Such an approach may send a project into a downward spiral of uncontrolled scope creep. To avoid this, we can consider a proactive approach. For example, let's try to control a flow of changes by first anticipating how certain requested change can impact the project scope and, second, act on such anticipation for the sake of "tightening our belt" around the scope. To make this happen, we can borrow from the proactive cycle of project control (see the "Be Proactive: Five Questions of the Proactive Cycle of Project Control" box in the "Jogging Line" section of Chapter 12) and ask these questions each time we receive a project change request:

● What will be the variance between the baseline scope and the actual scope caused by the requested change?

● What are the issues causing the variance, which, in turn, is causing the request?

● What is the current trend — that is, the predicted scope at completion if we accept the proposed change?

● What new risks may pop up in the future and how could they cause further change to the predicted scope at completion?

● What actions could we take to eliminate the need for the requested change and other possible changes in the future, so we deliver the project as close to the baseline scope as possible?

The rationale for this approach is not to ignore a need for change. Neither is it to engage in a futile exercise of proactive control where it has no place. Rather, it is to create a culture that cherishes the prediction of scope, catches early signals of the possible scope creep, and acts to prevent it. Just as that old adage goes, forewarned is forearmed.

Time to Develop. Assuming that an organization has the change control policy and rules in place, preparing a CCM is a relatively straightforward activity. It won't take more than an hour, because drafting a CMM will pretty much be reduced to a quick adaptation of the policy, followed by the recruitment of the change authority and the coordinator. This time requirement may significantly grow when an organization lacks the policy and the project team needs to build a CCM from scratch. Even in smaller projects where the project manager may be one-person band, 15 to 20 minutes may be necessary to clarify an informal CCM process with his or her manager.

Benefits. CCM's value is in bringing order to the project change process. Through a methodical prescription of the sequence and arrangement of tasks and people in the process, this order significantly diminishes the possibility of problems, including scope creep, budget overruns, and project slippage [6]. Making the change process how-to-do's known in advance also helps direct behaviors of project participants, eliminating notorious perplexity of people involved in project changes.

Advantages and Disadvantages. CCM's advantages are in its

▲ *Visual impact*. Pictorial appearance of CCM makes it user-friendly and easy to follow, building on a natural ability of humans to better process graphical than the narrative information.

▲ *Step-by-step nature*. The visual impact is further strengthened by presenting the CCM as a sequence of steps, a format that adds more transparency and user-friendliness to the process.

On the other hand, to those busy project managers struggling to get through their tight schedules, asking for a CCM may look like one more thing competing for their time. This feeling may grow even stronger in organizations without the change control policy, where building one from scratch may easily be a real time drain.

Customize the Change Coordination Matrix. The description of the CCM that we have offered here clearly reveals its true potential for customization. It can be developed in various ways, allowing us to find one that best fits our needs. The following ideas may help you customize the CCM.

Summary

The focus of this section was the Change Coordination Matrix (CCM), a tool that maps steps in the change control process, identifies actions to be taken, assigns responsibilities for the actions, and coordinates those responsible. CCM should be developed in each project that is subject to change, possibly in the early planning stage before the scope is defined. Its value is in bringing order to the project change process. Through a methodical prescription of the sequence and arrangement of tasks and people in the process, CCM considerably lessens problems such as scope creep, budget overruns, and project slippage. Following we sum up the key highlights in constructing a CCM.

Customization Action	*Examples of Customization Actions*
Define limits of use.	Use formal CCM in all larger projects that are subject to change.
	Understand and apply an informal CCM in smaller projects that are subject to change.
Modify a feature.	Require the existence of the change review board (in large projects).
	Identify the project manager as the change coordinator (usual in smaller projects).
	Show CCM process from left to right, not top-down as we did.

> **Change Coordination Matrix Check**
> Check to make sure your CCM is properly constructed. The matrix should
> - Be based on necessary information inputs.
> - Include crucial steps in the project change control process.
> - Involve major change players such as the change authority, originator, coordinator, and others who are impacted by the change.
> - Determine who does what in the process.
> - Be designed with as much of a proactive approach as possible.

Project Change Request
Performing an All-Angle Evaluation of a Possible Change

The impact of project changes on its scope, schedule, cost, quality, and other matters may easily surpass the awareness or expertise of their originator [7]. As a consequence, the project may severely suffer and in certain cases even collapse (see the boxes that follow, "Scope Creep by Design" and "Scope Creep by Lack of Design: A Quick Interview with a Project Manager"). For this reason, it is very important to ensure that each change is evaluated in a disciplined and professional manner before receiving a go-ahead. It is the task of the Project Change Request to help perform an all-angle evaluation of proposed changes (see Figure 11.3).

Developing a Project Change Request

While the request appears as a simple form (be sure to keep it to a one-page format, with as much supporting detail as necessary), it does require doing homework, which begins with preparing solid information inputs.

Prepare Information Inputs. For larger projects, major inputs to the application of the Project Change Request are the scope baseline with WBS and the change control plan (see the "Scope Statement" section in Chapter 5). For smaller projects, inputs usually involve verbal directions on issues involved in the plan.

> **Scope Creep by Design**
> Scope creep, or uncontrolled change of scope, is often perceived as a major threat to projects [2, 3]. But one company faced with highly uncertain semiconductor fab projects controls the creep one change at a time. At the time of defining the scope historically prone to many changes, the company identifies a bucket of money equal to 10 percent of the project budget that is called AFC (allowance for change). Its purpose is to pay for the scope items that can't be predicted. To bring control to the process, every time such an item pops up, it is treated as a scope change and the project manager has to formally approve it. This very successful practice has helped the company control scope creep by the AFC design.

PROJECT CHANGE REQUEST

Project Name: Cage 01   **Request #:** 121

Details of change requested and their impact or scope/quality:
Test plans in work package 4.03.03 have to be changed and finished with 1 week delay.
Consequently, the succeeding work package 4.04.01 "Unit Test" originally scheduled to start on 5/07/01 has to start on 5/12/01.
No impact on deliverables of the scope definition, baseline scope remains unchanged.
No impact on quality specs.

Change request originated by: George Best   **Date:** 4/17/01

Reason for request:
Errors discovered in the test plans. Change in the plans a must.

Emergency action requested (if any): No

Type of change: Major (requires replanning effort) _____   **Minor** ___✓___

                Cost of investigating change _$105_____

Impact on project schedule: **Schedule estimate ref:** CC014
Total project duration will be extended by one week and project completed on 6/18/01.

Impact on project cost: **Cost estimate ref:** CC002
The change budget is $1,840.

Customer funded? No   **If yes, give customer approval reference:**

Change review board instructions:

Authorized (for change review board): Paul McCartney   **Date:** 4/24/01

Figure 11.3 An example of the Project Change Request.

Scope Creep by Lack of Design: A Quick Interview with a Project Manager

Q: What kind of project do you manage?

A: I've been tasked with developing a product platform for BAC business unit.

Q: Any major problems?

A: Quite a few! After six months in the project, a corporate VP demanded that the platform be changed to include two more business units. A few months later, we were ordered to add the portable version of the product.

Q: How did these changes impact the project?

A: The original scope had an 18-month completion for $2M budget. Right now we are three years in the project, budget is up to $4M, and we need six more months to complete.

Q: But you have a great product platform, right?

A: Yeah. But the real problem is that some of our customers got tired of waiting for the new platform products and took their business to our major competitor.

When Do You Start Applying PCR? As the conventional wisdom goes, a change should be proposed as soon as it is needed. We would like to add a little bit more precision here. The early conceptual stage of scoping the project is the time when there is no use for PCR as a change control arm. With the scope being only roughed out, it is impractical to attempt to control changes. At a later stage in the progress of scope definition, however, it is practical to start using the PCR tool. For example, a new product development project that has not done any piece of design work or has not issued any drawing would not need to apply PCR. Rather, it would start using PCR for changes that would [7] either

▲ Constitute a departure from the agreed design specification.

▲ Affect specifications issued by the engineering department for purchasing or production planning.

Considering these as examples, an organization still needs to create its own policies.

When Do You Stop Applying PCR? Although the question may seem less than meaningful, there will come such time when any change may impede the progress pace, potentially leading to costly waste and rework. An effective action in such situations would be to impose a scope freeze, a mandate that no changes will be considered unless an overriding reason exists. An example may be a customer-funded requirement to add a new safety feature to a product. A good understanding of how these issues of the change control plan and timing will be handled is necessary before we start dealing with specific change requests.

Describe Requested Changes and Their Impact on Scope/Quality. This self-explanatory action assumes that the description of the requested change will be sufficiently precise to provide an easy understanding about which part, deliverable, or work package should be changed and in what manner. We don't rule out a need for longer descriptions when complex matters are involved. Generally, the language used for this purpose should be as laconic as possible.

Explain the Reason for Change. Why do we want to make this requested change? Motivation may vary. To make sure it comprehends this motivation and prevents motivations that harm it, an organization requires the originator of change to state whether the requested change is of a must or want type. In their corporate lingo, a must change is one without which the project might perish. Consequently, it draws the appropriate attention and approval. A want change, on the other hand, would typically provide more elegance but not substance to the project product. At the same time, the want change may easily lead to the replanning effort, thus making itself a primary target for disapproval. This well-aged must/want change concept may be a good messenger to those who ignore or underestimate risks of scope creep. We may or may not like this concept, but to be on the safe side, ensure that every proposed change has its rationale.

Identify Emergency Changes. One of the major problems with regular, formalities-oriented treatment of PCRs is that they are unnecessarily slow. To overcome this problem, you need a feature in the system that ensures fast responses to urgent PCRs, which is marked in the PCR as "emergency action required." Its task is to inform the change authority that the requested change needs an urgent consideration and approval. This may mean that the authority will have to act promptly. Depending on the policy rules, they may choose to interact in a face-to-face meeting, on the phone, or via an intranet-based program (see the box the follows, "An Intranet Helps Treat Customer-Funded Changes as a Royalty"). It is of utmost importance that emergency changes should not be an excuse to circumvent evaluation of matters related to quality, performance, reliability, safety, or any other aspect. Building safeguards that enforce appropriate consideration of the change may be a useful aid.

Explain Impact on Project Schedule. It is very difficult to evaluate the impact of changes on schedules without a good network diagram. Simply that is where dependencies between activities are shown, helping us analyze how a change to one deliverable and its activities will affect dependent activities down the road. Still, more often than not, that is what occurs—a schedule impact assessment is made on the basis of a gut feeling. To safeguard against risks related to such assessments, rely on network schedules to produce reliable estimates, even if you are dealing with a small project and have no time to document in detail. An informal but good network analysis will benefit the small project.

Evaluate Impact on Project Cost. Requiring a cost or resource estimate for the proposed change is a well-meaning strategy to prevent cost surprises. That humans have a tendency to underestimate the cost is well documented in many books and papers. Several decades ago as well as today, missing an estimate by 20+ percent has not been unusual. What is unusual is the failure on management's part to take this tendency into account when evaluating a change request. Asking for a databased, bottom-up estimate when the change request is proposed is a sound management safeguard against the tendency. If the change is major, it is possible to go further and request an estimate from an independent source to compare with the estimate of the change originator, a practice frequently used by Intel project managers under the name of "shadow estimating."

Identify the Type of Change. Some organizations have a habit of approving major changes to the scope without ever referring back to the original (often called baseline) scope. This phony practice results in an ever-changing scope, making it a moving target. The risk is that moving scope, schedule, cost, and quality may inflate to the point of representing an all-new project that needs planning and implementation different from the current one. To prevent this, screen all proposed changes by identifying how they will impact the scope and quality. Should the change have minor, if any, impact, it may be treated as a minor corrective action, not impacting baseline scope, quality, cost, and schedule.

An Intranet Helps Treat Customer-Funded Changes as a Royalty

Customers made it very clear for Oscope, Inc. (OI): "We are not willing to put up with your long turnaround time of our major change requests!" Keen to retain customers, OI redesigned its change procedure, adding three vital changes. First, a rule was made that turnaround time for major customer requests will be 48 hours. These changes required a significant evaluation effort, including involvement of design, tooling, and manufacturing engineers, as well as marketing and purchasing experts. In addition, these people and their representatives were not collocated, so communication among them was time-consuming and slow. Second, in response, OI built an intranet site that significantly sped up the communication. Third, instead of using the consensus decision making in the change board, typically a slow method, one of the board members was nominated the approver while others were considered reviewers only. The system redesign led to a drastic improvement, helping OI retain the customers.

In contrast, we may identify a major change to scope of work, funding, or schedule requirements. That may warrant a replanning (or rebaselining) effort, including changes to the scope statement, WBS, schedules, budget, and resource allocation. To make all involved fully aware of such consequences, identify whether the requested change is a major or minor one. This is, of course, possible only after we assessed the impact of the change on scope, quality, cost, and schedule in the upper part of PCR.

Still another type of change may be identified and reviewed under this section of PCR. For time pressures in our busy schedules, some organizations require prioritization of change requests, imposing shorter turnaround times to more important changes. One organization has installed a change system with four layers of priority, where the change authority has to respond to a first priority change in 24 hours and to the fourth one within 30 days. This is a fine example to provide responsiveness where needed and discourage change when unwanted.

Miscellaneous Issues. Part of PCR are also such requirements as the change board decision on the request or the need for supporting detailed information. Since they are rather self-explanatory, their more detailed treatment here may be a redundancy.

Follow Up and Close Down. Approving a change request is no more than a prelude to the change action. And as with any other project action, this one will demand planning, very meticulous for major changes and less so for minor ones. Organizing for the change action and closely monitoring its implementation are prerequisites to its successful closure. To get there, however, make sure that all conflicts are resolved—and those may occur in the course of the change action—and finish all changes resulting from the change action before proclaiming it is finished.

Utilizing the Project Change Request

When to Use. Each project should use a PCR to screen and rationally evaluate the proposed changes [8]. In larger projects the PCR use needs to be as documented as possible to leave a trail of recorded changes. Because of their small budgets pressure, smaller projects typically cannot afford the documented PCR, warranting a unique, informal approach to PCR. First, all issues present in documented PCR will be on the table of small project changes as well. However, these issues will be discussed, established, shared, and applied verbally in any considered change. The sheer presence of such discipline surely cannot make up for the rigor of the documented process but can easily accommodate the lack of resources and hectic pace typical of a small project environment. In short, let small projects live with what is possible.

Time to Use. A few minutes—that is all you need to complete a PCR. But that is only the technicality part, which must be preceded by a substantial analysis that is a function of the size and complexity of the proposed change. In a small-scale change, like in the example from Figure 11.3, figuring out its scope, cost, and schedule consequences may take 15 to 30 minutes. At the other end of the spectrum are major changes to major projects, where a group of experts may spend a week or two to fully assess the requested change's impact on the project's business purpose and goals—the triple constraint of scope, time, and quality.

Benefits. Because of its process, structure, and content, PCR brings to management the value of making conscious decisions instead of letting everything happen by default. Naturally, then, as a result of management and project team members working together on a PCR, the decisions tend to be of a higher quality. When the decisions are based on or coordinated with other tools necessary to assess the PCR impact, project scope, cost, and schedule are kept in check and agreement with each other. Also, the apparent outcomes are documentation of changes, reduced project participants' confusion, better-controlled scope change, lower total cost, and fewer delays.

Advantages and Disadvantages. Major advantages of PCR are its:

▲ *Clarity.* PCR projects an image of clarity: "this is the change we propose to make and here is how it impacts the project's fundamentals." To be able to fathom the essence of a change this clarity is of huge value.

▲ *Brevity.* PCR intentionally excludes any detail beyond the fundamentals. If there is a need for more details to support the case of change, they will be relegated to an attachment. In that manner, the very essence of the change, its crux, will not be compromised by excessive detail.

Customization Action	*Examples of Customization Actions*
Define limits of use.	Use PCR in every major project for every proposed change.
	Apply a "verbal" PCR in small projects.
Add a new feature.	Include prioritization of PCR with the turnaround time (works for time-to-market or customer-oriented companies).
	Add information about major risks and risk responses (favored by high-tech projects).
Leave out a feature.	Leave out "emergency action required" feature if prioritization of PCR is required.
Modify a feature.	If in a time-to-market project, perform a very tentative estimate of cost impact of the requested change (so-called order of magnitude estimate).

The very best advantages will be even more appreciated when disadvantages of PCR's time-consuming nature are revealed. In large contractual projects, the number of changes may run in the thousands (the author's first-hand experience). Even with a well-designed and applied PCR system, this number of changes exacts a sizable chunk of PM time, translating into major expense. Although this expense gives a solid return on investment when it comes to controlling scope, cost, and schedule, many will enjoy a sigh of relief knowing that small projects are usually spared from too many time-consuming changes.

Customize the Project Change Request. Different types of projects may need different approaches to PCR. To opt for a generic PCR such as the one described here is to ask for trouble, which can be avoided by customizing PCR for specific project needs. The preceding hints may give you some ideas on how to customize it.

Summary
In this section we covered the Project Change Request—a tool that helps perform an all-angle evaluation of a possible change. Each project should use a PCR to screen and rationally evaluate the proposed changes. In that case, the process, structure, and content of the PCR help businesses make decisions of a higher quality, keeping project scope, cost, and schedule in check and agreement with each other. Also, the apparent outcomes are documentation of changes, reduced project participants' confusion, better-controlled scope change, lower total cost, and fewer delays. Finally, the following box highlights key points in developing the request.

Project Change Request Check

Check to make sure you developed a proper project change request. The request should

- Be based on information inputs from the written (for larger projects) or verbal (for smaller projects) change control plan.
- Describe the proposed change, its reason and impact on scope/quality.
- Identify when the change request needs an emergency treatment.
- Assess the impact of the proposed change on project schedule and cost.
- Identify the type of change.

Project Change Log
What Is a Project Change Log?

Project changes may not come in small numbers; rather, they may proliferate. This creates a need for recording, numbering, and coordinating the flow of project changes (see Figure 11.4). Monitoring the change process in this manner is enabled by the Project Change Log. Administered by a coordinator, a PCL records each change request and assigns it a number, making sure the decision about it—whether it has been approved or rejected by the change authority—is recorded as well. When a request is approved and the change implemented, that information becomes part of a PCL.

PROJECT CHANGE LOG

Project Name: Aquarius

Sheet Number: 1 of 5

Project Change Request No.	Submitted By	Brief Description of the Requested Change	Date of Submission	Approved?	Issue Date	Complete?	Change's Cost/Delay	Project Cost/Completion
10	Peter Tan	Printer upgrade	4/27/01	Yes	5/04/01	Not Yet	$1,846 1 week	$312,000 8/01/01

Figure 11.4 An example of the Project Change Log.

Keeping a Project Change Log

Forms such as PCLs have a major advantage in that they look simple to complete. Their appearance of a simple spreadsheet (see the box that follows, "It Only Takes a Spreadsheet, Even If It Is Big"), however, fails to inform us that it takes some information and energy spent in an orderly sequence of steps to produce a meaningful entry to the PCL. Following is a description of these steps.

Prepare Information Inputs. There are four major inputs to a disciplined and sound process of running a PCL:

▲ The change control plan (see the "Scope Statement" section in Chapter 5) provides a full understanding of the project change rules, which are necessary to direct the log.

▲ An appropriate coordinator must be appointed to administer the PCL.

▲ The Change Coordination Matrix supplies a procedure of steps that a change goes though, something no PCL can function without.

▲ The Project Change Request is a subject of administration in a PCL.

Record the Submission of the PCR. When the coordinator receives a PCR from the originator, his or her job is to issue a serial number for PCR and record the name of the originator, with a brief description of the proposed change.

Enter the Submission Date. Recording the date when the PCR was registered in the PCL helps maintain the schedule of change coordination. Reckoning from the date, we know how many days are expected for subsequent steps in processing the request (see the "Change Coordination Matrix" section earlier in this chapter).

It Only Takes a Spreadsheet, Even If It Is Big

Final accounting is a painful part of any contractual project, especially if there were lots of approved project changes. It is even more painful when there is no project change log, as was the case in a project that approved several hundred changes over a yearlong project execution. Because of several changes in the project manager position, the log was never established. Then, at the end of the project, it took the project team several months and thousands of dollars of their time to track down all requested, approved, and rejected changes to include them into a negotiated final accounting with the owner. Eventually, the team learned the hard way a very simple lesson: it only takes a spreadsheet to have a change log. Even if it is a long one because of many changes, it is still just a spreadsheet.

What Do You Do with a Change That Was Implemented but Never Requested?

Alan DeFazio was a computer engineer with no prior experience in contractual project work. When he learned that their project's computer vendor went belly up, he simply ordered better and more expensive equipment from another, more reputable vendor. After all, that's what he has done many times for his own company's internal needs. "This will make everyone happy. Everybody likes better computer stuff," he thought to himself. Four months later when the equipment was delivered, everyone on the project team was more than mad at Alan. His project manager was screaming, "Alan, why did you change computer specs without going through the change request procedure?" Even worse, the owner's project manager refused to pay the price differential between the original and new equipment, saying "I have no money in the budget and to get it I have to go beg my chief financial officer. No way!" The epilogue? After several months of frustration, the owner approved the change, helping Alan and his project manager save their jobs. The moral to the story: Having a change procedure in place means nothing unless you train people to use it.

Record If the PCR Is Approved and When Issued for Execution. The decision about the request may go both ways. If it is not approved, write "No" in the "Approved" column of the PCL and cross out the entry, making sure it is still readable. In the opposite case, when PCR is approved, write "Yes" in the column and fill in the column "Issue Date," recognizing that the change order is now officially issued for execution.

Track the Change Implementation. Once they get to be executed, changes may fall through the cracks and disappear from the project radar screen. To avoid this, the coordinator will deem any change incomplete (therefore, "Not Yet" in the "Complete" column in Figure 11.4). Once complete, "Yes" will be entered in the column. Although it sounds paradoxical, sloppy change coordination may easily fail to catch a change that was never requested but implemented (see the box above, "What Do You Do with a Change That Was Implemented but Never Requested?"). Should this happen and if it doesn't hurt the project completion, put that change in the change coordination loop as soon as possible.

Keep a Running Total. Certain changes will impact the project cost and completion date. To know the true project budget and completion date, we need to add cost and completion date details for all approved changes to keep a running total.

Utilizing the Project Change Log
When to Use. PCL should be used in each project that is subject to change. The size and complexity of a project, again, will have a say in how PCL is administered. Unlike the emphasis on a disciplined recording of all requested changes in larger projects, smaller projects are likely to build an informal PCL approach (see, for example, the box that follows, "Small Projects: Running a Change Log on a Palm Pilot"). When changes come in small numbers, a PCL may not be of much help.

> **Small Projects: Running a Change Log on a Palm Pilot**
>
> Kelvin Peak loved his job of managing seven or eight information systems projects simultaneously. He didn't care that it involved 12-hour days. He simply loved the fast pace and lots of changes on the go. What he didn't love was tracking all those changes. Not because he is an anti-documentation type of guy, but because his busy schedule was so packed with action items that it couldn't accommodate any extra time for change log. One day he ran across a Palm Pilot. This was a big sigh of relief, since Kelvin could now use it to track changes, spending a minute or two of his daily time. When a colleague showed him how to connect the Pilot to Excel and instantly turn his change log into a spreadsheet accessible to his team members, Kelvin was sure, "That's it! That's what I need!"
>
> (We are not specifically stating that this is the best way to run a change log. But it is one way, and it works.)

Time to Use. Keeping a tally of requested and implemented changes may not be a favorite task in projects, perhaps because PCL's appearance of a spreadsheet may not be able to stir excitement. Balancing the act may be a consolation that it takes a few minutes to make an entry in the PCL. Adding up the minutes for many entries in larger projects will not be a significant time burden when spread over the course of the project.

Benefits. In projects fraught with changes, PCL's role of a repository of changes provides a good oversight of all requested, rejected, approved, ongoing, and complete changes. The value of this information is in its potential to prevent both cost losses and project delays, by indicating to the decision makers the impact of major changes on the project budget and schedule. Such clarity is bound to trim down the confusion often seen in situations when PCL is absent.

Advantages and Disadvantages. Advantages of a PCL is its

▲ *Clarity.* PCL looks like a model of transparency, offering the fundamental but brief information on all changes in the project in one place. For such clear oversight of changes to be maintained, any unnecessary detail is left out.

▲ *Simplicity.* A glance at a PCL is perhaps enough to be able to interpret it, and even possibly keep it. To busy project people, simplicity is a productivity booster.

The red-tape impression that PCL's appearance creates in some project participants is probably its major disadvantage. As a result a resentment toward maintaining a PCL is not unusual and may be a major threat to good and cost-effective oversight of project changes.

Customization Action	Examples of Customization Actions
Define limits of use.	Use PCL in all projects subject to change.
Add a new feature.	Add a column indicating the incremental cost of a requested change (useful for cost-conscious organizations).
	Add a column for keeping a running total of the project price (used in contractual projects).
Modify a feature.	Highlight major changes impacting the total project cost and completion date.

Customize the Project Change Log. Although a PCL's simplicity may not induce a flurry of creativity when it comes to its customization, we still should consider some changes to the standard format described. These changes should help better reflect your own needs for overseeing project changes by means of a PCL. The ideas below are designed to help you customize the scope and direction of a PCL.

Summary
Presented in this section was the Project Change Log, a tool with the purpose of recording, numbering, and coordinating the flow of project changes. A PCL should be used in each project that is subject to change. Unlike the emphasis on a disciplined recording of all requested changes in larger projects, smaller projects are likely to build an informal PCL approach. In both cases, a PCL provides a good oversight of all requested, rejected, approved, ongoing, and complete changes. Such clarity is bound to decrease the confusion often present when PCL does not exist. Tailoring PCL to specific project needs adds more value to its users. To sum up, we highlight key points in establishing the log.

Project Change Log Check
Check to make sure your project change log is properly configured. The log should

- Be based on necessary information inputs.
- Number, briefly describe, and date the requested changes
- Identify whether the change request is approved or not.
- Record the date the change request is approved and issued for execution.
- Record when the change was executed.
- Keep a running total of the project cost and completion date.

Concluding Remarks

The three tools reviewed in this chapter—Change Coordination Matrix, Project Change Request, and Project Change Log—are building blocks of the scope control (see the summary comparison that follows). Each is designed to meet a different, unique need in preventing the scope creep. Building a roadmap that defines the scope control methodology with a proactive focus is the essence of the Change Coordination Matrix. A Project Change Request, a key component of the methodology, secures control of each requested change. Finally, a Project Change Log leaves a trail of the requests. As complementary as they are, each one can be used without the other two, formally in larger projects or informally in small projects.

A Summary Comparison of Scope Control Tools

Situation	Favoring Change Coordination Matrix	Favoring Project Change Request	Favoring Project Change Log
Provide roadmap for managing changes	√		
Provide proactive approach	√		
Control individual change requests		√	
Track status of individual change requests			√
Small projects	√ Informally	√ Informally	√ Informally
Large and complex projects	√	√	√
Projects with lots of changes	√	√	√
Projects with few changes		√	
Need to use the tool informally	√	√	
Need to use the tool formally	√	√	√
Take little time in small projects, longer time in large projects	√	√	√

References

1. Harrison, F. L. 1983. *Advanced Project Management*. Hunts, U.K.: Gower Publishing Company.

2. Cleland, D. I. and D. F. Kocaoglu. 1983. *Engineering Management*. New York: McGraw-Hill.

3. Meredith, J. R. and S. J. Mantel. 1989. *Project Management*. 2d ed. New York: John Wiley & Sons.

4. Ra, J. W. and J. R. Hemsath. 2001. "Web-Based, Real Time Protocol for Management of Change in a North Slope Oil Exploration and Production Operation" at Portland International Conference on Management of Engineering and Technology. Portland, Oregon.

5. Kerzner, H. 2000. *Applied Project Management*. New York: John Wiley & Sons.

6. Hed, S.R. 1973. *Project Control Manual*. Geneva: Sven R. Hed.

7. Lock, D. 1990. *Project Planner*. Hunts, U.K.: Gower Publishing Company.

8. Klien, R. L. and A. H. Institute. 1986. *The Secrets of Successful Project Management*. New York: John Wiley & Sons.

Schedule
Control

There can't be a crisis next week. My schedule is already full.

Henry Kissinger

Major topics in this chapter are schedule control tools:

- ▲ Jogging Line
- ▲ B-C-F (Baseline-Current-Future) Analysis
- ▲ Milestone Prediction Chart
- ▲ Slip Chart
- ▲ Buffer Chart
- ▲ Schedule Crashing

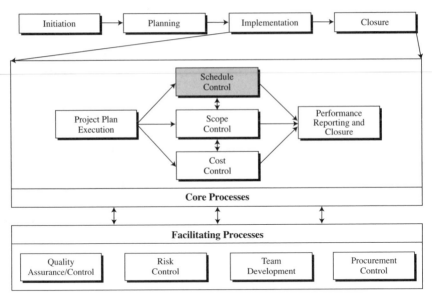

Figure 12.1 The role of schedule control tools in standardized project management process.

With these tools we will step into schedule control, doing our best to keep in check project schedule updates caused by the project plan execution. Coordinated with tools of scope and cost control, the updates become an important input into the performance reporting and closure of the project (see Figure 12.1). In that effort, a significant role belongs to facilitating processes of the project implementation phase. The purpose of this chapter is to help practicing and prospective project managers accomplish the following:

▲ Learn how to use various schedule control tools in the proactive cycle of project control.

▲ Choose schedule control tools that match their project situation.

▲ Customize the tools of their choice.

Mastering these skills is crucial to project planning and building a standardized PM process.

The Jogging Line
What Is the Jogging Line?

In a traditional sense, the Jogging Line spells out the amount of time each project activity is ahead or behind the baseline schedule. In that manner, the line indicates the fraction of completion for each activity to its left, and what remains to be completed to its right (see Figure 12.2). In recent, more innovative applications, the line is viewed as a step in the proactive management of the schedule. In particular, the amount of time each activity is ahead or behind the baseline schedule is used to predict the project completion date and map corrective actions necessary to eradicate any potential delay.

Constructing a Jogging Line

Prepare Information Inputs. To build a Jogging Line, you need quality information, including:

- ▲ The baseline schedule, whether a Gantt Chart or Time-scaled Arrow Diagram (TAD)
- ▲ Performance reports or verbal information about actual performance
- ▲ Change requests.

The first two components provide information on the planned progress (baseline) and the actual progress (report or verbal information) that the Jogging Line compares to establish where the project stands. Since the changes in the course of project implementation may occur such that they extend or accelerate the baseline, it is beneficial to take them into account when considering schedule progress and predicting completion date.

Work Packages/Tasks	Timeline												
	JAN	FEB	MAR	APR	MAY	JUN	JUL	AUG	SEP	OCT	NOV	DEC	JAN
1.01 Select Concept			▮										
1.02 Design Beta PC				▮									
1.03 Produce Beta PC				▮									
1.04 Develop Test Plans				▮									
1.05 Test Beta PC						▮							
2.01 Design Production PC								▮					
2.02 Outsource Mold Design							▮						
2.03 Design Tooling							▮						
2.04 Purchase Tool Machines										▮			
2.05 Manufacture Molds										▮			
2.06 Test Molds										▮			
2.07 Certify PC											▮		
3.01 Ramp Up											▮		

Figure 12.2 An example of the Jogging Line for a project.

Be Proactive: Five Questions of the Proactive Cycle of Project Control

Schedule control is no more than an application of the proactive cycle of project control (PCPC). It can be performed by answering the following five questions:

1. What is the variance between the baseline and the actual project schedule?

2. What are the issues causing the variance?

3. What is current trend—the preliminary predicted completion date if we continue with our current performance?

4. What new risks may pop up in the future and how could they change the preliminary predicted completion date?

5. What actions should we take to prevent the predicted completion dates from happening and deliver on the baseline?

The purpose of the first question is to establish the variance—that is, whether the project or a specific activity is ahead of or behind the time plan (for more details about the schedule variance, see Chapter 14). Understanding what issues cause the variance is what the second question is after. Given current variance and productivity and current issues, the third question strives to identify where the project (activity) schedule may end up in the future. But since the future is unknown, it is better to explore now what new risks may pop up in the future and further endanger the schedule completion. This is the focus of question number four. Finally, question five nails down the most important step: act, act, act! Or in other words, it specifies how to act in order to crash all variances, disturbances, and derailments and come out on top, reaching the schedule goal.

Every review of schedule progress, no matter what tool is used, should be driven by these five questions of schedule-oriented PCPC. Do it every time, whether reviewing the schedule for an activity, milestone, or entire project. Doing it this way is about reaching the heart of schedule control—being proactive, always looking into the future.

Do a Rehearsal with Activity Owners. Like in the days of managing-by-walking-around, we suggest the project manager first see each activity (or work package) owner privately, one-on-one. Ask about

▲ The progress of tasks that the activity comprises (for more details, see the box "When Assessing the Actual Status, Go for Satisficing" later in this chapter)

▲ Whether the progress differs from the plan, and if so, what causes and issues led to it

▲ When they anticipate finishing the tasks

▲ What can be done to finish them as planned; what help you can offer

All these questions are about the constituent tasks of an activity (work package). Now is the time to ask how the answers to the task questions will translate into the activity progress—ahead or behind the plan, predicted completion date, and major corrective actions. Essentially, a meeting like this is a rehearsal for the project progress meeting that will draw the Jogging Line. Its fundamental purpose is to train and prepare the activity owners for the project progress meetings by using a proactive cycle of project control (see the box on page 384, "Be Proactive: Five Questions of the Proactive Cycle of Project Control"). Sure, this looks like a time-consuming rehearsal for people with a busy work schedule. But probably one or two rehearsals will be necessary for the activity owners to learn how to prepare for the progress meetingAfter that, the meeting may be a few minutes long or it can be done over the phone or via e-mail. Or, if the activity (work package) managers learned how to prepare, no rehearsals would be necessary.

Call the Progress Meeting. Progress meetings instill the discipline and regularity in reviewing strides that you make in the project. They should be called on a regular basis—once a month for a long project or once a week for a short project, for example. The meetings may be formal, sit-down meetings with an official scribe, in which case, a clear and timed agenda should be communicated beforehand to the attendees. Or, in a high-tech organization, long on workload and short on time, it is perfectly appropriate to stage short, on-the-go, stand-up meetings. Again, the key to it is regularity and meeting discipline.

Proactive Cycle of Project Control and Deming's PDSA Cycle

The five questions of PCPC are in perfect harmony with Deming's plan-do-study-act, a never-ending circular approach to improvement (see the box titled "How Are Quality Improvement Map and PDSA Cycle Related?" in Chapter 14). Once the schedule baseline is established in the plan step and project work is being carried out in the do step, PCPC questions one, two, three, and four come into play in the study step (see Figure 12.3). Here the schedule variance is established, its cause determined, and trend forecasted based on current issues and future risks. Then, in the act step, question five leads to the identification of connective actions, which will be planned for and implemented in the plan-do steps of the next PDSA cycle.

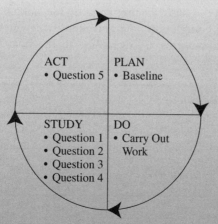

Figure 12.3 Proactive cycle of project control is part of the plan–do–study–act cycle.

Ask about the Activity's Actual Progress. The same five questions of PCPC that were asked in the rehearsal with the activity owners should be used again to assess their progress. For clear understanding of the relationship between PCPC and Deming's Plan-Do-Study-Act cycle, see the box on page 385, "Proactive Cycle of Project Control and Deming's PDSA Cycle."

Where there are interfaces and dependencies between activity owners, focusing on them is crucial, since that is where the potential seams may appear on the project. Depending on the culture of the organization, activity owners can bring written reports or just verbal information to the meeting.

Draw the Jogging Line. Drawing the line includes several steps:

- Begin with the project baseline schedule.
- Mark on the schedule's calendar the date of the progress meeting, whether it's called data date [3] or, more informally and also more traditionally, the reporting date.
- From that date, draw a line vertically down until reaching the first activity in the schedule. The activity owner will tell how many days the activity is ahead or behind the baseline. It is of vital importance that their information is reliable. If that's not the case, see what may happen in the box that follows, "Garbage In, Garbage Out."
- Draw a horizontal line to the left off the data date for as many days as the actual schedule is behind. Or, if you are ahead of the schedule, draw the horizontal line to the right off the data date for as many days.
- At that point, draw a vertical line crossing the first activity.
- Repeat the whole exercise with all other activities.
- When vertically crossing the last activity, draw the line horizontally back to the data date and then turn vertically down. In this way, the Jogging Line begins from and ends at the data date.

Predict Project Completion Date. Knowing how much each activity is ahead of or behind the schedule prepares you for making an educated forecast of the project completion date and identifying early actions to finish on time. Details about this are described in the box "Forgetting Trend Analysis: Déjà Vu?" later in this chapter.

Garbage In, Garbage Out

"In a progress meeting, an activity owner told me his deliverable was three weeks behind the schedule. A week later, he reported that he caught up with the deliverable and was right on schedule. Knowing that he was the only person working on the activity, I polled several experts: 'how many hours would it take to complete the catch up work?' The answer was 'about 150 hours.' You think he could do that? No! He is not on top of things, and submits inaccurate reports. Tell me, then, how can I be proactive and predict the project completion date?" This is a story we heard from a project manager in a leading software firm. The moral to the story: poor activity reports, poor trend forecast.

Utilizing the Jogging Line

When to Use. Large and small as well as complex and simple projects are a fertile ground for the Jogging Line's use. In such situations, when it comes to tracking progress only, the line is equally a nice fit with both the Gantt Chart and TAD. When you are relying on the Jogging Line for its innovative forecasting use in larger and complex projects, it is easier to apply it in a TAD (see the box that follows, "Using a Jogging Line in a Project War Room is a Blast"). Because TAD includes dependencies between activities, it is easier and more reliable to translate the amount of time each project activity is ahead or behind the baseline schedule into the projected completion date for the project. This is not the case when working on small and simple projects. There, having the dependencies is not an issue, and the Gantt Chart will suffice. At any rate, the Jogging Line may be used in all progress reviews, whether formal or informal.

Time to Develop. A reasonably skilled and prepared project team can prepare a Jogging Line for a 25-activity Gantt Chart or TAD in 15 to 30 minutes. As the number of team members grows, so does the necessary time.

Benefits. The value of the Jogging Line is in both its historic and forecasting power. The former means that it accurately tells an activity owner and the project team the history of the activity's progress. Using the history, then, of course, they can forecast future schedule trend and strategize actions to deliver the project as planned. For more details about trending see the box titled "Forgetting Trend Analysis: Déjà Vu?" later in this chapter.

Advantages and Disadvantages. The major advantages of the Jogging Line is that it is

- ▲ *Visual*. It vividly indicates the work done, supplying the project team with an invaluable communication means.
- ▲ *Simple*. In a matter of minutes, almost any project participant can read and draw the Jogging Line.
- ▲ *Proactive*. When used for predicting trends, the Jogging Line helps build an anticipatory mindset, equipping the project team to act in advance to combat an expected difficulty.

> **Using a Jogging Line in a Project War Room Is a Blast**
>
> There is something ritual about major project progress meetings in Southern Systems. A huge TAD is sitting on the wall of the project war room, ornamented with strips of adhesive tape, all of them differently colored. No secrets here, each tape is a Jogging Line, visually displaying progress in a certain week. Well armed with information, activity owners report on where their activities are compared to the baseline, filling the air with their passionate voices, interacting with the users of their activity outputs, and taping a piece of Jogging Line for their activities. Like fortune-tellers, they anticipate their activity future to help visualize when their project will end. Obviously, they are happy to be needed by their company, working on this product development project.

The major shortcoming of the Jogging Line is that it may be

- *Time-consuming.* It may take time to prepare good information to draw the line and predict a trend.
- *Demanding when forecasting completion.* The Jogging Line shows how far behind or ahead the activity currently is. It does not automatically forecast how far ahead or behind it will be at completion. Rather, the project team has to make that forecast. B-C-F Analysis displays this more effectively.

Variations. There are several Gantt Chart-based tools for schedule control that project teams usually consider before discovering the value of the Jogging Line (see Figure 12.4):

- Shaded Bar Method
- Plan vs. Actual Bar Method
- % Complete with Plan vs. Actual Bar Method

The Shaded Bar Method uses shading to indicate the portion of the activity that has been completed. While very visual, that shaded portion cannot show when the implementation of the activity started, how much of the activity scope was really completed, and how much the actual implementation is behind or ahead of the schedule. Only one of these shortcomings is resolved by the Plan vs. Actual Bar Method, which visually compares the plan bar with the actual bar—we see when the implementation of the activity started. % Complete with Plan vs. Actual Bar Method also relies on the plan and actual bar but adds percent complete for each bar. The strongest of these three methods, % Complete with Plan vs. Actual Bar Method, suffers from an inability to reveal how much the actual implementation of an activity is behind or ahead of the schedule. By adding actual bars, it also increases the number of bars on the Gantt Chart, making it more complex. Faced with this increased complexity, many would see the Jogging Line as more effective.

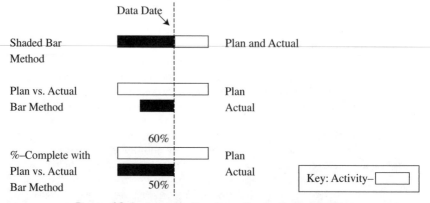

Figure 12.4 Less effective alternatives to the Jogging Line.

Customization Action	Examples of Customization Actions
Define limits of use	Use Jogging Lines on all projects, for controlling schedule of either individual projects or a group of multiple projects.
	Each activity owner reports verbally in the meeting where the Jogging Line is drawn.
	No Jogging Line progress meeting will exceed 30 minutes.
	Project managers will do several rehearsals with all new activity owners.
Add a new feature	Add percent complete next to where the Jogging Line crosses an activity.

The Jogging Line often goes by the name of time line [9]. A variation of the Jogging Line is called the zigzag line. The zigzag line connects data date on the calendar with the point of actual fraction of completion on the first activity to the point of actual fraction of completion on the second activity and continuing through points of actual fraction of completion of other activities. Perhaps the single most important variation is the Jogging Line that replaces each activity with one project, becoming a great tool for multiproject managers eager to track the schedule of their multiple projects. On a single sheet, you can portray the Jogging Line for 20+ projects.

Customize the Jogging Line. A focused adaptation of the Jogging Line accounting for a project's specifics is a must and provides the project with the best-possible benefits. Above are a few examples of adaptations.

Summary

This section featured the Jogging Line, a tool that spells out the amount of time each project activity is ahead or behind the baseline schedule. Large and small as well as complex and simple projects can use the Jogging Line, and it meshes equally well with both the Gantt Chart and TAD. The value of the Jogging Line is in both its historic and forecasting power. Because it shows the history of the activity's progress, it can help forecast a future schedule trend and strategize actions to deliver the project as planned. To sum up, the following box highlights the key points in developing the Jogging Line.

Jogging Line Check

Check to make sure you developed a good Jogging Line

- The line should be continuous.
- It should start at the data date.
- It should cross each activity (or project) to indicate its time variance—that is, the amount of time each activity (project) is ahead of or behind the baseline schedule.
- It should end at the data date.
- It should be drawn on an appropriate baseline schedule in the form of a TAD or Gantt Chart.

B-C-F Analysis
What Is B-C-F Analysis?

The B-C-F Analysis, which stands for Baseline-Current-Future Analysis, compares the baseline project schedule with two predicted schedules—the first one based on the current schedule performance and the second one derived from the worst-case future scenario (see Figure 12.5). As a result, we detect schedule trend, or in other words, where our schedule is going. Most importantly, if the trend is unfavorable, it forces us to design actions to prevent it, and to many users this is the ultimate purpose of all of these proactive schedule control tools.

The three schedules can be either in the format of the Gantt Chart or TAD. Whatever the format, the B-C-F schedule is no more than an application of the old tools with a new twist of anticipatory attitude. Simply, this tool is much more about a novel mind set than about a novel tool design.

Performing the B-C-F Analysis

Prerequisites to an effective B-C-F Analysis are

- ▲ The baseline schedule, whether a Gantt Chart or TAD
- ▲ Performance reports or verbal information about actual performance
- ▲ Change requests

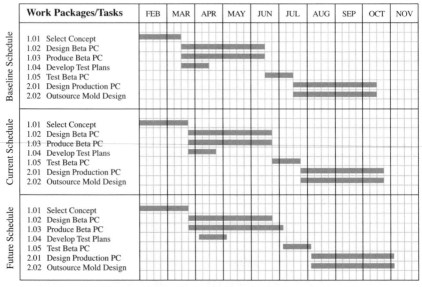

Figure 12.5 An example of B-C-F Analysis.

The baseline schedule—the first prerequisite—is the first part of the analysis. Performance reports or verbal information about actual performance help establish where the project stands, thus serving as an input in predicting the current schedule. Another input into the current schedule building is information on change requests.

Draw a Jogging Line. The B-C-F Analysis begins with the steps that we described in developing the Jogging Line. In that process the goal is to perform a rehearsal with activity owners, call the progress meeting, ask about activities' actual status (see the box that follows, "When Assessing the Actual Status, Go for Satisficing"), and draw a Jogging Line over the project baseline schedule. This should tell clearly how much each activity is ahead of or behind schedule.

Prepare the Current Schedule. Armed with the knowledge of how much each activity is ahead of or behind the baseline schedule, what current issues are, and what remains to be done, forecast when each of the activities will be completed. This will produce a new duration for each activity, when it was started, and when it will be finished. Using this information, a new schedule will be drawn (a Gantt Chart or TAD). This is the current schedule, reflecting how the project is expected to unfold in the future given the activity performance of the past.

When Assessing the Actual Status, Go for Satisficing

To respond to question one of PCPC—What is the schedule variance between the baseline and the actual project schedule?—you must first assess the "actual" schedule status. A productive way to do this is to apply schedule variance (SV) or schedule performance index (SPI) discussed in Chapter 13 on Earned Value Analysis. Each of them aims at establishing the actual amount of completed scope of work for each point in time on the schedule and finding the variance by comparing it with the baseline. Which method should you use?

The majority of managers use the satisficing approach when making complex and unprogrammed decisions [4]. They don't search for a perfect solution, since it requires time and resources they may not have on their hands. So, they don't seek for being perfectly rational. Rather, their goal is to find a solution that is good enough and fits in with how much time and resources they have. This means they are boundedly rational, or *satisficing*, a word combined from two words, satisfactory and sufficing.

Then, how does a project manager go about a satisficing approach when assessing the actual status? Take, for example, a small and simple project, one of five the manager is concurrently managing. She is not satisficing if she spends hours to find out that the project is 7.2 days late. The cost of acquiring this information is too high for her, although it appears like a perfectly rational approach. Instead, she may spend half an hour with her team in an informal PCPC-based meeting to establish that the project is approximately a week to ten days behind. When satisficing, the crux of the matter is to match the size and complexity of the project with the possible solution to a decision problem and your available time. Generally, most of the project manager's project decisions could benefit from such an approach.

Develop a Future Schedule. To develop this schedule, you should ask, "What risks may pop up in the future and sink our ship?" This is a worst-case scenario type of planning, in which the axe of the Murphy's law—if things might go wrong, chances are they will—is hanging above the project's head (see the box titled "Forgetting Trend Analysis: Déjà Vu?" later in this chapter). Visualizing threats, dangers, and risks that the project may encounter in its march toward the end helps in developing the future schedule. If it doesn't work, asking seasoned project managers about their worst experiences with the same type of project is another alternative. For example, one of them may say her technology vendor went belly up during her project and set her back by six months. Brainstorming with team members to identify such risks is helpful as well, especially if the organization has a risk checklist (see the box that follows, "Issue Database: A Checklist for Future Schedules") [10].

Once the risks have been identified, the next step is to figure out how each of them can impact future schedules (see the box titled "Issues and Risks" later in this chapter). For this, the project team needs to look closely at the dependencies between activities in the current schedule. Will the risk impact activities on the critical or noncritical path? If the impacted activities are on a noncritical path, will the whole total float be eaten up? How much could it push out the completion of impacted activities? If the critical path is impacted, how much will the project completion be pushed out? When the team develops answers to these questions, take the current schedule and extend activity durations accordingly to obtain the future schedule.

Action, Action, and Action. The major purpose of all preceding steps is to equip the project with an early warning signal. A signal that says, "take action to resolve the issues causing the current schedule variance and mitigate risks to future schedule." If the issues cannot be put to a stop, there is a need to rechart the future schedule and find alternatives that would lead the project team to deliver as expected per the baseline schedule. Two approaches are handy here: fast-tracking and crashing, or their combination.

Issue Database: A Checklist for Future Schedules

When developing the future schedule in the B–C–F Analysis, a project team in a high-tech organization first goes to its issue database as a checklist (see Chapter 15 for information about issue databases). These issues that already happened in past projects may hit their project in the future. Therefore, they are potential risks to their project. For each issue in the database, the project team asks: "Could this issue become a risk in our project? Would its impact on our project be different from the impact it had on the past projects? Would actions taken in the past projects work in our project, or would we have to use different actions to defend against such risks?" The database is built over time to include crucial lessons learned about major issues and their impact in past projects. So, it is both experiential and realistic. Using this tactic saves a lot of time in meetings, helps build future schedules of better quality, and uses past experience in mitigating risks.

For fast-tracking:

▲ Go back to the future schedule and focus on hard and soft dependencies between activities.

▲ Turn any of the sequential activities with hard dependencies into overlapping activities as much as possible, while still observing the hard dependencies.

▲ Reexamine all soft dependencies in order to overlap as many activities as much as possible.

▲ Also, given the soft dependencies, pick activities that can be performed out of the sequence established in the schedule.

Although all of these changes can help accelerate the project, you need to search for more opportunities. Crashing the schedule by throwing in more resources to reduce duration of activities on the critical path can provide additional time savings (see the "Schedule Crashing" section later in this chapter). Fast-tracking and crashing, however, may substantially increase the number of critical activities and paths, putting more pressure on project time management. In situations like this, you must live with an attitude of "if it is to be, it's up to me." In other words, get to work (see the box that follows, "Tips for B-C-F Analysis).

Utilizing B-C-F Analysis

When to Use. B-C-F Analysis is very beneficial to those using Gantt Charts or TADs. In projects with Gantt Charts, applying B-C-F Analysis requires knowing well the dependencies between activities, even if they are not shown on the Gantt. When dependencies are formally identified via TAD, the stage is well set for the B-C-F Analysis. In both applications, the bottom line is to be proactive (see the box that follows, "This Is Exactly What I Need") and insist on applying the B-C-F Analysis consistently in progress reviews, both formal and informal ones. The analysis is more applicable to smaller and medium-sized projects than to large and complex projects.

Time to Perform. A reasonably skilled and prepared project team of smaller size can prepare a B-C-F Analysis for a 25-activity Gantt Chart or TAD in 45 to 60 minutes. The necessary time will expand as the team size is increased.

Tips for B-C-F Analysis

● Give rehearsals for the B-C-F progress meetings ritual significance. The better the meetings are, the better the analysis.

● Ask for good enough, not perfect, current and future schedules. The focus is on trend, not precision.

● Insist on maximum interaction among activity owners in progress meetings to help them understand how they impact each other and the project.

● Observe which activity owners tend to be too optimistic or pessimistic when forecasting completion times. They may need personal coaching to overcome these tendencies.

This Is Exactly What I Need

"I run five-six projects at a time in a matrix environment filled with a constant pressure to deliver faster. Typically, my teams use Gantt charts with about ten activities and meet once a week for 45 minutes to review schedule progress. When the meeting is over, I know no more than how much project activities are delayed. Reporting this to my boss is embarrassing, because he always tells me to be more proactive, asking me when the project will be completed. I don't know what to do," confided a project manager. When introduced to B-C-F Analysis, his reaction was, "This is exactly what I need."

Benefits. B-C-F Analysis is one of those simple tools that helps reach the ultimate in project management: predictability of the project schedule. Through consistent forecasting of the activity and project completion date in progress reviews, combined with deep understanding of the root causes of the reasons for schedule delay and reinforced with remedial actions, B-C-F Analysis prepares the project team to see and tell the future of their project schedule.

Advantages and Disadvantages. The advantages of B-C-F Analysis are that it is

- ▲ *Visual.* It provides a graphic representation of the project schedules that is conducive to dynamic and productive communication between the project team and management and within the project team.
- ▲ *Proactive.* It helps prevent project fires by acting now instead of later when the fires are burning and are difficult to extinguish.

On the other hand, B-C-F- Analysis may

- ▲ *Pose a behavioral resistance.* Many project participants are not used to being proactive. Rather, they prefer to respond to unfavorable project situations when they occur. Asking them to use the analysis may mean fighting their resistance, an uneasy proposition to many project managers.
- ▲ *Be complex.* In projects with large schedules with many dependencies, the application of B-C-F Analysis may be too challenging and time-consuming.

Variations. Multiproject managers tend to use B-C-F Analysis as a tool for schedule control of multiple projects they manage concurrently. In that case, each line in three schedules represents not a project activity but a project. Essentially, the analysis produces the baseline, current, and future schedules for a group of projects—for example, four to seven projects.

Customization Action	Examples of Customization
Define limits of use.	Use B-C-F- Analysis as a cornerstone of the formal progress reporting.
	Use B-C-F Analysis as a substitute for formal progress reports, if the organization does not require such reports.
	Use B-C-F Analysis with time-scaled diagrams on smaller and medium-sized projects.
Amend a feature.	Replace the Jogging Line with percent complete for each activity.

Customize the B-C-F Analysis. What we have described so far is a generic analysis. It may be good for any project, but more often than not, you need to adapt it to specifics of the project situation to get the most out of it. The examples above illustrate the adaptation.

Summary

B-C-F Analysis was the central point in this section. This tool compares the baseline project schedule with two predicted schedules—the first one based on the current schedule performance and the second one derived from the worst-case future scenario. Whether this is done with Gantt Charts or TADs, B-C-F Analysis offers a proactive scheduling approach. In particular, by means of consistent forecasting of the schedule, B-C-F Analysis prepares the project team to visualize the future of their project schedule and devise actions necessary to get there. The analysis is more applicable in smaller and medium-sized projects than in large and complex ones. The following box recaps the key points in performing B-C-F- Analysis.

B-C-F Analysis Check

Check to make sure you performed the analysis effectively. The analysis should include the following:

- Baseline schedule, Gantt Chart, or TAD
- Data date
- Current schedule, developed on the basis of performance reports or verbal information about actual performance
- Future schedule, based on current schedule and information about risks
- Specified actions that could help deliver the project per baseline schedule, if the future schedule indicates a negative trend.

Milestone Prediction Chart
What Is the Milestone Prediction Chart?

Like other proactive schedule control tools, the Milestone Prediction Chart anticipates the expected rate of future project progress. Unlike other proactive schedule control tools, it focuses in those predictions on major project events—milestones and project completion. Note in Figure 12.6 that a vertical axis shows the team's predicted completion date for a specific milestone, while the horizontal axis shows the actual date the prediction was made. Obviously, the beginning point on the horizontal axis is the time when the schedule baseline is prepared, and its milestone dates are marked on the vertical axis. After the beginning point, the project work is kicked off and the horizontal axis represents the actual project time line; the team reviews progress regularly and makes milestone predictions. By connecting all predictions for a particular milestone into a line, we can obtain the milestone trend line. If the line approaches the completion line going upward, we would know that there would be the milestone slip. Delivering the milestone right on time would produce a line approaching the completion line horizontally. If we estimate an early milestone completion, the completion line would be approached downward. Although it is effective in predicting milestone progress, the chart is even more effective if used to develop actions required to eliminate any potential deviation from the baseline milestone schedule.

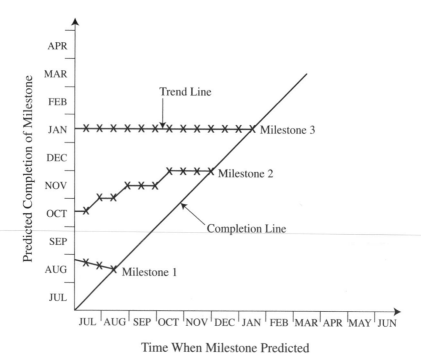

Figure 12.6 An example of the Milestone Prediction Chart.

Constructing the Milestone Prediction Chart

Prepare Information Inputs. The following quality inputs are necessary to build the chart:

- ▲ The baseline schedule, preferably with dependencies
- ▲ Performance reports or verbal information about actual performance
- ▲ Change requests

The first baseline's purpose here is to provide the planned milestones to be entered on the vertical axis. Also, the schedule reveals activity dependencies necessary for the milestone impact analysis. Future predictions of milestones (for importance of predictions, see the box that follows, "Forgetting Trend Analysis: Déjà Vu?") will be based on performance reports or verbal information, as well as on change requests that are being reviewed or are in preparation.

Forgetting Trend Analysis: Déjà Vu?

Western corporations invented quality management, forgot it, and relearned it in hard battles with the Japanese counterparts [1]. Are we seeing the same phenomenon of forgetting when it comes to trend analysis, questions three and four of our PCPC? Our experience shows that the majority of projects in their schedule control and progress reports place an emphasis on the history part, leaving out the trend forecast of the completion date. As experts wrote a long time ago, the purpose of the project control is "No sudden shocks, please!" [5]. In other words, we predict trend because it is much better if we could know in good time what is going to happen, "rather than just watch it happen" [5]. This makes trend analysis—where we are going—the single most important piece of information in project control.

Our managers, especially those in the time-to-market business, hate to suddenly hear that the schedule is going to slip. To them, it is much more meaningful to learn that ahead of time, if the project is going to be delayed indeed. Having a trend projection, then, provides the managers with "early warning signals" so they can act while it is still possible to reverse unfavorable trends [5]. Are these comments belittling the role and value of the historic information in schedule control, question one and two in PCPC? Not really. What we are saying is that history, of course, can't be influenced and the future can, so it is more important for the project team to use that historic information to forecast a future schedule trend and strategize actions to deliver the project as planned. It is for this reason that trend analysis should be the central piece of the schedule control.

Do a Rehearsal with Milestone Owners. Good preparation for progress meetings that are chartered with developing the Milestone Prediction Charts has a tremendous impact on how well you control a project. This is the reason we say that rehearsal with milestone owners is indispensable. Because the rehearsal with milestone owners is essentially the same as the one with activity owners, you can refer to the "Do a Rehearsal with Activity Owners" subsection in the "Jogging Line" section in this chapter. That will help you design an appropriate process for the rehearsal with the milestone owners, which would also include their rehearsal with owners of activities that constitute the milestones.

Call the Progress Meeting. In terms of substance, the "Call the progress meeting" subsection in the "Jogging Line" section of this chapter holds true here as well. Where the difference may occur is the meeting attendance roster. The project manager typically needs milestone owners to attend, and they may want their activity owners to attend. All of them need disciplined and regular meetings with a clear and timed agenda, where invaluable information is exchanged for the benefit of proactive milestone schedule control. It is an imperative to use a proactive control approach (see the box titled "Be Proactive: Five Questions of the Proactive Cycle of Project Control" earlier in this chapter).

Ask about the Milestone's Actual Progress. Formally or informally, based on verbal or written information, milestone owners need to provide answers to the rehearsal questions again in the progress meeting. Since cracks on the project usually appear during interfaces between milestones or their constituent activities, the project team has the benefit of scanning and dissecting the interfaces and their related dependencies. In this process, the understanding of the interfaces and the impact they may have on milestone progress that the team wants to predict will be enhanced. Face-to-face, enriched exchange of information between milestone owners and others involved has no par (see the box that follows, "It is *the* Tool").

Draw the Milestone Prediction Chart. To draw the chart:

- ▲ Mark milestones from the baseline schedule on the vertical axis of the chart—these are "as planned" milestones.
- ▲ Next, draw the completion line. Because the vertical and horizontal scales use the same project schedule as the basis, the line has a 45°angle relative to both scales.

Then the first progress meeting is held. The process is as follows:

- ▲ Well prepared, the owner of the first milestone crisply describes its actual progress (for more details see the box titled "When Assessing the Actual Status, Go for Satisficing" earlier in this chapter), potential variance from the baseline, as well as current issues causing the variance, and gives a preliminary prediction of the milestone completion. Relying on a schedule indicating the dependencies between milestones/activities, the owner opines how the actual status of the milestone can impact the progress of other dependent milestones. A pointed discourse unfolds.

▲ Owners of dependent milestones ask for more information and share their opinion about their actual progress, variance, and current issues. They also give a preliminary prediction of their milestone completion, analyze their actual impact on dependent milestones, and review possible future risks that may further derail the milestones.

▲ At this point, preliminary predictions of milestone completions are made; however, the job is not done yet. Measures must be taken to prevent possible slips. Everybody goes back to the drawing board, analyzing dependencies between milestones/activities. Given the hard and soft dependencies, which activities can be overlapped? Which need to be compressed? The resulting agreement produces corrective actions and final predicted milestone completion dates. Those are marked on the Milestone Prediction Chart at the date of the prediction.

This exercise should be repeated every time the progress is reviewed. Once a milestone is completed, mark it on the completion line.

Utilizing the Milestone Prediction Chart

When to Use. The Milestone Prediction Chart is primarily designed to project the completion date for major milestones [11]. It includes no more than six to seven milestones on the highest level of the project, small or large. Despite this, quite a few project managers apply it with minor milestones, numbering in the tens. They seem to not be hindered by what to us appears as an overcrowded and ineffective chart. Whether built on major or minor milestones, using the chart proactively calls for good understanding of dependencies between milestones. Geared with such understanding, the project team can build reasonably good charts in regular progress reviews, both formal and informal ones.

Time to Develop. To a smaller, well-versed project team, building a prediction chart with six or seven milestones should take no longer than 30 minutes. Adding more team members will inevitably increase the building time.

It Is *the* Tool

It took about 30 minutes. That's how much a group of senior program managers from high-tech organizations allowed for a presentation titled– "A Few Good Project Control Tools." They looked at the project control acumen of the tools in terms of their ability to track progress, reveal trend, and instill a proactive approach. Jogging Line, B–C–F Analysis, Slip Chart, Earned Value Analysis, Milestone Analysis, and Milestone Prediction Chart were briefly reviewed. At the end, the program managers were asked to pick one of these tools they believed had the highest acumen for a high, summary level of the project control. The majority of them selected the Milestone Prediction Chart as their first choice.

Benefits. The basic value of the Milestone Prediction Chart is in its ability to create a sense of predictability of the major project events, or milestones. Through an environment of disciplined progress reviews, the chart helps predict completion dates for major milestones on a regular basis, helping identify trends and leading to actions to correct possible negative trends. Strongly based on the knowledge of the internal dynamics of the project, this proactive approach forces the project team to look into the schedule of milestones on the higher level of the project.

Advantages and Disadvantages. The Milestone Prediction Chart features the following:

▲ *Graphic appeal.* The clarity of direction of the line connecting milestone predictions is unmatched and is likely a major reason for the affinity that higher-level managers have for this tool. As an expert commented, "These graphic representations are nearly infallible in improving schedule predictions over trying to use PERT or Gantt Charts alone" [12].

▲ *Simplicity.* It may take no more than several minutes to master reading or interpreting the prediction chart, even by laypeople.

▲ *Potential for improvement.* Openly charting milestone predictions will make project teams better estimators of their progress [12].

Disadvantages are as follows:

▲ *In corporate cultures that are not in a relentless pursuit of excellence, the chart may be used as a means of catching people when their schedule indicates a trend of slipping.* This may be viewed as a threat among staff and deserves consideration before you use the chart.

▲ *A chart graphing a larger number of milestones*—for example, 15—may become too crowded and convoluted to comprehend.

▲ *Not all milestone owners understand* fully the dependencies of previous milestones to their own. This can lead to inaccurate estimates of milestone completion.

Customization Action	Examples of Customization Actions
Define limits of use.	Use the prediction chart to graph both projects and milestones.
	Include no more than six milestones or projects in the chart.
	Use only for major milestones on the summary level of the project.
Amend an existing feature.	Move the completion line to the right (keeping it at 45⁰). For example, the horizontal axis begins from January, and the vertical one six months later. This is good when graphing project completions that take a longer time.

Variations. A very popular variation leaves out other milestones and predicts and tracks project completion date only. Although ten or so projects can be included in the chart, be aware that too many projects may make it very confusing. Mark predicted completion dates for each project, then connect predictions for an individual project to get a line indicating the trend for the project. Similarly to the Milestone Prediction Chart, projects right on schedule approach the completion line horizontally. Those finishing earlier than scheduled bend downward, while poorly behaved ones approach upward. This variation is used by multiproject managers as well as their bosses.

Customize the Milestone Prediction Chart. Since we presented a generic type of the Milestone Prediction Chart, you should adapt it to satisfy needs of your projects in order to get the best value out of it. In the box above is an illustration how that can be done.

Summary
Described in this section was the Milestone Prediction Chart. Through a disciplined approach, the chart helps predict completion dates for major milestones on a regular basis, helping identify trends and leading to actions to correct possible negative trends. The Milestone Prediction Chart is primarily valued for its ability to create a sense of predictability of the major project milestones. Adapting the chart to specific project needs additionally increases its benefits. In the box above we summarize the key points in developing the Milestone Prediction Chart.

Slip Chart
What Is the Slip Chart?
The Slip Chart tracks progress and signals the trend of the project schedule. In the former purpose, the chart estimates how much time the project is ahead of or behind the baseline schedule at the time of reporting (see Figure 12.7). When consecutive estimates are linked, the latter purpose is revealed in the form of the trend line. As a result of more practitioners subscribing to proactive project control, the Slip Chart took on another purpose: helping predict the project completion date and plan for corrective actions to fight potential completion date slips. While this does not alter the basic design of this tool, it does require an innovative frame of mind to use it.

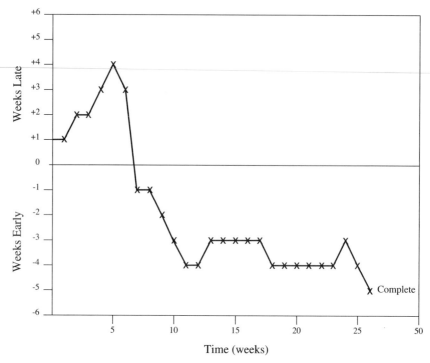

Figure 12.7 An example of the Slip Chart.

Developing the Slip Chart

Prepare Information Inputs. The probability of having a good Slip Chart rises with the quality of the following inputs:

▲ The baseline schedule, preferably a Critical Path Diagram, although a Gantt Chart can be used

▲ Performance reports or verbal information about actual performance

▲ Change requests

While serving as a benchmark to establish the magnitude of the slip, the schedule's dependencies and the critical path help fathom and chart the corrective actions as well. Against the benchmark will be compared the actual progress information from performance reports. Change requests in preparation, and their impact on the schedule, are another critical ingredient to figure out the magnitude of the slip.

Issues and Risks

Remember questions two and four in PCPC? Here they are:

- What are the issues causing the variance?
- What new risks may pop up in the future, and how could they change the preliminary predicted completion date?

The former uses the term *issues*, the latter *risks*. What is the difference between issues and risks? Without a pretense to deal with the semantics of the difference, let's take a look at their use in the industry. According to PMBOK, "Reports commonly used to monitor and control *risks* include *Issues Logs...*" [3]—the two terms can be used interchangeably. Some project managers believe that risks and issues define different concerns and should, therefore, be defined into different categories that need different managerial responses [7]. This is how we use them here.

An issue has already happened, and it is the difference between what should be and what is. Its time horizon includes the past and the present. For example, a loss of a team member is an issue that led to a month delay (variance). In contrast, risk can be characterized as what could happen to the detriment of the project [8]. For example, "a possibility of losing the project manager could cause a late completion of the project." Apparently, risks are in the future. Consequently, while we strive to resolve an issue, our managerial response is to prevent or mitigate a risk (see Chapter 9).

Back to our question "What are the issues causing the variance?" The aim is to identify what's happened that caused the schedule variance and needs to be treated. On the other hand, answers to "What new risks may pop up in the future and how could they change the preliminary predicted completion date?" seek to find future candidate events that need to be responded to in order to mitigate their impact on the project schedule.

Do a Rehearsal with Activity Owners. Doing the homework to professionally prepare for progress meetings to develop a Slip Chart is half of the success. The mechanism for this—rehearsals with activity owners—is well described in "Do a Rehearsal with Activity Owners" in the "Jogging Line" section. Since the project team deploys the mechanism, do not forget that not all activities are created equal. Activities on the critical path are the hard core of the project assignment. Their deviation from the baseline schedule and mutual impact will be the primary determinants of the overall project slip. To exercise proper management caution, do not ignore near-critical activities and paths, the secondary determinants. They may easily slip to become critical.

Call the Progress Meeting. The information for this action has been described in "Call the Progress Meeting" in the "Jogging Line" section of this chapter. Use it with an open mind, adapting it to the specific situation.

Ask about the Activity's Actual Progress. This is the time when the activity owners share the substance they prepared in the rehearsal. The eyes of the project team are primarily on those who own critical activities, and secondarily on near-critical activities owners. With a vivid portrayal of activities' actual progress, variance from the baseline, current issues causing the variance (see the box on page 403, "Issues and Risks"), one after another, the owners deconstruct activity dependencies to assess their impact on the progress of subsequent activities. Other owners join the impact analysis, receiving and giving more information with the purpose of calculating how slips of individual activities on the critical path combine to establish how much the slip is. More precisely, how much is the critical path—that is, the whole project—late (behind the baseline) or early (ahead of the baseline). For this, the satisficing approach should be applied (see the box titled "When Assessing the Actual Status, Go for Satisficing" earlier in this chapter).

Draw the Slip Chart. Several actions are involved in drawing the chart:

- ▲ Find on the Slip Chart form the time of the first progress meeting, or the *data date*.

- ▲ If late, go up the vertical axis as much as you are late to mark your slip point. When early, do the same, only go downward.

- ▲ Connect point zero on the horizontal axis with the slip point.

- ▲ Repeat at each data date, creating a line consisting of connected slip points. For example, as Figure 12.7 indicates, the slip is one week at the end of week 1 and four weeks at the end of week 4. With a strong corrective action, the project is brought to be one week ahead of schedule at the end of week 7. Having the line connecting slip point is usually where the conventional application of the Slip Chart ends.

Project teams with an anticipatory mind set take the Slip Chart use further—into the prediction land.

In the Prediction Land. Here the owners pin down current issues and future risks, making the preliminary predictions of project completions (see the box titled "Be Proactive: Five Questions of the Proactive Cycle of Project Control" earlier in this chapter). To prevent slips that are possible to prevent, they scrutinize dependencies between activities, soft and hard. Looking for ways to fast-track or crash the schedule, at the meeting the team comes up with required remedial actions and a final predicted project completion date (for an example of inappropriate prediction, see the box that follows, "The Window May Be Closed"). Perhaps one of the most important project rituals, future-oriented progress reviews are destined to make or break a project. Rooting them in the trend analysis is instrumental in making them effective.

The Window May Be Closed

Soon after the start of a hardware development project, its Slip Chart showed a three-week slip. The team added the slip to the baseline completion date, predicting the project would be three weeks late. Here is an example of how this extrapolation may be a risky business. One of the project's later critical activities, one-week-long rapid prototyping, is subcontracted to a vendor, who accepted the activity's start date with the comment, "If you come to us a week later, that's fine. If you come later than that, our window will be closed. At that time, add seven extra weeks to the planned delivery date for the prototype, for we will be having another project at the time." Apparently, the extrapolation is misleading. The predicted completion date is not three but at least seven weeks late. The moral to the story: Do not extrapolate; be proactive! Follow the cycle of the proactive project control, or the window will close on you.

Utilizing the Slip Chart

When to Use. Small and simple projects can benefit from the Slip Chart. So can large and complex projects. When applied to track progress in these projects, the chart can work off both the Gantt Chart and network diagrams (except the Critical Chain Schedule). Working off the latter is easier and more accurate because the amount of time each activity is behind or ahead of the baseline schedule can smoothly be translated into the amount of time the critical path is ahead of or behind the baseline schedule. As we know, the amount of time the critical path is ahead of or behind the schedule is equal to the overall project's time ahead of or behind the baseline. In contrast, with the absence of dependencies between activities, the Gantt Chart may pose a challenge when you are converting each activity's amount of time behind or ahead of the baseline schedule into an overall project's time behind or ahead of the schedule. It is this same reason that makes the prediction of the project completion date off the Slip Chart based on network diagrams easier than based on the Gantt Chart.

Time to Develop. A smaller, knowledgeable team may be able to develop a Slip Chart for a 25-activity Gantt Chart or Time-scaled Arrow Diagram in about 30 minutes. The growing team size is likely to increase the time.

Benefits. The Slip Chart's value is primarily in its ability to record the history of project progress and thus reveal the historic trend [13]. For this value to be further enhanced, an extra step, typically part of the proactive approach arsenal, needs to be taken—use the historic trend to forecast future schedule trend and organize actions to deliver the project as planned. Misusing the chart is bound to reduce its value (see the box that follows, "Three Errors").

Three Errors

Executives in a vehicle manufacturing company love Slip Charts. Consequently, all major product development projects they oversee are required to submit the chart. Says a project manager, "My team gets together on a regular basis and the major emphasis is on activity owners reporting the progress status. I, then, sit down all by myself, prepare the Slip Chart, and send it to the execs. They review it and tell us what corrective actions to take." There are at least three errors here related to the use of the chart. First, the team doesn't use the chart to predict the future trend. Second, the team doesn't identify the corrective actions. Finally, the execs identify the actions without having full understanding of these issues involved, robbing the team of their ownership. The lesson here is to follow the proactive cycle of project control when using the clip chart.

Advantages and Disadvantages. The advantages of Slip Charts are that they are

- ▲ *Visual.* The chart graphically communicates to the project participants the amount of time the project is ahead of or behind the baseline schedule and reveals the historic trend.

- ▲ *Simple.* Even to a person lacking project sophistication, it doesn't take more than a few minutes to figure out the Slip Chart.

- ▲ *Proactive.* Because it is only when used for forecasting future trends, the Slip Chart instills an anticipatory way of thinking. This helps the project team detect early warning signals of a problem and act to prevent it from happening.

On the other hand, more often than not, the Slip Chart is also

- ▲ *Time-consuming.* It takes time to prepare sound progress information and turn it into a project slip time.

- ▲ *Reactive.* Unfortunately, too many users see the chart as "project historian," not even attempting to use it for trend prediction and other steps in the proactive cycle of project control.

Customization Action	*Examples of Customization Actions*
Define limits of use.	Use the Slip Chart for all projects with the network diagram (except the Critical Chain Schedule).
Add a new feature.	Use the Slip Chart for predictions of the project completion date and other steps in the proactive cycle of project control.

Slip Chart Check

Check to make sure you developed a complete Slip Chart. It should show:

- The data dates
- The amount of slip at each data date
- A continuous historic trend line, connecting all points indicating slips at the time of data date

Variations. This tool can be used as the multiproject Slip Chart, with a separate line for each project. Having more than five or six projects can make it overly busy and less effective. Some companies use a version of the tool to show days of the critical path on the x-axis and a number of days of positive or negative total project float on the y-axis. For example, at one-half of the critical path completed (x-axis), the project may have a negative float of five days (y-axis), meaning it is five days behind the baseline schedule.

Customize the Slip Chart. The generic Slip Chart that we have described needs to be adapted in order to account for the specific situation of a company's projects. Consider the examples above as a primer when thinking how to adapt the chart.

Summary

We presented the Slip Chart in this section. This tool tracks how much time the project is ahead of or behind the baseline schedule and signals the historic trend of the project schedule. Using the historic trend, you can forecast the future schedule trend and organize actions to deliver the project as planned. Small and simple projects can benefit from the Slip Chart, as can large and complex ones. The chart can work off both the Gantt Chart and network diagrams (except the Critical Chain Schedule). In the box above are key points in developing the Slip Chart.

Buffer Chart
What Is the Buffer Chart?

A Buffer Chart measures the status of buffers established by the Critical Chain Schedule (CCS) to provide an early warning system in order to protect the project's due date. First, the Buffer Chart takes an instantaneous snapshot of buffers' percent consumed relative to the percentage of the work completed on the Critical Chain (see Figure 12.8). Consecutive snapshots taken at regular periodic intervals are then linked on the chart to obtain a line indicating the trend. For example, in Figure 12.8 the line suggests that the buffer is being consumed at a faster pace than the pace of progress in completing the Critical Chain. In other words the line answers the question "How are we doing today?" providing information to make a proactive decision to impact the value of the buffer. In Figure 12.8 that decision might be to take an action in order to recover the project buffer.

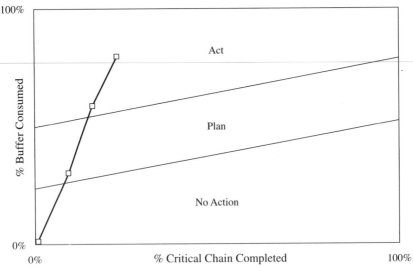

Figure 12.8 An example of the Buffer Chart.

Developing the Buffer Chart

Prepare Information Inputs. The preparation of the following inputs is of critical value in effectively using the Buffer Chart:

- ▲ The CCS
- ▲ Performance reports or verbal information about actual performance
- ▲ Change requests

The first two inputs provide a baseline and the actual status information to calculate the values for the x and y axes in the chart, percent complete of the critical chain, and percent of the buffer used. The inclusion of change requests in the analysis is necessary to visualize the extent of change of the values.

Do a Rehearsal with Activity Owners. Professionals do homework to prepare for a progress meeting to develop a buffer chart. A good mechanism to check on the activity owner's preparation is a rehearsal. Conceptually, the rehearsal is similar to what is described in the "Jogging Line" section in "Do A Rehearsal with Activity Owners." The practical questions, though, differ and will be addressed in "Ask about the Activity's Actual Progress."

Call the Progress Meeting. Essentially, calling a progress meeting to build a Buffer Chart is very much like progress meetings we described in the "Jogging Line" section of this chapter. In a nutshell, regularity (usually weekly but at least monthly), combined with substantive issues and a disciplined approach, is what drives meaningful progress meetings.

Ask about the Activity's Actual Progress. In the meeting's spotlight are those owning activities that are being executed, primarily on the CC (Critical Chain) and secondarily on the subordinate merging paths. With their activity status information prepared and tested in the rehearsals (see the box titled "Be Proactive: Five Questions of the Proactive Cycle of Project Control" earlier in this chapter), the activity owners clearly answer the crucial question of "How many days are remaining on this project activity?" While measuring the project health in this fashion is beneficial for overall good control of the project, it also helps with the next step— monitoring the buffers. Actual progress assessment benefits from the satisficing approach (see the box titled "When Assessing the Actual Status, Go for Satisficing" earlier in this chapter).

Monitor the PB and each CCFB. Knowing how many days remain on activities underway indicates the completion date for the activity. The date, then, provides background information to answer the next question, "What percentage of each buffer is consumed?" The emphasis in this monitoring is on all buffers, PB (project buffer), and CCFBs (critical chain feeding buffers) on the merging paths. Typically, the monitoring occurs in a climate without pressure or focus on the estimated activity completions [14]. Rather, there is a realistic expectation that the estimates may vary on a daily basis and may even go beyond their baseline duration estimates. As long as the activity owners continue sticking with CC principles of work and behavior, actual durations of their activities are of no concern.

Monitor the completion of the CC. While the consumption of a buffer is of utmost significance, its consequences can only be understood in the context of our performance on the activity chain associated with the buffer. In particular, in the meeting the team needs to estimate the percentage of work that is completed on CC and other activity chains. When this information is available, the team can compare the percentage of each buffer consumed with the percent complete of the activity chain associated with the buffer and be able to establish the project's status or health at any given time. This comparison is conveniently illustrated on the Buffer Chart (Figure 12.8).

Draw the Buffer Chart. To develop a chart as shown in Figure 12.8, begin with marking the percent CC (or activity chain associated with the buffer) completed on the horizontal axis at the time of the progress meeting. Go vertically upward until reaching a point equal to the percent of buffer consumed, and connect this status point with point zero on the horizontal axis. Repeat drawing status points in each progress meeting, creating a line consisting of connected consecutive status points.

Use the Buffer Chart in a Proactive Manner. The purpose of the chart is to provide an anticipatory tool with clear decision criteria. In the box that follows, "Using the Buffers," we explain the decision criteria and possible actions as proposed by Goldratt, who pioneered the chart. Following his criteria, for the chart to be beneficial, it should be updated as frequently as one-third of the total buffer time. The reason is simple—the decision criteria are based on thirds of the buffer length. For example, whether a buffer is less or more than a third of the total buffer late (less or more than 5 days for a 15-day buffer) determines the type of action we take. In contrast, the chart in Figure 12.8 uses different decision criteria. See the box titled "You Need to Experiment" for an explanation.

Utilizing the Buffer Chart

When to Use. Being an integral part of the CCS, the Buffer Chart's use is closely linked to how the CCS is used. Accordingly, the most appropriate application of the Buffer Chart is in a dedicated project team, destined to significantly slash the project cycle time. Geared with all necessary resources, the team is part of a company with a strong performance culture, relentlessly working to exceed its customer expectations, delivering maximum value to its shareholders, and providing strong growth opportunities to its employees.

Using the Buffers

Buffers—expressed in time units—are used to measure activity chain performance. The crucial point is to establish explicit action levels for decisions expressed in terms of the buffer size, measured in days (see Figure 12.9). Goldratt, the developer of the CC scheduling method, proposes the following decision criteria [2]. If a buffer is negative—for example, the latest activity on the chain is late compared to its original completion date—and you penetrated the first third of the buffer, take "No action" (see the graph). Should you start consuming the middle third, it is time to assess the problem and develop a "Plan" of action. Once you are within the final third, you need to "Act." Note that these hold true for both the project buffer (PB) and critical chain feeding buffers (CCFBs). If these decision criteria were used, the lines separating "No action," "Plan," and "Act" zones in Figure 12.8 would be horizontal.

	1st Third	2nd Third	3rd Third
Project Buffer	No Action	Plan	Act
CCFB-1	No Action	Plan	Act
CCFB-2	No Action	Plan	Act

Figure 12.9   Using buffers.

Time to Develop. A well-trained and prepared, smaller project team can develop the Buffer Chart for a 100-activity Critical Chain Schedule in one to two hours. As the number of team members grows, so does the necessary time.

Benefits. The value of the Buffer Chart is in its "proactive rather than reactive" use. It prompts the project team to first take an "out the windshield" look at the project health and second, if that health is not good enough, to act now rather than later to secure steps necessary to deliver a project as expected [6].

Advantages and Disadvantages. The Buffer Chart has advantages in that it is

▲ *Visual.* Its graphical message is clear to a project user: This is the percentage complete for both the activity chain and the associated buffer. And, here is the line revealing the historic trend.

▲ *Simple.* Reading off the chart poses no challenge to any project participant. The chart's words such as "plan" or "act" (Figure 12.8) are understandable even to a layperson.

▲ *Partially proactive.* The chart is designed to act as an early warning system, prompting the project team to take different actions in situations with different levels of buffer consumption. The fact that the real actions are taken only after a significant portion of the buffer is consumed makes the approach partially proactive.

You Need to Experiment

The Buffer Chart is a new tool based on a distinctive philosophy of CC scheduling. To really comprehend its potential and put it to best-possible use, you need to experiment with it and find the comfort zone. For example, some companies modified the original criteria for using the buffers. Rather than relying on thirds of the buffers as decision levels given by the originator of the tool, they chose decision levels that change as the consumption of CC changes. Look at the chart in Figure 12.8. Decision levels are borderlines between zones of "No action," "Plan," and "Act, " mandating which action type to use. And, the higher the percent of CC completed, the higher are decision levels. This, of course, makes sense in general— the more work the project team completes, the more buffer consumption the project team can tolerate. But an exact amount of "More" should be picked by the company to fit its business purpose and nature of projects [6]. The moral to the story: Experiment to find the decision levels that best fit the company's projects.

Customization Action	Examples of Customization Actions
Define limits of use.	Use the Buffer Chart for projects with strategic importance that have a dedicated team.
Modify a feature.	Determine decision levels for using the buffers and the associated type of action.

On the other hand, the Buffer Chart can be described as

▲ *Time consuming.* It takes time to prepare dependable status information and translate it into a Buffer Chart.

▲ *A new tool based on a new paradigm of CC;* therefore, learning and becoming comfortable with it may take time.

Customize the Buffer Chart. Modifying the generic Buffer Chart described in this section helps fit it to the project needs of the company. The examples above as well as the "You Need to Experiment" box illustrate the modification.

Summary

This section dealt with the Buffer Chart, which measures the status of buffers established by the Critical Chain Schedule to provide an early warning system in order to protect the project's due date. This nature of the warning system comes from the Buffer Charts "proactive rather than reactive" use. In particular, it requires the project team to take a look at the schedule health, and when that health is not good enough, to act now rather than later to secure steps necessary to deliver a project as expected. The following box summarizes the key points in building the Buffer Chart.

Buffer Chart Check

Check to make sure you developed a dependable and complete Buffer Chart, which should show at progress status points:

● The percentage of CC completed

● The percentage of buffer consumed

● Whether the status point is in the "No action," "Plan," or "Act" zone

● A continuous historic trend line, connecting all status points.

Schedule Crashing

What Is Schedule Crashing?

Schedule Crashing is a method of shortening the total project duration without changing the project logic, which means that the sequence of dependencies between project activities remains the same [15]. To compress the duration, the project usually deploys more resources in performing activities. As a consequence, the total project cost grows (see Figure 12.10).

Performing Schedule Crashing

Schedule Crashing requires a process of disciplined and patient steps outlined here, which we will follow in crashing the schedule example from seven to four days (Figure 12.10).

Prepare Information Inputs. The result of Schedule Crashing heavily hinges on quality inputs, including:

- ▲ The baseline schedule in any network format except Critical Chain
- ▲ Performance reports or verbal information about actual performance
- ▲ Change requests
- ▲ Resource and cost information

The network diagram provides the job logic and the baseline, which is called the "normal schedule." When the baseline is compared with the actual status information, the schedule variance is identified. Further increase of the variance is possible because of the project changes, prompting us to include change requests in the analysis. To figure out the optimum way of the crashing, we rely on resource and cost information.

Develop a Normal, Cost-Loaded Schedule. This is the baseline schedule developed in the phase of project planning. Here resources are assigned to project activities, and their costs are calculated. Without this resource and cost information, Schedule Crashing the way we define it here is not possible. In Figure 12.10 we give an example of a normal schedule (starting position) with duration and cost for each activity in the table.

Develop a Crash, Cost-Loaded Schedule. Preserving the project sequence of dependencies between activities:

- ▲ Estimate for each activity the shortest possible time to complete it.
- ▲ Ask the activity owners and team, "What resources do we need to complete each activity in this way? How much does it cost?" This is sometimes a painstaking exercise, requiring a lot of quick and good information. Also, it may take multiple iterations to develop

these estimates, called crash durations (time) and crash cost. In that process, various challenges will be encountered. For example, some activities cannot be completed in shorter time than the normal schedule shows, or say you need to rent a small piece of testing equipment, but only the big one (and more costly) is available. Not surprisingly, some resources that the project needs may not be available, although there is a budget for them. Generally, these estimates will be developed following the rules of time and cost estimating, which are heavily dependent on the knowledge of project technology and productivity of resources. Once crash durations and crash cost are prepared, the crash schedule is ready. In Figure 12.10 duration and cost for each activity in the crash schedule is given in the table.

Activity	Duration (Days)		Cost ($)		Cost/Time Slope*
	Normal	Crash	Normal	Crash	
A	3	2	30	50	20
B	2	1	40	60	20
C	2	1	20	80	60
D	3	1	30	50	10
E	1	1	40	40	0
F	1	1	40	40	0
	Total: 7 days		Total: $200		

* Cost/Time Slope = (Crash cost − Normal cost) / (Normal duration − Crash duration)

Starting Postion:
– Critical Path A-D-F
– Total (normal) duration: 7 days
– Total (normal) cost: $200

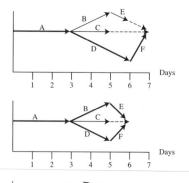

Step 1:
– Cut D by one day
– Total duration: 6 days
– Total cost: $200 + $10 = $210
– Critical Path A-D-F and A-B-E

Step 2:
– Cut A by one day
– Total duration: 5 days
– Total cost: $210 + $20 = $230
– Critical Path A-D-F and A-B-E

Step 3:
– Cut D and B by one day
– Total duration: 4 days
– Total cost: $230 + $30 = $260
– Critical Path A-D-F, A-B-E, and A-C

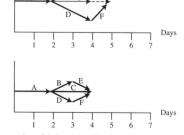

Figure 12.10 An example of Schedule Crashing.

Compute Cost/Time Slope. All activities are not created equal. Some of them are more costly to shorten than the others. Calculating the cost/time slope will show the cost of reducing duration of each activity by one day. Use the following formula to compute the slope for each activity (see table in Figure 12.10):

Cost/time slope = (crash cost − normal cost)/(normal time − crash time)

This has created the basis to identify the cheapest activities to crash. But don't start crashing yet. First, the sequence of crashing activities needs to be figured out.

Crash the Critical Path Only. The critical path is the longest path in the network schedule, composed of activities whose float is zero. The duration of the critical path is the minimum time to complete all project activities. Therefore, the duration of the critical path is equal to the total project duration. The only way to shorten the project duration, then, is to shorten duration of activities on the critical path (see the box that follows, "Wasting Money By Crashing Noncritical Activities"). Simply, crashing the critical path duration by a certain number of days will translate into reducing the total project duration by the same number of time units of the schedule (we will use days). Now look at the network diagram and identify the critical path (in Figure 12.10 we showed critical path for each schedule with a bold line). Working off a TAD (time-scaled diagram) is the easiest way to crash the schedule because it shows activities with and without float visually, which is why we use it in our example in Figure 12.10.

First Crash Critical Activities That Are Cheapest to Crash. When crashing, we want to do it with the minimum cost increase. For this reason, we don't choose to crash just about any activity on the critical path. Rather, we focus on the cheapest one to shorten first, by selecting a critical activity with the least cost/time slope. In our example in Figure 12.10, activity D has the least cost/time slope on the critical path, $10/day. Cut it by one day. Now the schedule is one day shorter, or six days long, and its cost equals the normal cost of the schedule plus the cost/time slope of the activity that was cut, which makes $210 (see Step 1 in Figure 12.10). Continue with cutting critical activities one day at a time, first those with the least cost/time slope—see Steps 2 and 3—until reaching the desired schedule duration of four days and its cost ($260).

Wasting Money By Crashing Noncritical Activities

How often does a project schedule slip? In our experience, project delays are widespread and many of those projects take actions to accelerate and catch up with the baseline schedule. A frequent action of this type is to throw in additional resources indiscriminately and shorten duration of project activities. Since all too often the schedule does not show dependencies, and accordingly the critical path, both critical and noncritical activities. are crashed. Crashing noncritical activities increases the total project cost without compressing the schedule. What a waste of money! The only way to reduce a project's duration without changing the project logic is to crash critical activities.

When There Are Multiple Critical Paths, Crash All of Them Simultaneously. It is a rare privilege to have a single critical path. More often than not, as we crash activities on the original critical path (activities A-D-F in Figure 12.10), new critical paths appear. After cutting D in the first step, there are two critical paths—A-D-F and A-B-E. When this happens, to shorten the total project duration, shorten duration of all, in our case both, critical paths at the same time. This is why in Step 2, we cut activity A, which is on both critical paths, and in Step 3, we cut D on one critical path and B on another critical path.

Failing to shorten one of them will leave the total project duration unchanged; simply, the longest path(s) determines the total duration. As multiple critical paths are crashed simultaneously, follow the rule of first crashing the least slope activity on each of the paths. Having multiple paths is why we enforce the rule of "crash one day at a time?" Because when the original critical path is crashed by one day, a new critical path may emerge. This new critical path would not be discerned if we cut the original critical path by two or more days at a time. In that case the shortened schedule would not have the lowest cost, since we might have missed shortening cheaper activity on the new critical path than the one crashed on the original critical path.

Utilizing Schedule Crashing

When to Use. Schedule Crashing primarily is a method involving two project scenarios. In the first, the project is in the planning stage, the execution has not started yet, and the project team proposes a schedule for management approval. Management finds the schedule too slow and demands it be shortened. To accomplish this, the team goes back to the drawing board, employing Schedule Crashing. The second scenario occurs when the project is underway and its schedule slips [16]. To catch up, the team may use the Schedule Crashing procedure. While for both scenarios the team can apply Schedule Crashing alone, many teams combine it with fast-tracking. Remember, fast-tracking changes the project logic, altering the dependencies between project activities. Because many smaller projects do not employ network diagrams but stick with Gantt and Milestone Charts, Schedule Crashing is generally a choice for larger projects.

Time to Use. Crashing a 250-activity network diagram in a half-day to one-day time is realistic for a skilled, smaller project team. Expect this time to increase with the increase of the team size, since larger teams need more time for internal communications. Crashing off a time-scaled diagram may help reduce the time because of the diagram's visual appearance.

Benefits. The value of Schedule Crashing lies in its capacity to provide a roadmap for shaving an amount of time off the project schedule. Step-by-step, it shows which activities to crash, what resources it takes, and how much it costs. To all organizations cherishing time-to-market speed, or more generally, fast cycle times, this capacity is a significant benefit (see the box on page 417, "Five Golden Rules Of Schedule Crashing").

> **Five Golden Rules Of Schedule Crashing**
> - Crash only activities on the critical path.
> - Crash by one time unit of the schedule at a time. (e.g., one day at a time)
> - When there are multiple critical paths, crash all of them simultaneously.
> - First crash critical activities that are cheapest to crash (the least cost/time slope).
> - Don't crash noncritical activities.

Advantages and Disadvantages. The strong points of Schedule Crashing come from its internal mechanism:

- ▲ *Clear steps.* The procedure lays down an easy-to-understand and stable sequence of crashing steps. With a very limited training measured in tens of minutes, a person familiar with the network diagram can internalize these steps.
- ▲ *Cost minimization.* To organizations focusing on cost management, Schedule Crashing offers an opportunity to reduce duration of their projects while minimizing the cost of the reduction.

Schedule Crashing poses challenges that make it

- ▲ *Complex for larger projects.* Because of the inherent complexity of larger networks with multiple critical paths, crashing them may be an overwhelming task for many (computers significantly reduce the complexity).
- ▲ *Time-consuming.* Just as developing a sizeable network diagram takes time, so does crashing it, especially so when multiple scenarios are tested.

Variations. Companies competing on fast cycle times tend to apply Schedule Crashing without cost analysis. They do not calculate and apply the least cost/time slope as a guide in determining the sequence of activities to crash. Rather, they throw in more resources that they need to shorten duration of critical activities, and in the end may or may not calculate the total cost. A typical driver of such behavior is their perceived lack of time and need to follow crashing procedure oriented toward cost minimization.

Customize Schedule Crashing. To get the best value out of Schedule Crashing, we advise adapting its generic procedure to account for the project situation. Examples that follow illustrate possible adaptation.

Customization Action	*Examples of Customization Actions*
Define limits of use.	Schedule Crashing in larger projects needs to be based on a network diagram.
	Schedule Crashing will be used in combination with fast-tracking.

> **Schedule Crashing Check**
> Check to make sure you crashed your scheduled properly. The crashed schedule should have
>
> - Unchanged dependencies between project activities
> - The duration you targeted
> - The lowest cost of crashing

Summary
This section focused on Schedule Crashing, a tool for shortening the total project duration without changing the project logic. Used primarily in larger projects, Schedule Crashing helps when a project needs to be accelerated. Step-by-step, Schedule Crashing indicates which activities to crash, what resources it takes, and how much it costs. In the box above, we offer the key points in Schedule Crashing.

Concluding Remarks

The six tools reviewed in this chapter are designed for different project situations. In most projects the project manager can afford to use one, possibly two, of them, facing the problem of eliminating the others. To help in this effort, the table that follows lists situations and each tool's suitability for them. If these are not sufficient to characterize the project, add more of them and mark how they favor the tools used. The tool with the highest number of marks is probably the most appropriate to deploy. Using more than one is also a viable option, considering that some of them finely complement and reinforce each other, as indicated in the chapter material. Whatever tool is selected, its real essence is the application in the proactive cycle of project control mode.

A Summary Comparison of Schedule Control Tools

Situation	Favoring Jogging Line	Favoring B-C-F Method	Favoring Milestone Prediction Chart	Favoring Slip Chart	Favoring Buffer Chart	Favoring Schedule Crashing
Small and simple projects	✓		✓			
Formal progress reviews	✓	✓	✓	✓	✓	
Informal progress reviews	✓	✓	✓	✓		
Short time to train how to use the tool	✓	✓	✓	✓		
Focus on highly important events			✓			
Increase goal orientation			✓			
Large, complex, and cross-functional projects	✓		✓	✓	✓	✓
Fast projects					✓	✓
Projects of strategic importance					✓	
Focus on top-priority activities				✓	✓	✓
Summary detail needed			✓	✓		
Schedule control of multiple projects	✓	✓	✓	✓	✓	
Display trend		✓	✓	✓	✓	
Provide built-in proactive approach		✓	✓		✓	
Little time available for schedule control	✓					
Make up for the project delay						✓

References

1. Summers, D. C. S. 2000. *Quality*. 2d ed. Upper Saddle River, N.J.: Prentice Hall.

2. Goldratt, E.M. 1997. *Critical Chain*. Great Barrington, Mass.: North River Press.

3. Project Management Institute. 2000. *A Guide to The Project Management Body of Knowledge*. Drexell Hill, Pa.: Project Management Institute.

4. Simon, H. 1987. "Making Management Decisions: The Role of Intuition and Emotion" *Academy of Management Executive*, 1: 57–64.

5. Kharbanda, O. P., E. A. Stalworthy, and L. F. Williams. 1980. *Project Cost Control in Action*. Farnborough, U.K.: Gower Publishers.

6. Kania, E. 2000. "Measurements for Product Development Organizations, A Perspective from Theory of Constraints." *Visions* 24(2): 17–20.

7. Githens, G. D. and R. J. Peterson. 2001. "Using Risk Management in the Front End of Projects." at The Project Management Institute Annual Seminars and Symposium. Nashville.

8. Wideman, M. 1992. *Project and Program Risk Management*. Newton Square, Pa.: Project Management Institute.

9. Hed, S. R. 1973. *Project Control Manual*. Geneva: Sven Hed.

10. Lientz, B. P. and K. P. Rea. 1999. *Breakthrough Technology Project Management*. San Diego: Academic Press.

11. Weisflogg, U. 1998. "Servant Leadership as the Means for Collective Action in Project Management." at International Conference on Management of Technology. Orlando, Florida.

12. Silverberg, E. C. 1991. "Predicting Project Completion." *Research-Technology Management* 34(3): 46–47.

13. Rosenau, M. D., et al. 1996. *The PDMA Handbook of New Product Development*. New York: John Wiley & Sons.

14. Leach, L. P. 1999. "Critical Chain Project Management Improves Project Performance." *Project Management Journal*. 30(2): 39–51.

15. Kerzner, H. 2001. Project Management: *A Systems Approach to Planning, Scheduling, and Controlling*. New York: John Wiley & Sons.

16. Meredith, J. R. and S. J. Mantel. 2000. *Project Management: A Managerial Approach*. 4th ed. New York: John Wiley & Sons.

Cost Control

There is no terror in a bang, only in the anticipation of it.

Alfred Hitchcock

Major topics in this chapter are cost control tools:

- ▲ Earned Value Analysis
- ▲ Milestone Analysis

These tools are designed to help you successfully perform cost control of the project. More precisely, they help you get a handle on revised cost estimates, budget updates, and forecasts of final cost (see Figure 13.1). In the process, you will rely on cost baseline and information about project plan execution, while synchronizing cost control tools with the tools of scope and schedule control and facilitating processes. Together, these tools should enable you to report performance in the ongoing march toward project closure. The objective of this chapter is to offer practicing and prospective project managers the following learning opportunities:

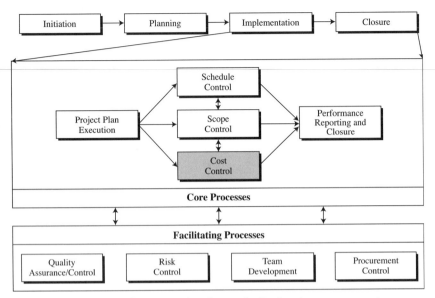

Figure 13.1 The role of cost control in the standardized project management process.

▲ To become familiar with cost control tools in the proactive cycle of project control

▲ To choose a cost control tool that matches their project situation

▲ To customize the tool of their choice

These skills are vitally important in implementing a project and building the standardized PM process.

Earned Value Analysis
What Is Earned Value Analysis (EVA)?

Earned Value Analysis periodically records the past of a project in order to forecast its future (see Figure 13.2). During progress statusing, EVA measures a project's schedule and cost performance to find out whether they are ahead or behind the plan (schedule and cost variances) and why. Then, final project costs (estimate at completion) and completion date (schedule at completion) are predicted. While the practical elegance of such an approach comes from EVA's seamless integration of project scope, cost, and time, its special value is in the proactive, predictive approach. In particular, these predictions warn us of possible problems, creating opportunities to fix them in a timely manner and keep the project on its planned course. In summary, EVA strives to establish the accurate measurement of physical performance against a plan to enable the reliable forecast of final project costs and completion date [1]. In this chapter, we describe the deliberate sequence of several steps to help explain the conceptual simplicity of performing EVA. For further understanding, refer to the box that follows, "Basic Earned Value Analysis Terminology."

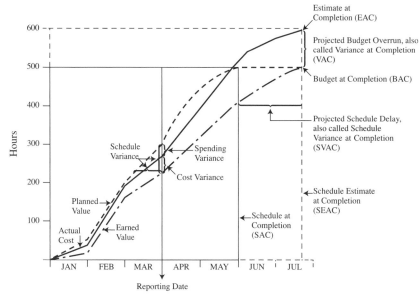

Figure 13.2 An Earned Value Analysis chart.

Basic Earned Value Analysis Terminology

Planned value (PV) = Budget = Planned standards = Scheduled work = Budgeted cost of work scheduled (BCWS); expressed in hours, dollars, units, and so on

Actual cost (AC) = Actuals = Actual cost of work performed (ACWP); expressed in hours, dollars, units, and so on

Earned value (EV) = Accomplished = Earned standards = Budgeted cost of work performed (BCWP); expressed in hours, dollars, units, and so on

Cost variance (CV) = EV – AC; expressed in hours, dollars, units, and so on

Schedule variance (SV) = EV – PV; expressed in hours, dollars, units, and so on; with a different formula, it can also be expressed in time units (e.g., days)

Spending variance = PV – AC; expressed in hours, dollars, units, and so on

Cost performance index (CPI) = EV/AC

Schedule performance index (SPI) = EV/PV

Budget at completion (BAC) = Original total budget to complete the project; expressed in hours, dollars, units, and so on

Estimate at completion (EAC) = (AC/EV) x BAC; projected budget at the end of the project; expressed in hours, dollars, units, and so on

Schedule at completion (SAC) = (PV/EV) x Original schedule; projected duration at the end of project; expressed in time units (days, weeks, months)

Projected budget overrun = Budget variance at completion (VAC); expressed in hours, dollars, units, and so on

Projected schedule delay = Schedule variance at completion (SVAC); expressed in time units (days, weeks, months)

Reporting date = Data date = Today = Point in time at which EVA is performed

Performing Earned Value Analysis

First conceptualized by the industrial engineers of the late nineteenth century, EVA grew to its comprehensive and dominant form under the auspices of the government PM [3]. In the process of growth, the original simple terminology of the engineers yielded to a confusing terminology. Although EVA has become an effective tool, primarily in large government projects, it has not been able to attract a large following in the private sector. The business world nowadays is generally made up of small and medium projects, often managed by companies that have started using the PM approach recently [4]. These companies need EVA in a simpler form, built on a terminology as simple as that of the industrial engineers and often based on resource hours as much as on dollars. Stellar, although rare examples of private organizations pursuing such EVA can be found in the industry [5]. Fully valuing both approaches, for large government and smaller private projects, we will first focus on a comprehensive approach and then in the customization section offer suggestions for the establishment of simpler forms of EVA.

Collect Necessary Inputs. To be effective, EVA has to be set on some firm foundations. These are essentially inputs such as

- ▲ Fully defined project scope
- ▲ Project schedule
- ▲ Time-phased budget.

Fully defining the project scope is, of course, not an easy task, especially when you are dealing with a new job fraught with unknowns. Among the available tools for scope definition, it is our belief that the logic of systematic decomposition of project work into successive levels of manageable chunks of work, as in WBS, provides a sufficient degree of confidence that the scope will be fully defined, including all work to do in the project. This is why one of the golden rules of the WBS structuring is to show all project work. That this is critical becomes clear when we know that EVA may require an estimate of the percent of work completed. If the estimate is 20 percent complete and the project scope is not fully defined, the estimate is apparently inaccurate, because it does not refer to the full scope of work. A disciplined and appropriate application of WBS can help create a reasonable representation of a fully defined project scope.

WBS provides a basis for scheduling the project scope. Each task will be carefully analyzed to determine when in the project time line it will be executed. Details about beginning and ending points of work tasks, as well as their durations, will be determined in the schedule. Such information, along with the approved budgets for tasks, essentially defines scheduled work or planned value. As the project implementation unfolds, physically completed work is evaluated and earned value determined. Both the plan value and earned value are derived from the project schedule information and are critical for successful EVA. For that reason, the project schedule is a critical input to EVA.

A fully defined project scope that is scheduled for implementation must be based on careful resource estimates. Specifically, resources to complete each WBS element need to be identified and allocated for certain time periods on the schedule. This creates a time-phased budget of resources. These resource estimates along with the scheduled work help generate the planned value. Also, these estimates combined with the completed work constitute the earned value. Apparently, the time-phased budget is a critical input to EVA. In summary, EVA requires a fully defined project scope integrated with allocated resources, all translated into a sound project schedule for performance. Often these three inputs are termed a "bottom-up project baseline plan" [6].

Set Up a Performance Measurement Baseline (PMB). You will establish a PMB in order to determine how much of the planned work you have accomplished as of any point in the project time line. Establishing a PMB involves three tasks:

- Determining points of management control and who is responsible for them
- Selecting a method for measurement of earned value
- Setting up a PMB

The foundation for the tasks is the project baseline plan, which fully defines project scope, integrating it with allocated resources and translating them into a project schedule for performance, all within the framework of WBS. Given that WBS has elements on multiple levels, you have to decide which elements (on which level) will be management control points. These points are called *control account plans* (CAP). Although at first sight this might seem a confusing term, in actuality its concept is simple—a CAP is a basic building block of EVA, a point at which we measure and monitor performance. The makeup of a CAP is defined in the box that follows, "Key Components of a Control Account Plan."

Key Components of a Control Account Plan
- Narrative scope definition
- Location in WBS; that is, which level (e.g., level 1 in a WBS with levels 0 for project, 1 for CAP, and 2 for work packages)
- Constituent work elements (e.g., level 2 for work packages)
- Timeline (e.g., begin/end dates of each work package)
- Budget (resource hours, dollars, or units for each work package)
- Owner; the person responsible for CAP (e.g., marketing VP)
- Type of effort (e.g., nonrecurring or recurring)
- Methods to measure EVA performance (e.g., weighted milestones)

CAPs may be located on a selected level of a WBS—at level 1, or 2, or 3 (the project is level 0 of WBS), or all the way down to whatever is chosen as a lowest level to exercise management control. The essence is that a CAP is a homogeneous grouping of work elements that is manageable, which brings us to the issue of its size. How large or how small should a CAP be? According to the current trends in private industry, the size of CAPs is on the increase [1]. One reason is that, not surprisingly, project managers want to concentrate on CAPs that includes larger work elements, which are typically on higher levels of the WBS. Also, they include into a CAP all organizational units responsible for its constituent work elements. The desired result of these trends is to enable project managers to focus their attention on fewer but more vital control points of their projects, making EVA significantly easier to use and much more time-efficient. Such a CAP, then, has a clearly defined narrative scope, location in the WBS, constituent work elements, timeline, and budget. Although the budgets are often expressed in dollars (usually in large projects), they come in all forms—from resource hours to units to standards. Because so many project managers manage only resource hour budgets, we will use the hours in our examples. To ensure accountability for the budgets, each CAP should be assigned to a person responsible for its performance. Figure 13.3 displays an example of a CAP.

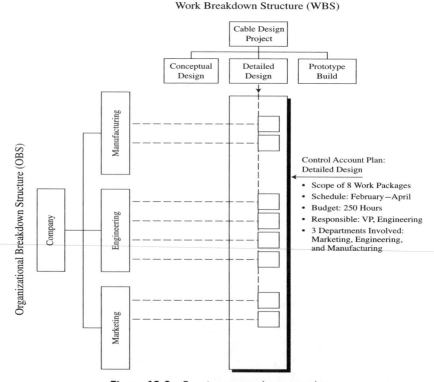

Figure 13.3 Forming a control account plan.

Measurement of a CAP's performance, the cornerstone of the whole EVA, calls for well-defined methods of measurement. While we review several such methods (see Table 13.1), hard-and-fast rules for selecting the appropriate one do not exist. Rather, the choice you have to make is a personal one, often arbitrary, and may vary on a project-by-project basis. In the selection process, the project team and CAP managers should focus on the ease and accuracy of measurements that can be consistently applied to appropriately support their specific project needs.

The *percent complete* method uses a periodic—e.g., monthly or weekly—estimate of the percentage of completion of a work package, expressed as a cumulative value (e.g., 65 percent) against the full 100 percent value of the work package. Hailed as a simple and fast method, which perhaps explains its wide popularity, the method has also been viewed as being overly subjective. Defining work packages' scope well and checking on accuracy of the estimates helps make the subjectivity reasonable.

Table 13.1: **Fundamentals of Major Earned Value Measurement Methods**

Type of the Method (1)	When to Use (2)	Major Advantage (3)	Major Disadvantage (4)
% complete	Well-defined work packages; management reviews in place; nonrecurring tasks	The easiest method to administer	Made purely on the subjective basis
Fixed formula	Work packages are detailed and short-span; nonrecurring tasks	Easy to understand	Rather subjective
Weighted milestones	Work packages run two or more performance periods; nonrecurring tasks	Perhaps the most objective method	Difficult to plan and administer
% complete with milestone Gates	Works in any industry, on any type of project; nonrecurring tasks	Both easy and objective	Requires time and energy to define meaningful milestones
Earned standards	Preestablished standards of performance; nonrecurring or recurring tasks	Perhaps most sophisticated of all methods	Requires the most discipline
Equivalent units	Long performance periods; nonrecurring or recurring tasks	Simple and effective	Requires a detailed bottom-up estimate

50/50 Formula Offers Reasonable Accuracy

If the work package size is appropriately set, the package completion estimates by means of the 50/50 formula will still provide reasonably accurate overall project performance evaluation. 50/50 formula means that when a work package is started, 50 percent of the package's budget is earned, while the completion of the package earns another 50 percent.

If reporting is weekly in a year-long project budgeted at $1,040,000, with a week-long average work package, 520 packages overall, 10 packages per week, assuming current week's packages estimated at 50 percent in error and all off in the same direction, Brandon found this maximum error using the 50/50 formula [2]:

Maximum error = (Average Packages Per Week x Average Cost Per Package x 0.5)/Total Cost

= (10 x $2,000 x 0.5)/$1,040,000 = 0.009—i.e., less than 1 percent, which is reasonably accurate

Fixed formula by work package includes various options: 25/75, 50/50, 75/25, and so on. For example, 25/75 formula means that when a work package is started, 25 percent of the package's budget is earned, while the completion of the package earns another 75 percent. Apparently, any combination that adds up to 100 percent is possible. This is a quick way of estimating, applicable in situations where work packages are short-span and performed in a cascade type of time frame. It can also be accurate (see the box above, "50/50 Formula Offers Reasonable Accuracy."

Weighted milestones is a method of dividing a long-span work package into a several milestones, each one assigned a specific budgeted value, which is earned when the milestone is accomplished. As objective as it is, the method's success hinges heavily on the ability to define meaningful milestones that are clearly tangible, budgeted, and scheduled.

Percent complete with milestone gates strives to balance the ease of percent complete estimates with the accuracy of tangible milestones. A work package of, say, 600 hours is broken down into three sequential milestones, each budgeted at 200 hours and placed as a performance gate. You are allowed to estimate the first milestone's earned value by percent complete up to 200 hours. To go beyond the point of 200, you need to meet predefined completion criteria for the first milestone. This procedure is repeated for subsequent milestones [1].

Earned standards is a method often applied by industrial engineers to establish planned standards for performance of work packages, which are then used as the basis for budgeting the packages and subsequently measuring their earned value. For example, the planned standard for producing a cup of lemonade at $0.20/cup is used to budget the work package including the production of 1,000 cups for $200. When 500 cups are produced, regardless of the actual cost the earned value is 500 cups X $0.20/cup = $100. Widely applied in repetitive types of project work, the method's foundations are the planned standards developed from historical cost data, time, and motion studies [1].

In *equivalent completed units*, a planned work package is earned when it is fully completed. Similarly, a planned portion of it is earned when completed. For example, a work package to build 5 miles (five units) of freeway is estimated at $3M/mile for a total of $15M. It is fully earned when all five miles are finished. Also, the completion of half a mile will earn $1.5M. Based on detailed bottom-up estimates, the method is favored by the construction industry without ever having been called its real name—the earned value.

After this short review of the six methods, two things need to be mentioned in closing comments about the task of measuring EVA performance. First, note that the work package is the place where the measurement is taken, while measurement for a CAP is a summation of work packages' measurements. Second, there is no single best way to measure earned value for any type of project task. This contingency rule means that different types of tasks will use different methods, and perhaps the most appropriate method is to combine multiple methods, relying on CAP managers to collectively estimate the earned value of individual work packages. For an example of a project where multiple methods are used, see Figure 13.5a. A design project consists of three CAPs, essentially three phases on level 1 of WBS. Each of them applies a different method of EVA measurement—percent-complete, weighted milestones, and earned standards. Since each CAP consists of multiple work packages, this means that all work packages within the CAP are measured with the same method. As we remind you that the EVA measurement method is the last on the list of components of a CAP (see the previous box, titled "Key Components of a Control Account Plan), we move to establish the PMB.

The PMB is a time-phased sum of detailed and individually measurable CAPs. What is included into CAPs depends on how companies define cost management responsibilities of their project managers. Many companies allow their managers of internal projects to manage only direct labor hours, which is our focus here. In that case, their CAPs and PMB will include only direct labor hours. On the other end of the spectrum are project managers whose job is to manage all project costs, as well as management reserves and profit. Accordingly, their PMB will reflect this situation. Still other companies may select that their PMB and related cost responsibilities of project managers is somewhere between the two ends of the spectrum.

In projects with lower uncertainty, a firm PMB with detailed CAPs can be established before the project implementation begins. What if you have to start executing an uncertain project in which front-end CAPs are detailed out while the later ones cannot be planned for the lack of information (see Figure 13.4)? And, what if a CAP's scope starts changing? The answer for the first issue is the rolling wave approach; as you progress in executing the available detailed CAPs, you will generate more information that enables you to plan other CAPs [7]. As for the second issue of scope changes, you should establish a PMB change control procedure. By carefully handling all changes to the scope, you will be able to update and maintain the approved PMB, a prerequisite to successful EVA.

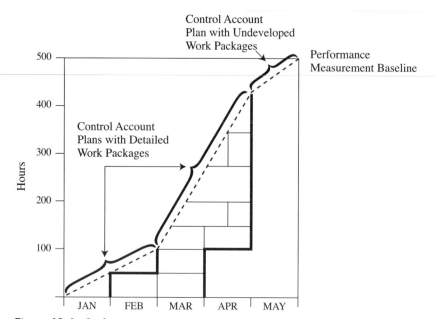

Figure 13.4 Performance measurement baseline: the sum of control account plans.

From *Earned Value Project Management*, Second Edition, Quentin W. Fleming and Joel M. Koppelman. Newtown Square, PA: Project Management Institute, Inc., 2000. All rights reserved.

For practical purposes of EVA, the time-phased PMB can be displayed as a cumulative performance curve representing the planned value over the project schedule. That is the curve shown in Figure 13.5b, developed for a design project whose performance data is given in Figure 13.5a. In summary of this step, a project's PMB is now in place, consisting of detailed CAPs, each one essentially a form of subproject.

Evaluate Project Results. This step compares the actual results of performing the project with its plan (PMB) following the proactive cycle of project control (see the box "Five Questions for PCPC for Cost Control" later in this chapter). While the very measurement of performance occurs within individual CAPs, you may monitor and periodically (weekly or monthly) evaluate performance results at three levels: within individual CAPs, at some intermediary summary level (either a WBS element beyond the CAP or an organizational breakdown structure level), and at the project level. This step includes the following:

▲ Focus on schedule area: evaluate schedule variance (SV) and schedule performance index (SPI).

▲ Focus on cost area: evaluate cost variance (CV) and cost performance index (CPI).

▲ Identify the cause of the variances, if any.

Our example in Figure 13.5c shows the comparison—the cumulative performance curve for actual values against the performance curve for planned values. To tell the truth, the majority of project managers, despite its potentially deceptive results, favors this traditional cost management approach. At the end of March in our example, the difference between the two curves—called the *spending variance*—only reflects whether the project stays within the approved budgeted hours. In does not in any way determine the project's true cost performance status. If used for establishing the project's true cost performance status in Figure 13.5c, the comparison would mislead us by indicating that the actual performance is under budget (300 hours – 270 hours = 30 hours), a positive development. This couldn't be further from the truth—the project is in cost trouble, as we will see soon, and that can't be discerned using this planned-versus-actual approach. The reason for this false finding is in comparing points on the planned and actual curves that include different scopes of work. In short, apples are compared with oranges. Another trouble with this two-dimensional traditional approach from Figure 13.5c is in that it relates to cost only. To get an insight into the project schedule performance, we need a separate planned versus actual schedule chart, and that one would not match the cost chart from Figure 13.5c. The remedy for these problems is a chart that integrates true cost and schedule performance. This is where the earned value performance curve steps in, as illustrated in Figure 13.5d.

A comparison of the earned and planned value at the end of March indicates the following:

$$\text{Schedule variance (SV)} = EV - PV = 225 \text{ hours} - 300 \text{ hours} = -75 \text{ hours}$$

This negative SV means that the project falls behind its planned work. A look back at Figure 13.5d reveals two manifestations of the same SVs—one drawn vertically is expressed in hours (budget units), the other horizontal in time units. Not surprisingly, you may prefer the one expressed in time units (days, weeks, months). It is generally easier to identify such time units of delay by means of the schedule performance index. Before we get there, it is worth mentioning that anytime SV is negative, the project is late to its planned work, and anytime SV is positive, the project is ahead of its planned work.

Another task in evaluating the schedule position is calculating SPI. At the end of March in Figure 13.5d, the situation is as follows:

$$SPI = EV/ PV = 225/300 = 0.75$$

SPI quantifies how much actual earned value was accomplished against the originally planned value. In other words, it represents how much of the originally scheduled work has been accomplished at a certain point of time [1]. An SPI equal to 1 means perfect schedule performance to its plan. Any SPI greater than 1 implies an ahead-of-schedule position to the original plan of work. SPI running below 1 reflects behind-schedule position to the originally scheduled work. Therefore, our SPI of 0.75 indicates that 75 percent of the originally planned work is accomplished. This means our project is way behind, more precisely, 25 percent $(1 - 0.75 = 0.25)$ behind the baseline plan of work. Since our reporting date at the end of March is the 90th day of the project, we can tell that our project is 22.5 days (25 percent of 90 days) behind the original plan of work.

Schedule analysis in EVA deserves a word of caution. Specifically, anytime you find a schedule delay condition that includes negative SV and SPI that is less than 1, you should know that EVA schedule variance is not based on the critical path information and may be deceptive. Poor schedule performance of some work packages or tasks may be balanced by schedule performance of other work packages or tasks. Use your critical path schedule and risk analysis in conjunction with EVA schedule analysis [8]. If the late work packages/tasks are on the critical path or are highly risky to the project, complete the work packages/tasks at the earliest possible date [1].

(a) Data on multiple methods of earned value measurement for a project

CAP	EV Method	Measure	Jan	Feb	Mar	Apr	May
Conceptual Design	Percent-Complete Estimate	Planned	45	55			
		Earned	20	30	50		
		Actual	35	45	50		
Detailed Design	Weighted Milestones	Planned		100	100	50	
		Earned		100	0		
		Actual		115	0		
Prototype	Earned Standards	Planned			100	50	
		Earned			25		
		Actual			25		
Total Project	Planned	Inc.	45	155	100	150	50
		Cum.	45	200	300	450	500
	Earned	Inc.	20	130	75		
		Cum.	20	150	225		
	Actual	Inc.	35	160	75		
		Cum.	35	195	270		

Key: CAP–Control Account Plan
　　 Inc.–Incremental
　　 Cum.–Cumulative

Reporting Date

(b) Cumulative performance curve for the planned value

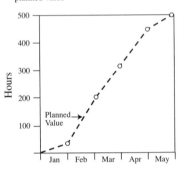

(c) Cumulative performance curves for the planned value and actual cost

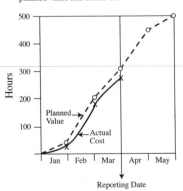

Reporting Date

(d) Cumulative performance curves for the planned value, actual cost, and earned value

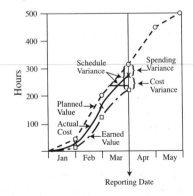

Reporting Date

Figure 13.5 Performing Earned Value Analysis.

Now we can move to our second area of interest in this step—calculate cost variance (CV) and cost performance index (CPI). At the end of March in Figure 13.5d, CV is as follows:

$$\text{Cost Variance (CV)} = EV - AC = 225 \text{ hours} - 270 \text{ hours} = -45 \text{ hours}$$

and

$$CPI = EV/AC = 225/270 = 0.83$$

The purpose of CV is to indicate the differential between the earned value for the physically accomplished work and the actual cost to accomplish the work. Therefore, the positive CV means that the project is running under budget, while negative CV signals the project is in the red, overrunning the budget. Our case is, apparently, experiencing the latter, consuming 45 more hours than we have allocated for the amount of accomplished work.

CPI is a cost efficiency factor. By relating the physically accomplished work to the actual cost to accomplish the work, CPI establishes the cumulative cost performance position. When CPI is equal to 1, that means perfect cost performance to its original budget. Values of CPI exceeding 1 indicate under original budget position, while those less than 1 spell the trouble of over original budget position. In our example, the CPI's reading of 0.83 is that earned value for the physically accomplished work is only 83 percent of the actual cost to accomplish the work. Put it differently, 17 percent $(1 - 0.83 = 0.17)$ of actually consumed hours are a budget overrun. Why are there variances in our example? Identifying and understanding the root cause of the variance are the last task in this step of evaluating project results. The variances are so important that a careful dissection of the project is a necessary strategy to uncover the cause and, later, develop corrective actions.

Both CPI and SPI cumulative curves enable a very effective tracking of a project, as illustrated in Figure 13.6. Note that both rate and trend of the indices are crucial here. Key to it is using the cumulative data, rather than incremental data (weekly or monthly). Unlike the incremental data, which is prone to fluctuations, the cumulative data tends to smooth out the fluctuations and is very effective in forecasting the final project results, the focus of our next step.

Forecast the Final Project Results. If there is one single, most compelling reason to use EVA, it is for its proactive ability—the ability to reasonably forecast the final project results most of the time during project execution. We say most of the time because it is only somewhere at the 15 percent completion point and beyond that a sound, statistically reliable forecast becomes feasible by completing the following tasks:

▲ Forecast project's completion date.
▲ Forecast project's cost at completion of the project.
▲ Take corrective actions, if necessary.

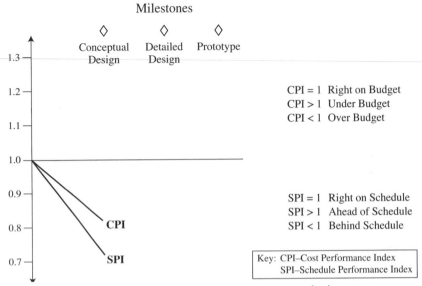

Figure 13.6 Tracking cumulative Schedule Performance Index (SPI) and Cost Performance Index (CPI).

A quick forecast of the completion date for our example in Figure 13.5d at the end of March is as follows:

Schedule at completion (SAC) = original schedule /SPI = 150 days/0.75 = 200 days

This quick method may unfortunately turn risky. As mentioned earlier, an EVA schedule delay condition that like in our example has a negative schedule variance and SPI that is less than 1 is not based on the critical path information and may be deceptive. Therefore, a better solution is to predict the completion date based on results of the critical path analysis in combination with EVA schedule variance.

Out of 20+ available formulas to estimate a project's cost at completion of the project [1], we will only look at two that are frequently used. Here is the low-end formula and forecast:

Estimate at completion (EAC) = Budget at completion (BAC)/CPI = 500 hours/0.83 = 600 hours

This means that at the end of the project we would need 600 hours to get this project done, a whole 100 hours variance at completion (VAC) over the original budget. Clearly, this method—called *constant cost efficiency rate*—relies on to-date cost overrun and projects it to the end of the project. Figure 13.2 includes the prediction of the final results developed by means of the quick forecast for SAC and low-end formula for estimate at completion (EAC).

A more rigorous method, based on our forecast on both the cost overrun and schedule slippage to date, is called *constant cost and schedule efficiency rate*:

$$EAC = (BAC)/SPI \times CPI = 500 \text{ hours}/0.625 = 800 \text{ hours}$$

With this method, we get variance at completion of 300 hours. Some researchers found that the low-end forecast is a reliable measure of the "minimum" hours, while the high-end method produces a forecast of the "maximum" hours we may need [1]. Their claim that the high-end method is the most appropriate forecasting method should be contrasted with a recent study's finding that the low-end method is the most accurate [9]. With such differing views using both methods to develop a range of final cost projections, in our case between 600 and 800 hours, makes sense. This is the absolute essence of prediction—produce a sanity check of the trend and final direction of the project (for major factors impacting the final results, see the box that follows, "Three Factors Influencing the Final Project Results"). In our example, the prediction is not good; actually, it is very bad, but its ultimate value purely depends on the willingness of management to act or not to act. If the option is to not act, the whole EVA is meaningless—it has no value whatsoever. Choosing to act by developing corrective actions rooted in the root causes of the problems is what EVA is designed for.

Utilizing EVA
When to Use. EVA is a choice for any project, regardless of the industry and size. With the amount of resources at stake in large projects, a full-scale EVA can be easily justified. Simplified versions of EVA such as cost and achievement analysis (see the "Variations" section coming up in the chapter) are a right fit for smaller projects. In either case, a good measure of customization is recommended.

Three Factors Influencing the Final Project Results
- Sound project baseline. Only when the scope is well defined, the schedule is realistic, and the budget is accurate can we expect a realistic forecast of the final project results.
- Actual status of the project. The actual status of the project, as quantified by SPI and CPI, will be a vital factor in determining with what final results the project will end up. Better SPI and CPI rates and trends indicate probably better final results.
- Corrective actions. What will management do if the forecast is poor? Not believe it and do nothing? Or believe it and aggressively pursue corrective actions to alter the forecast? This is the moment of truth for management that will critically influence the final results.

Time to Perform. Any project in need of EVA must establish the scope, budget resources, and schedule tasks. When that is available, small projects are ready to deploy simple versions of EVA. In such an environment, a project with ten tasks, 300 to 500 hours in budget, with some overlap of tasks may have several tasks to evaluate weekly. This would probably need no more than an hour to accomplish. The situation changes in large projects, where tens of hours may easily be expended to run a regular, full-scale EVA.

Benefits. A disciplined employment of EVA that is based on a clear understanding of what a company wants to accomplish with it offers multiple benefits. (see the box that follows, "Five Benefits of Earned Value Analysis"). We begin with how EVA can help handle a fundamental question in today's project business: Is the project on schedule, behind, or ahead of schedule? Using the schedule variance and SPI in conjunction with the critical path method can reliably answer this question. In a similar manner, the cost variance and CPI play the crucial role in establishing the true cost position of the project by finding whether it is on, over, or under budget.

Each true schedule and cost position may be viewed as a significant step, telling what happened in the project at a specific point of time of its history. Knowing that they cannot change the history, proactive project managers use the position to look into the future and impact it. In particular, research of the past use of EVA indicates that cumulative cost performance index for larger projects becomes very stable at the 15 percent completion point in the project [10]. Simply, this means that early in the project the cost performance index exhibits a consistent pattern, enabling reliable forecasts of the project cost at completion. Similarly, the schedule performance index combined with the Critical Path Method can be used for predicting the final completion date. Hence, you can periodically ask, "Given my current performance, what will be my final costs and completion date?" The answer offers trend performance and, if the trend differs from the baseline plan, an early-warning signal. This may be the highest value of EVA—providing project managers with the early-warning signal about possible problems in the future and serving them an opportunity to devise and take needed corrective actions while there is still time to fix such project problems. Most importantly, this works for smaller projects as well. According to some experts, smaller projects' accurate cost performance index readings necessary for trend performance and early-warning signals could become available at 10 percent completion in the project life cycle, even earlier than for larger projects [1].

The true credibility and, eventually, the value of EVA are rooted in its integration of project scope, schedule, and cost. WBS provides the vehicle for the integration of all project work through its hierarchical tree of deliverables called *work elements*. For each element with its scope of work, resources are allocated, schedule determined, and cost estimated. By measuring current and predicting future performance for each work element, and aggregating them up the WBS hierarchy, we can arrive at the measurement of the current performance and prediction of the future performance for the total project. This integrated and consistent manner of performance measurement and prediction [11] is a vital improvement compared to the traditional separate schedule performance and cost performance charts, a fairly typical approach in the industry. These are fraught with risks of having nonintegrated, multiple, and often conflicting measurements of performance.

Project managers focus on tasks that matter. To separate tasks that matter in controlling their projects from those that don't, they can use EVA, highlighting tasks that have the highest variances from the approved project baseline. These tasks become critical in ensuring that the project ends in line with the baseline. For that reason, project managers focus on such critical tasks, devoting them most of their attention and time. By doing that, they are managing by exception, putting their expertise to work where it is most needed. This certainly becomes easier with the establishment of thresholds for tasks—specific levels of schedule or cost variances that, if exceeded, trigger management action.

The assessment and improvement of efficiency and effectiveness of projects call for comparable, consistent, and transparent information about their results. This is not possible if this information comes from multiple systems, where one system is used for large projects and another for small projects, with project managers interpreting results in one way, senior mangers in another. These incompatible systems rid a company of the great opportunity to compare all its projects and opt to go for the most efficient and effective ones. A true alternative is a system in which the information comes from a single, universal yardstick for measurement of current performance and prediction of the future performance for all projects and all management levels. EVA provides such a universal system. For example, a comparison of the scheduled work with accomplished work produces a true indicator of whether a project is meeting time expectations set forth by management [1].

Advantages and Disadvantages. Advantages are that EVA is

▲ *Conceptually simple.* Although at first sight it may appear complex because of its three-dimensional makeup (planned, actual, earned), in essence, EVA is conceptually simple.

▲ *Relatively easy to learn.* Realistically, it does not take extensive training to get a handle on the EVA fundamentals. This is an advantage that should propel otherwise the low use of the tool.

Five Benefits of Earned Value Analysis

A disciplined use of EVA offers the following benefits [1]:

● Assesses true schedule and cost position.
● Monitors trend performance and generates an early-warning signal.
● Integrates scope, cost, and schedule.
● Focuses on exceptions.
● Provides a single control system for all management levels in all projects.

Disadvantages are primarily in EVA's

▲ *Low use*. Some experts estimate that perhaps less than 1 percent of all projects, primarily including the major systems acquisitions by governments, deploy EVA [1]. The reasons may lie in its past history of application, plagued by confusing terminology and many rules and interpretations that governments as major users prescribe. This red-tape mentality might have scared many potential users away, creating an image of EVA as a government type of tool that is not useful for the private sector. With such a myth in circulation, many good companies missed the opportunity to customize the tool for their own project needs and enjoy numerous benefits.

Variations. There are several cost control tools that are conceptually founded on earned value, although they are not, at least explicitly, referred to as what they really are—a simplification of EVA. Two such tools that enjoy a high level of popularity are Milestone Analysis and cost and achievement analysis [12]. Overall, their appeal is in that they use simple terminology and straightforward procedure, which is perhaps why they are so time-efficient. Because of a perception in the PM community that the Milestone Analysis is a tool of its own, it is described as a separate tool in the next section in this chapter. The cost and achievement analysis that is briefly covered in the following is illustrated in Figure 13.7.

Based on the scope and schedule for a task, its budget (same as the planned value in EVA) of resource hours is defined. Multiplying the budget by the percent complete will produce the achieved value (equivalent to the earned value in EVA). Actual consumed hours (equivalent to actual cost in EVA) to complete the scope defined by the achieved value are recorded as well. Values for the budget, achieved, and actual cost are then used to predict the final cost for the task. Doing this on a regular basis for each task, in cumulative terms, allows you to sum them to produce budget, achieved, and actual values for the whole project and predict the project's final cost. The approach offers a great way to be proactive in smaller projects.

COST AND ACHIEVEMENT ANALYSIS

Project Name: <u>Bull</u>   Sheet: <u>1 of 1</u>   Estimate Date: <u>May 02</u>

1	2	3	4	5	6	7
Task No.	Task Description	Budget (hours)	Percent Complete to Date	Achieved to Date (hours) (3) x (4)	Actual Cost to Date	Predicted Final Cost (hours) $(6) + \frac{(3) - (5)}{(5) \div (6)}$
12	Prepare Bill of Materials and Routing	8.0	40.0	3.2	5.0	12.5
	Total	312.0	36.4	113.6	118.0	206.1

Figure 13.7   Cost and achievement analysis.

Customization. If EVA is such a good project performance measure and a trend predictor, why is it not used more in the private industry? In addition to already mentioned confusing terminology, extremes of regulations, and strong red-tape association, we suspect there is a lack of understanding of how to customize EVA to be much simpler, user-friendly, and time-efficient. Examples of private companies that managed to do so seem to support our argument [8, 13]. It is for that purpose that we offer some ideas for customization of EVA in the following table.

Customization Action	Examples of Customization Actions
Define limits of use.	Use EVA for all projects, smaller and larger, but allow for different formats. For example, a full scale EVA can be used for larger projects, while simplified versions may be chosen for smaller ones.
Modify a feature.	Speak EVA in friendly terms, not in awkward terms. Friendly terms are the planned value, actual cost, earned value or the value of work completed, or other terms from the popular parlance of the corporate lingo. BSWS, ACWP, BCWP, etc. are awkward terms that seem to breed resistance to EVA.
	Set appropriate CAP size. For example, a small project can have only several CAPs on level 1 of WBS. Similarly, a large project can have 20-30 CAPs on level 2 of WBS (project is level 0Set appropriate work package size. This size enables proper tracking but doesn't generate an excessive amount of paperwork. For details, consult the "Work Breakdown Structure" section in Chapter 5.
	Be flexible with methods of EVA measurement. Letting CAP and work package managers select the method that best suits them is a flexible approach that has to be combined with a proper management review system.
	Rely on simpler methods of EVA measurement whenever possible. Simpler means that most of the time the 50/50 or a similar formula may work well, saving time to estimate the accomplishment.
	Utilize hours instead of dollars. If you manage a budget of hours or units, do EVA based on hours/units.
	Set up a feed-forward cost reporting system. Since most accounting departments are not project oriented, don't wait for their data. Set up hourly CAPs' budgets, track actual hours on project level, apply a simple measurement method (e.g. 50/50 rule). That will provide enough information for schedule and cost analysis in EVA.

continued

Customization Action	Examples of Customization Actions
	Rely on simple and straightforward software tools [2]. Using EVA in today's dominant project software programs is far from an easy task, lowering interest in EVA. Going for a database product or spreadsheet, manually or through interfacing, is relatively easy to do, making EVA easier and more attractive.
	Use EVA for measuring projects, not for individual performance evaluations. If used for the evaluations, it may not be acceptable to many project teams. Simply, to sell EVA use, we have to make it neutral and nonintrusive.

Summary

This was the section about Earned Value Analysis (EVA), a tool that strives to establish the accurate measurement of physical performance against the plan and enable the reliable forecast of final project costs and completion date. EVA is a choice for any project, regardless of the industry and size. While a full-scale EVA is necessary for larger projects, simplified versions are of a right fit for smaller projects. In either case, a good measure of customization is recommended. Perhaps the greatest benefit of applying EVA is in its providing a single control system for all management levels in all projects. To recap the whole section, we present the key points of EVA in the following box.

EVA Check

Check to make sure you performed a good EVA. It should begin with the following:

- Set a performance baseline in the form of planned (value) cumulative performance over project schedule and continue with collection of data to periodically evaluate:
- Actual (cost) cumulative performance, and
- Earned (value) cumulative performance against the baseline in order to establish
- Schedule and cost position, and root causes for them if they are unfavorable that will be used to
- Forecast final project schedule and cost results, and
- Develop corrective actions, if necessary.

Milestone Analysis
What Is Milestone Analysis?

Milestone Analysis compares the planned and actual cost performance for milestones to establish cost and schedule variances as measures of the project's progress (see Figure 13.8). A milestone's cost is planned and tracked on the y-axis, and its schedule on the x-axis The gap between the milestone's planned and actual cost provides the cost variance. Similarly, the schedule variance is obtained through the differential between the planned and actual schedule for the milestone. Both the planned and actual values are portrayed by cumulative curves. These two curves—as opposed to EVA's three curves of plan, actual, and earned values—are made possible by using milestones as a platform for the integration of scope, schedule, and budget. Although effective in tracking project progress, Milestone Analysis is far more effective when used proactively to predict the final project cost and completion date.

Performing Milestone Analysis

Milestone Analysis should be performed in complete harmony with the proactive cycle of project control (PCPC) that we discussed in Chapter 11.

Collect Necessary Inputs. Foundations of the Milestone Analysis are as follows:

- ▲ Fully defined project scope
- ▲ Project schedule
- ▲ Time-phased Budget

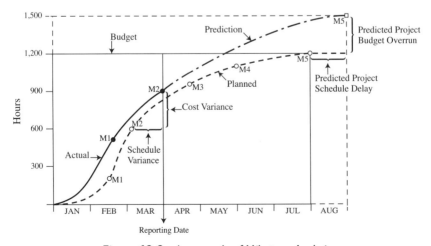

Figure 13.8   An example of Milestone Analysis.

Preparing these inputs, as described in the Earned Value Analysis section, is a prerequisite for the effective Milestone Analysis. Together, they form a bottom-up project baseline plan.

Set Up and Track Milestones. Using the Cost Baseline (Time-phased Budget), draw a planned cost performance curve that will be the baseline, annotating milestones on it (see Figure 13.8). This is a cumulative curve, typically expressed in resource hours or monetary units. A certain number of hours is budgeted for each milestone, and because of the cumulative nature of the curve, when a milestone is reached, the cumulative number of hours for the milestone and all preceding milestones should be consumed. As the project unfolds, actual cost data are collected and used to draw a cumulative actual cost curve, but what really matters is when a milestone is accomplished and marked on the actual curve. Hence, all performance is measured on the milestone level, following a fixed formula of 0/100, an approach of EVA: When the work on a milestone is started, 0 percent of the milestone's budget is earned, while the completion of the milestone earns a full 100 percent. Because of the cumulative nature of curves, the milestone acts as the culmination point of all previous project work, making its performance equate with project performance at that point.

Evaluate Project Results. This is the time to compare the actual results of performing the project against its plan. The goal is to establish the schedule variance (SV) and cost variance (CV), and to identify the cause of the variances, if any. A good way to prepare is to do a rehearsal with milestone owners, and then to implement the evaluation in a progress meeting (for details see the "Milestone Prediction Chart" section in Chapter 12). For both the rehearsal and meeting, an effective framework is the Proactive Cycle of Project Control (see the box that follows, "Five Questions for PCPC for Cost Control").

Five Questions for PCPC for Cost Control

Like schedule control, cost control should follow the questions of Proactive Cycle of Project Control (PCPC):

- What is the variance between cost performance baseline and the actual project cost?
- What are the issues causing the variance?
- What is current trend—the preliminary predicted cost estimates at completion if we continue with our current performance?
- What new risks may pop up in the future and how could they change the preliminary predicted cost estimates at completion?
- What actions should we take to prevent the predicted cost estimates at completion from happening and deliver on the baseline?

Both EVA and milestone analysis should follow the cycle, from the rehearsal with the activity/milestone owners to progress meetings to defining corrective actions. More details about this are available in Chapter 12, in the boxes titled "When Assessing the Actual Status, Go for Satisficing," and "Forgetting Trend Analysis: Déjà Vu?"

In our example from Figure 13.8 the variances are as follows:

Schedule Variance for Milestone 2 = Planned – Actual = 2 months – 3 months = – 1 month

Cost Variance for Milestone 2 = Planned – Actual = 600 hours – 900 hours = - 300 hours

While the negative variance indicates that the project falls behind its plan, the positive variance means the project is ahead of its plan. No variance implies the performance is right on plan. Therefore, in our example, the project is one month late and 300 hours over the budget.

Predict Final Results. Certainly the most important and also the toughest step is the prediction of the final results. It is important because it enables a proactive look at the direction and trend of the project—where are our final cost and completion date going to end up? The absence of formulas for prediction such as those used in EVA makes the prediction an intuitive, challenging assignment, typically performed in the progress meeting. Such an exercise is very similar to one described in detail in the "Milestone Prediction Chart" section in Chapter 12. In particular, as the owner of the milestone describes its actual progress, potential variance from the baseline, and issues causing the variance, owners of dependent milestones opine the ripple effect of the milestone on subsequent milestones. The ripple effect is analyzed in the context of the critical path schedule, indicating the dependencies between milestones and related tasks. As a result of the analysis, predictions of milestones' cost and completion dates are made, all the way to the end of the project. If the final results are not favorable, corrective actions are charted to alter the trend and send the project back on track.

Utilizing the Milestone Analysis
When to Use. Milestone Analysis is a good candidate for both smaller and larger projects. With its visual power and little time to develop, the analysis serves well the needs of projects with smaller budgets. In larger projects, its primary rationale for use is its ability to supply summary view of the project status to high-level managers, focusing on high-level milestones.

Time to Use. With a bottom-up project plan already in place, a well-versed project team should take no longer than 30 to 45 minutes to perform a Milestone Analysis that includes five or six milestones. As the number of milestones increases, so will the necessary time for the analysis.

Benefits. Crucial to the benefits of the Milestone Analysis is an understanding that it is a simplification of EVA. Using a milestone as a precisely defined scope of work, the analysis integrates cost and schedule with the scope, eliminating the need for the earned value curve. As a result, the Milestone Analysis includes only two curves, as opposed to EVA's three. This makes it more attractive and easier to use than EVA, while providing some of the values created by EVA. In particular, Milestone

Analysis establishes cost and schedule position, indicates performance trend and detects early-warning signals, integrates scope, cost, and schedule, and facilitates management by exception. Most often, these benefits are confined to smaller projects and larger projects that use the Milestone Analysis for higher-level milestones, its typical application areas. Since the Milestone Analysis with a large number of milestones tends to be convoluted and impractical, the analysis cannot be a single control system for all management levels in all projects, something that EVA is.

Advantages and Disadvantages. Two major advantages of the Milestone Analysis stand out:

- ▲ *Graphic appeal.* The ease of visually discerning the schedule and cost variances, along with the predicted line of future milestone performance, are bound to win the acceptance of mangers searching for a graphically appealing view of the project's summary status.

- ▲ *Simplicity.* Adding to the graphical appeal is the analysis' simplicity, enabling almost any project participants to grasp its message in several minutes of training.

Users should be aware of the Milestone Analysis's major disadvantages such as:

- ▲ *Potential for confusion.* When the analysis includes a large number of milestones, performing it may be confusing and challenging, although some companies experienced in its use apply it with hundreds of milestones.

- ▲ *Potential for misuse.* Some organizational cultures may misuse the cost and schedule variances, as well as final results predictions, primarily for performance evaluation of the project team. This may prompt teams to manipulate data in the analysis.

Variations. In the high-tech industry competing in the time-to-market race, there is a popular variation leaving out the cost variance and prediction. Consequently, the analysis is entirely focused on the schedule part, precisely the schedule variance and prediction of the final completion date of the project.

Customize the Milestone Analysis. The generic type of the Milestone Analysis may very well fit your projects. It is more likely, however, that certain adaptations could be made to better reflect your specific project situation. Following are a few examples to illustrate our point.

Customization Action	Examples of Customization Actions
Define limits of use.	Use the Milestone Analysis for all smaller projects.
	Use the Milestone Analysis only for milestones on the summary level of larger projects.
	Include no more than six-seven milestones in the analysis.
Amend an existing feature.	Use a formula for predicting the final project cost. For example, the final project cost = cumulative actual cost for the milestone + (project budget – cumulative actual cost for the milestone)/ (cumulative plan cost for the milestone/cumulative actual cost for the milestone). These predictions could be made each time a milestone is achieved.

Summary

This section was about Milestone Analysis, a simplification of EVA. When a milestone is used as a precisely defined scope of work, the analysis integrates cost and schedule with the scope, eliminating the need for the earned value curve. As a result, the Milestone Analysis includes only two curves as opposed to EVA's three. It also indicates performance trend and detects early-warning signals equally well in smaller and larger projects. The following box presents the key points about performing the Milestone Analysis.

Milestone Analysis Check

Check to make sure you performed a good Milestone Analysis. It should begin with

- Setting a performance baseline in the form of planned cumulative performance over project schedule
- Annotating planned milestones on it and continue with the collection of data to periodically evaluate
- Actual cumulative cost performance for accomplished milestones
- Accomplished milestones against the baseline in order to establish
- Schedule and cost variances, and causes for them if they are unfavorable

that can be used to:

- Forecast final project cost and completion date
- Develop corrective actions, if necessary.

Concluding Remarks

There are only two tools in this chapter: the Earned Value Analysis (EVA) and the Milestone Analysis. Both offer functionalities targeting certain applications. Still, project managers living under time pressures often ask, "Given my project situation, which one is more appropriate to use?" To decide, take a look at the set of project situations given in the following table, indicating how each situation favors the use of the two tools. First, identify the situations that correspond to your project. If these situations do not characterize the project well, think of more situations in addition to those listed, marking how each favors the tools. The tool with the higher number of marks for identified situations is probably a better fit for you (probably, because simplified versions of EVA may be an excellent choice in situations favoring either the Milestone Analysis or full-scale EVA). In any case, you should focus on a proactive method.

A Summary Comparison of Cost Control Tools

Situation	Favoring Earned Value Analysis	Favoring Milestone Analysis
Small and simple projects	√ (simplified version)	√
Large and complex projects	√	√ (higher level milestones)
Formal progress reviews	√	√
Informal progress reviews	√ (simplified version)	√
Short time to train how to use the tool		√
Focus on exceptions	√	
Provide early-warning signal	√	√
Integrates scope, cost, and schedule	√	√
Provide a single control system for all management levels in all projects	√	
Take little time to apply	√ (simplified version)	√
Use dollars or person-hours	√	√
Summary detail needed	√	√
Use two curves		√
Use three curves	√	
Display trend	√	√
Provide built-in proactive approach	√	
Need more accurate approach	√	
Little time available for schedule control		√

References

1. Fleming, Q. W. and J. M. Koppelman. 2000. *Earned Value Project Management*. 2d ed., Newton Square, Pa.: Project Management Institute.

2. Brandon, D. M. 1998. "Implementing Earned Value Easily and Effectively." *Project Management Journal* 29(2): 11–18.

3. Fleming, Q. W. and J. M. Koppelman. 2001. "Earned Value for the Masses." *PM Network*. 16(7): 29–32.

4. Dinsmore, P. C. 1999. *Winning in Business with Enterprise Project Management*. New York: AMACOM.

5. Hatfield, M. A. 1996. "The Case for Earned Value." *PM Network*. 10(12): 25–27.

6. Barr, Z. 1996. "Earned Value Analysis: A Case Study." *PM Network* 10(12): 31–37.

7. Harrison, F. L. 1983. *Advanced Project Management*. Hunts, U.K.: Gower Publishing Company.

8. Singletary, N. 1996. "What's the Value of Earned Value?" *PM Network* 10(12): 28–30.

9. Zwikael, O., S. Globerson, and T. Raz. 2000. "Evaluation of Models for Forecasting the Final Cost of a Project." *Project Management Journal* 31(1): 53–57.

10. Christensen, D.S. and S.R. Heise. 1993. "Cost Performance Index Stability." *National Contract Management Association Journal*. Vol. 25.

11. Beach, C.P. 1990. "A-12 Administrative Inquiry." Navy Memorandum.

12. Lock, D. 1990. *Project Planner*. Hunts, U.K.: Gower Publishing Company.

13. Ingram, T. 1996. "Client/Server, Imaging and Earned Value: A Success Story." *PM Network* 10(12): 21–25.

Quality Control

We are what we repeatedly do. Excellence, then, is not an act, but a habit.

Aristotle

In this chapter we will set our sights on tools for quality control in managing projects:

- ▲ Quality Improvement Map
- ▲ Pareto Chart
- ▲ Cause and Effect Diagram
- ▲ Control Charts

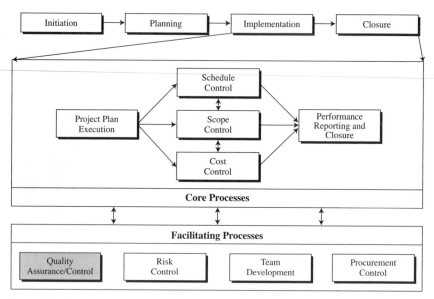

Figure 14.1 The role of quality control tools in standardized project management process.

These tools are designed for monitoring specific results of project implementation (see Figure 14.1). They are deployed to ensure that the project quality program happens in line with the project specification that is defined with the customer. The proper use of the tools will help you review the project work results, accepting the good ones and using the others to make corrections, adjust project processes, and work continuously to improve the desired quality outcomes of the project. To reach the outcomes, quality control will inevitably require a close coordination with scope, schedule, cost, risk, and other controls. The goal of this chapter is to help practicing and prospective project managers

- ▲ Learn how to use various quality control tools
- ▲ Select quality control tools that are appropriate for their project situation
- ▲ Customize the tools of their choice

These skills are vitally important in implementing a project and building a standardized PM process.

Quality Improvement Map
What Is Quality Improvement Map?

Quality Improvement Map (QIM) is a structured approach to problem solving and quality improvement in projects (see Figure 14.2). Leading a project quality team through a logical sequence of stages and steps, QIM enables a meticulous analysis of the project problem, its potential causes, and possible solutions [9]. Its foundation on data—rather than on opinion—makes it possible to direct the team to the core problems instead of peripheral ones. The combination of all these features presents QIM as a convenient tool for building a culture of continuous quality improvement in project management.

Deploying the Quality Improvement Map

While there are many types of QIMs, only a few of them are developed for PM. The QIM in this section is developed for PM and includes five stages: problem definition, cause analysis, corrective actions, results, and standardization (for a comparison with the widely known Plan-Do-Study-Act (PDSA) cycle, see the box that follows, "How Are the Quality Improvement Map and the Plan-Do-Study-Act Cycle Related?"). It will take us step-by-step through a disciplined effort of solving project quality problems.

Collect Information Inputs. The map draws heavily on an organization's

- ⬥ Quality policy
- ⬥ Strategic plans

As the highest source of quality principles and objectives, quality policy defines the execution of quality improvement actions in the organization (see the box that follows, "Breaking Through to Unprecedented Levels of Project Quality"). On the other hand, strategic plans provide major criteria for selecting quality improvement projects with the highest impact. Together, the policy and plans set a direction for the deployment of the map. As such, they are vital inputs to the map deployment.

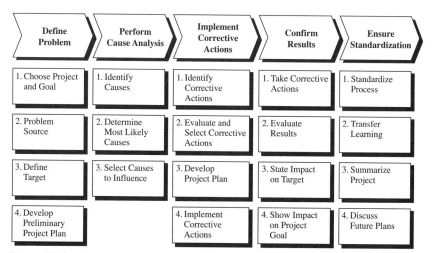

Figure 14.2 An example of the Quality Improvement Map in a high-tech manufacturing company.

How Are the Quality Improvement Map and the Plan-Do-Study-Act Cycle Related?

The major purpose of the Quality Improvement Map (QIM) is to facilitate quality improvement. W. Edward Deming's Plan-Do-Study-Act (PDSA) cycle also aims at the improvement (see Figure 14.3). The "Plan" step focuses on planning a change (i.e., an improvement project), which is then carried out in the "Do" step, possibly on a small scale. Studying the results of the implemented project to understand what has been learned is central to the "Study" step. And adopting the results, abandoning them, or repeating the cycle is the subject of the "Act" step [6]. Obviously, much of PDSA's essence is instilled in QIM. Still, while PDSA cycle is a more general and broader model of quality improvement, QIM is a more specific and practical one [9].

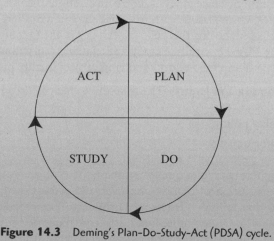

Figure 14.3 Deming's Plan-Do-Study-Act (PDSA) cycle.

Define the Problem. In many ways our success to improve quality profoundly hinges on our ability to identify and solve problems. But what are the quality problems that we want to solve? A problem is the deviation between what should be happening and what is actually happening, assuming that the deviation is sufficiently important to merit an action to solve it [10]. Problem solving, then, is an action of turning the state of what is actually happening into what should be happening.

Major steps in this stage are choosing the project, defining problem sources, defining the target, and developing the preliminary project plan. The action begins with selecting a problem to solve. This can be done, for example, by brainstorming multiple problems, where each problem may be treated as a potential quality improvement project. Comparing and evaluating them is a usual step before zeroing in on one of them. Examples of criteria for the evaluation include the basis of their necessity and urgency and alignment with the organization's policies and plans. Through methods such as forced ranking, voting, and group consensus, you can rank the projects and select one for implementation. Naming the project is important for giving it a focus, for example, "Increase (How) milestones accomplishment (What) in internal software projects (Where)." Naturally, the selection of the project needs to be justified in terms of benefits of correcting the problem, consequences of inaction, and so on.

The effort continues with data collection to grasp the current state of the process in order to clarify past trends and present level of control. Presenting data in a graphical form—by means of histograms, for example—can help show the problem's seriousness and variation with time, as well as further describe the situation. Selecting the project ends with defining its goal. "Increase the percent of all project milestones accomplished (How) in the third quarter (When) by 20 percent (How Much)" may be an example.

Breaking Through to Unprecedented Levels of Project Quality

Making improvements in the quality of PM is simply good business. According to a quality guru Joseph Juran, these quality improvements are no more than the process of breaking through to unprecedented levels of performance. This breakthrough sequence unfolds through several phases:

- Proof of the need. Managers of project managers speak the language of money. If project managers can prove that their quality improvement will make money for their managers, they are in; managers will approve project managers' improvement effort. To prove the need, collect information on poor quality and productivity and translate it into the language of money.

- Project identification. There is only one effective way of accomplishing breakthroughs: project by project. This means that each improvement effort is organized as a stand-alone project, usually small and short. A constant stream of such small projects/improvements is the road to breakthroughs.

- Organization for breakthrough. Who is in charge of guiding improvement projects? Whether it is the PM office on a company level or a functional department or just a few participants in a large project pursuing improvements in their project, the responsibility has to be clearly established. When that is the case, it is known that those responsible will define quality improvement project goals, scope project work, and perform it.

- Diagnostic journey. This is the time in an improvement project to collect data and use statistics and other problem solving tools, all with the purpose of detecting root causes of the quality problem that the project is trying to solve.

- Remedial journey. With the diagnosis in hand, the project team will identify alternative solutions, select one, and implement it, fighting the resistance to change along the way.

- Holding the gains. The implemented solution should become the standard. Training people to live by the standard and building controls to make sure it does not die over time are your top priorities in this phase.

The Quality Improvement Map is used as a vehicle for the breakthrough journey to PM quality.

Finding an answer to the question "What are sources of the problem?" is the next target of the quality improvement project. Here the idea is to dissect the problem into its components and understand the true nature of the problem. Therefore, the problem is broken down into its component sources, which have a measurable impact on the problem. For example, the percent of milestones accomplished may have components (sources) in large, medium, and small projects. Viewed from another angle, the components may be milestones including senior management in milestone reviews and those that do not include them. These source data should be collected and stratified to see their contribution to the total projects (e.g., "70 percent of the missed milestones are in large projects including senior management milestones reviews") and determine the focus of the project by targeting a certain problem source (e.g., "Focus on milestones including management in large projects"). One good tool for stratifying data in this way is the Pareto Chart.

Based on the selected problem source, the team will now develop a statement explaining the level of desired improvement. We call that statement the *target*, including what is to be improved ("Increase the percent of accomplishment of milestones reviewed by senior management in large projects "), when the target will be achieved ("in the third quarter") by how much ("by 50 percent"). Since the team may select several problem sources, you may need to identify several targets that combine to help achieve the project goal defined earlier. For each target, it is useful to show how it impacts the goal.

The first stage of the quality improvement project—define the problem—ends with the development of the preliminary project plan. It is simple and its main parts are a Gantt Chart and responsibility matrix, covering all steps in the remaining stages of the quality improvement. Typically, it requires management approval.

Perform Cause Analysis. There are three steps in this stage. First, the team needs to identify causes of the problem under investigation for improvement. Brainstorming and visually representing the problem by means of the Cause and Effect Diagram is a convenient way to identify its major causes. In this process, the role of the team's experience and knowledge is crucial, as is in the second step—determining the most likely causes.

Collecting evidence to support the listed causes is important in acting on data, not on the gut feeling. Check sheets, scatter diagrams, and histograms can be helpful in finding evidence that may be a basis to group or combine related causes. When there are causes that are not clear or specific, make sure they are clarified. To move further, the team needs to use the collected evidence to quantify the magnitude of the causes. Those with the highest magnitude are the most likely causes of the problem whose solution is searched for.

The third step—selecting the causes that the team can influence—recognizes that there is no logic in attacking all causes at once. Rather, a more focused approach is needed, one that looks at selecting the causes that the team can control and are likely to make the greatest contribution to solving the problem. If there is a need, this is the time to update the preliminary project plan and report to management on the progress.

Implement Corrective Actions. There probably are multiple ways to eliminate the causes of the quality problem the team selected to address. These ways are actually the corrective actions that the team needs to identify, evaluate, and prioritize in the first step of this stage. Given the multitude of corrective actions, a natural step is to select for implementation those that have the highest potential for impact. In that sense, a pro/con analysis can help by listing positive and negative effects of each corrective action. Another, more specific way is to select a corrective action based on

- Effect. Is it effective in achieving the targets?
- Feasibility. Is it possible from a technical point of view?
- Economics. How expensive is it to implement it?

Still another way of making the selection is rooted in the cost/benefit analysis of each corrective action. When an action is chosen, a good planning tool is to estimate its effects in attaining the targets, making sure it does not adversely affect the project goal.

In the final step in this stage, the team needs to develop a detailed project plan, including the implementation of all selected corrective actions as well as actions from subsequent stages. A well-defined responsibility matrix for the actions, along with their completion dates presented in a Gantt Chart, is the heart of the plan. When the plan gets a go ahead from management, it can serve as a baseline for the progress report.

Confirm Results. In a quality improvement project, this stage is the time to act. It begins with a step of undertaking the corrective actions. Then, collecting data on their impact follows. The essence here is to confirm the results of the improvement. For that reason, the team will first evaluate results by comparing the present versus past magnitude of the problem as defined in the first stage of problem definition. This can be done graphically, for example, by updating the Pareto Chart(s) developed in the first stage of defining the problem. Also, the team will estimate cost savings generated by the improvement.

The third step focuses on making a statement concerning both positive and negative impacts of the improvement on sources of the problem. Additionally, a statement assessing the impact on the target is developed. What is the impact of the corrective actions on the project goal? This issue is tackled in Step 4, for example, by comparing graphically the situation before and after the actions were taken and annotating the graph with information regarding significant and unusual events during the actions. The culmination of this step is a statement about the impact of the actions on the project goal. A progress report to management is another regular activity.

Ensure Standardization. Not all of the corrective actions will prove to be effective. Those found to be effective need to become a standard part of the process where the treated problem is. Some ways to do this include

▲ Rewriting the process description and redoing its Flowchart

▲ Revising checklists

▲ Informing and training those involved in the process.

The purpose of this step is to show how the process was changed so that the problem does not recur. Often, other areas of the organization may have faced the same problem. This is why there is a need for the second step of transferring what was learned to other parts of the project or other projects. The intent is to complete the standardization by ensuring that all PM processes that might experience this problem are corrected and standardized.

Because quality improvement projects are a method of spreading learning throughout the organization, there is a need for summarizing what was learned. Step 3 does this by preparing the final project report showing lessons learned, benefits, low points, and high points of the project.

As recognition of the participants is made, the team moves on to the fourth step to discuss future plans. One possibility is that the project problem was not totally resolved, offering an opportunity to the team to set a new target and continue with the problem solving described in this QIM. The problem may have been eliminated, however, and the team may state intentions to select a new project in the never-ending race of quality improvement.

Utilizing the Quality Improvement Map

When to Use. The team is a pillar of the quality improvement. Therefore, relying on teams when solving project quality problems is the most effective way of deploying the QIM. This does not rule out the opportunities for individual project participants to use it for solving their individual quality problems that do not call for skills of other participants. However, quality teams should be aware of differences in applying the map in large and small projects. Typically, large projects are long enough to accommodate quality projects—which can be treated as subprojects of the large projects—while the large project work is going on [11]. For example, within a two-year product development project there is enough time to launch and finish several quality projects, which usually span from several weeks to several months (see the box that follows, "The Solution Is the Short-Term Outlook Schedule"). It is not an easy case for a small project to host quality projects as its subprojects. Simply, the time crunch and lack of resources may make it next to impossible. The real opportunity for the deployment of the quality map is in organizations that have continuous streams of projects, whether small or large. Such quality projects would primarily be targeting quality problems that cut across organizational projects and PM processes, perfecting the processes over time (see, for example, the box "Improving the Project Product Definition" coming up in the chapter).

Time to Use. The completion of a quality improvement project per QIM is a significant, time-consuming effort. Since it is primarily team-based, the project requires the involvement of multiple participants, where a membership of five can be a good size. Considering that the core philosophy of quality improvement is delivering improvements in small, incremental doses, QIP can be deployed in quality projects in a short time, typically spanning from a few weeks to a few months. Pressured by their regular daily tasks, the team members may set aside a certain hourly budget to work on the quality project. It is not unusual for management to limit the size of the budget to four to six hours per month. With a five-hour time budget per member per month, a five-member team pursuing a three-month quality project can easily spend in excess of 75 hours deploying QIM.

Benefits. The overall value of the QIM is in providing focus in facilitating quality improvement of project processes [9]. This is made possible by QIM's organization, logic, and thoroughness. Clearly pointing where, when, and how certain quality tools can be used in the process, QIM brings the use of data instead of opinions in the forefront. Its equally strong contribution to the effective quality team dynamics cannot be overlooked. QIM leads quality project team members to listen to each other more carefully, heightening mutual respect at the same time. Perhaps because of its data-based nature, QIM helps team members learn more about each other and their organization, which assists in lowering territorial fences of the team members, promoting cooperation, communication, and trust [9]. By helping induce all these new behaviors, QIM eases a significant cultural change, a cornerstone of project quality improvement.

The Solution Is the Short-Term Outlook Schedule

In its 50+ year history, AtlasCom has excelled in manufacturing state-of-the-art specialized construction equipment. Recently, management has noticed that manufacturing productivity and quality was falling behind the competitors'. In response, several projects were launched, the first one focusing on reengineering the factory layout. Management tasked a two-tier team to get the job done. The first tier was the core project team, a cross-functional group of middle managers responsible to manage the project effort. The second tier— the extended team of manufacturing specialists—was in charge of the doing the project work. Faced with the lack of a formal PM process and experience, the core team received basic PM training before hiring a consultant who helped develop a detailed CPM schedule. Gearing to launch the execution, the core team explained the CPM schedule to the extended team members, asking them to get the work started and to report progress in a week. The problem was, the extended team members commented, that because of CPM's complexity, they were not able to use it as a basis for planning, organizing, and reporting. A quality improvement project (QIP) including members from both teams was chartered to find a solution to the problem following AtlasCom's Quality Improvement Map. In a week, QIP developed the solution: a short-term outlook schedule. It extracted activities from CPM that were owned by a team member and were to be done in the coming two weeks, presenting them in a Gantt Chart format. Familiar with the Gantt Chart, each member thus had a user-friendly two-week plan that they started using for organizing work, reporting weekly, and updating on a weekly basis.

> **Improving the Project Product Definition**
>
> SoftCo Inc. prided itself on the experience of its senior project managers, capable of developing good software products on time, most of the time. Then, in a surprise move a large competitor recruited two of them. To fill the vacancies, SoftCo promoted two of its best junior project managers to senior project managers. At first, everything functioned well—the new guys took over projects in progress left by the "defectors" and finished them as planned. As they kicked off new projects, the new guys seemed unable to get their hands around the product definition, which was changing almost every day, threatening to delay projects. Concerned by the problem, management directed them to follow SoftCo's PM process when developing the definition. A look at the process revealed that it did provide a flexible framework of project phases, milestones, and deliverables, leaving experienced project managers to deal with the details, among them the product definition. Now that their experienced predecessors were gone, the new guys were struggling. Quickly, a QIP was launched, including experienced members from engineering and marketing. In two weeks a new process for product definition development, compatible with the PM process, was ready. Its cornerstones were description of the intended users, product concept and benefits, positioning, and functionalities. The new guys immediately moved to apply the new process, under the careful eye of management.

Advantages and Disadvantages. Structure and simplicity are perhaps the top two advantages of QIM.

- ▲ *Structured.* Through a linear sequence of stages and steps, QIM offers a structured approach that doesn't leave unknowns when it comes to deploying QIM.

- ▲ *Simplicity.* This same linearity and straightforward quality tools that enable its data-based nature make QIM simple to deploy, so simple that even people with little education quickly become skilled users.

A successful deployment of QIM calls for cultural change and discipline. These requirements are the source of a major disadvantage of QIM. Specifically, the ability to change culture and exercise a disciplined approach are not faculties in abundance. Therefore, asking for them in deploying QIM can be overly demanding in today's project world.

> **Quality Improvement Map Check**
>
> Check to make sure that the quality improvement map is appropriately structured. It should include the following:
>
> - ● Components that enable you to understand the project quality problem, find facts, analyze problem sources and causes, generate improvement ideas and solutions, and implement them
> - ● A manageable number of stages and their steps
> - ● Quality improvement tools.

Customization Action	Examples of Customization Actions
Define limits of use.	Use QIM for improvements of PM processes across projects (for organizations with smaller projects); don't use it within a project.
	Use QIM for improvements of both PM processes across projects and within a project (for organizations with large projects).
Add a new feature.	Include other quality tools in QIP, such as the Control Chart.
Modify a feature.	Organize QIM in seven stages: define the project, study the current situation, analyze the potential causes, implement a solution, check the results, standardize the improvement, and establish future plans [9].

Variations. The seven-step method and problem solving process for quality improvement are just two of many variations of QIM [9]. While some are simpler, others are more complex [12]. The sequence of the steps may also vary, and the toolbox may be configured to include either quality improvement tools or both quality planning and quality improvement tools. Their purpose, however, is identical to QIM's.

Customize the QIM. QIM that we described is a generic tool that may serve general needs of a variety of organizations managing projects. Tailoring the tool to one's specific needs may be an option of higher value for the users. Preceding are a few ideas for doing so.

Summary

The focus of this section was on Quality Improvement Map (QIM), a structured approach to problem solving and quality improvement in projects. The major value of the QIM is facilitating quality improvement of PM processes. Also, it contributes to the effective team dynamics. Finally, QIM helps team members learn more about each other, promoting cooperation, communication, and trust. Customizing the map to your own project situation may even further increase QIM's benefits. In the following checklist, we emphasize the key points about structuring the map.

Pareto Chart

What Is the Pareto Chart?

A Pareto Chart is a histogram of the data, including a series of bars that indicate frequency of occurrence of problems or causes (see Figure 14.4). Arranged in ranking order from left to right, the problems represented by the taller bars are thus relatively more important than those on the right. Adding a cumulative frequency curve on the chart is a visual way to clearly indicate the relative magnitude of the problems and identify where opportunities for improvements are (see the box that follows, "The Vital Few and the Trivial Many").

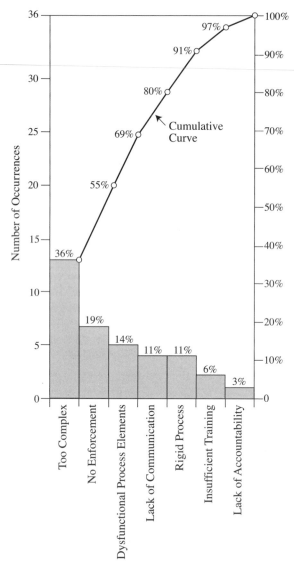

Figure 14.4 Pareto Chart of problems in a new product development process.

Developing a Pareto Chart

Select a Quality Problem. What quality problems have the highest impact on our company's market performance? This is a question that the CEO of Hypex Corp. posed to its executive committee after glancing at results of customer satisfaction surveys and quality audits. Soon, strategic prioritization of the quality problems singled out the inefficient new product development (NPD) process as one of the major culprits. In a quick action the committee assembled a team of experts to tackle the problem. Although practices in organizations vary, at the heart of selecting a quality problem is its impact on an organization's performance. The example from Hypex Corp. exemplifies this approach.

The Vital Few and the Trivial Many

While he was studying the distribution of income, an Italian born Swiss economist by the name of Vilfredo Pareto learned in 1898 that 85 percent of the wealth in the Italian city of Milan was owned by only 15 percent of the people. Surprising? Not really. Many less learned people observed similar patterns in their environments. What is surprising, however, is that this valuable principle was forgotten for years, until it was rediscovered and applied to business situations recently. Specifically, the principle was expressed in numerical terms and became a well-known 80:20 rule, stating that 80 percent of the trouble comes from 20 percent of the causes. Following are examples:

The Vital Few (typically 20%)		Most (typically 80%)
Line items	Account for	Project cost budget
Types of field failures	Account for	Failures of car engines
Consumers	Account for	Beer consumption
Major projects	Account for	Work time
Products	Account for	Sales and profits

Although the percentages may never be that accurate, the point is that most trouble comes from only a few causes. Quality guru Joseph Juran named it the Pareto rule and expressed its essence: "Vital few versus trivial many" [8]. Therefore, if we follow the rule when problem solving, we should concentrate on the vital few items. Eliminating them will give us the most improvement. Similarly, ignoring the trivial many won't hurt us. In summary, the vital few—the highest bars in the chart—are the top priority.

Identify the Causes of the Quality Problem. There are two major ways to find out which causes or problems drive the quality trouble and therefore need to be monitored, compared, and rank-ordered. One is based on the existing data; the other is brainstorming. Hypex Corp.'s project quality improvement team (QIT) relied on both with a zeal for uncovering the real causes. For example, a microscopic look at the database indicated that new products were spending too much time in the conceptual design stage, reaching the production release point with a significant delay compared to the customer-agreed delivery schedule. The data analysis helped uncover major areas for further search of more specific causes. At this time, the QIT enlisted the help of a larger, cross-functional group of NPD specialists, chartering them with brainstorming to identify the causes. In a four-hour brainstorming session, the group came up with over 100 causes, which through grouping by similarity were reduced to seven major categories of causes (listed on the x-axis in Figure 14.4).

Pick the Most Meaningful Method of Measurement. Earlier, we stated that the Pareto Chart includes a series of bars that indicate the frequency of occurrence of problems or causes. This implies that the most meaningful way of measuring the causes is the frequency. The QIT's initial intention was to take an alternative route that would assess the cost of the causes to the company's NPD or their impact in terms of the lost profit. While this sounded very attractive to the executives, the QIT found that there were too many obstacles to obtaining dependable assessment of the cost/profit impact. Rather than getting obscure measurements of this type, QIT opted for the frequency of occurrence of problems or causes. In situations where measuring the cost impact reliably is possible, there is a lot of value in using cost as a method of measurement.

Collect Data on Causes. This is possible to do either by reviewing historical files or gathering data in real time. If historical files contain good-quality information, they may be an excellent and probably fast source of compiling data about frequencies of occurrence of the causes. Often organizations lack good written records, and it may be more practical to collect real-time data about the causes during the predetermined time frame, say, three weeks. Take, for example, the case of an organization searching for causes of low proposal success rate in its projects. Using a real-time approach was out of the question because it would require a period of perhaps a year to submit a number of proposals sufficient to collect data on the success rate. Rather, the company opted to review its past proposal records and was able to extract the necessary data.

In the case of QIT, neither approach offered much practical promise. Historic records did not contain necessary data, and the real-time approach would take too long. Instead, the team asked the cross-functional group of NPD specialists to assess the frequency of the seven major categories of causes relying on their experience. Responses were tabulated into a check sheet (see Table 14.1), often the easiest way to collect data.

Table 14.1: Check Sheet for Categories of Causes

Cause/Problem Category	Incremental Frequency	%	Cumulative Frequency	%
Too complex	13	36	13	36
No enforcement	7	19	20	55
Dysfunctional process	5	14	25	69
Lack of communication	4	11	29	80
Rigid process	4	11	33	91
Insufficient training	2	6	35	97
Lack of accountability	1	3	36	100
Total	36	100	36	100

Incremental frequency in the check sheet is the total number of occurrences for an individual category. A category's portion of the total of all frequencies is its incremental percentage. To figure out cumulative frequency, add the total for the highest category to the next highest category's total—this is the cumulative frequency for the two highest categories. Add this number to the next-highest frequency to obtain the cumulative frequency for the first three categories, and continue with this procedure for the remaining categories.

Repeat this sequence of steps for cumulative percentages of categories.

Draw the Pareto Chart. The horizontal axis of the chart represents the categories of causes or problems in descending order from left to right. Using numbers from the check sheet:

- ▲ Draw bars above each category that represent their incremental frequency, with units of frequency recorded on the left vertical axis.

- ▲ Draw the cumulative percentage curve. For this purpose, enter "100%" on the right vertical axis opposite the point on the left vertical axis that represents the total sum of incremental frequencies for the categories. Add the remaining percentages in 10 percent intervals.

- ▲ Draw a dot at the upper right-hand corner of the bar for the highest cause/problem category. Then draw above the remaining bars dots that represent cumulative percentages at that point until 100 percent is reached. When all the dots are connected, the curve is ready for interpretation.

QIT followed these steps to arrive at its Pareto Chart in Figure 14.4.

Which Stocks to Invest In?

Picture causes of a quality problem as stocks in the stock market. If we gave you $1K, in which stocks (causes) would you invest (put effort to eliminate them) to earn the best return (achieve the greatest impact in quality improvement)? This is how Bud-Mud, a subsidiary of a large multinational organization, directs its quality improvement teams when ranking causes (an equivalent of measuring frequencies) for Pareto Charts. While some may appreciate this practice for its entertaining value in the otherwise monotonous work of quality projects, Bud-Mud takes it very seriously. To the company's culture of "the bottom line is the bottom line," this is a strategy to focus employees on making project decisions on the basis of return on investment.

Interpret the Chart. Clearly, the highest bars represent the vital few, those causes that contribute most to the overall problem, which in QIT's case is an inefficient NPD process. They are the indication for action—it makes common sense to take them on first to improve quality (for a different way of ranking causes, see the box on page 463, "Which Stocks to Invest In?"). As we ponder what action to take, one clarification may be important: What is most frequent is not necessarily the most important. Understanding what has the strongest influence on value that our projects provide to customers is crucial in deciding what causes to go after.

Using a Pareto Chart

When to Use. The Pareto Chart can be used throughout the process of solving quality problems. While early in the project it can help you decide which problem to study, later its contribution is in selecting which causes of the problem to tackle first. Here the Pareto Chart is primarily a planning tool. Once the causes are selected, solutions—sometimes called countermeasures—will be deployed to eliminate or reduce the impact of causes. At that time, we can use a Pareto Chart as a control tool to measure and prioritize the impact of the causes after the solutions were implemented. A comparison of two Pareto Charts, a planning and a control tool, will clearly reveal how much of the causes of the problem our action eliminated (for details see the box that follows, "NPD Process Before and After").

Time to Develop. When the frequency data for causes or problems is available, the time for a skilled team to draw a Pareto Chart can be measured in minutes.

Benefits. What makes a Pareto Chart valuable is its ability to help project teams concentrate on those causes of the problem that if removed will have the largest improvement impact [13]. Progress of eliminating the causes is measured and displayed with a visual impact that motivates the team to go for even more improvement. In this effort, the implemented solutions may eliminate some causes and make others worse, tempting the team to shift its attention to other problems. This is where the Pareto Chart steps in to prevent the shifting and provide a consistent focus for the team's quality team improvement effort.

Advantages. Pareto Chart's advantages are in its

- ▲ *Simplicity and visual impact.* These are the product of Pareto Chart's ability to display the relative importance of problems in a simple, visual, and easy-to-interpret way. To busy and often overworked team members, this is a productivity booster.

- ▲ *Less intimidating appearance.* Realistically, some team members searching for ways to eliminate project quality problems have a fear of concepts of statistics and probabilities that are often offered in quality improvement efforts. With its benign appearance, a Pareto Chart offers them a significantly less intimidating alternative.

NPD Process Before and After

After analyzing the major causes of NPD process inefficiency, Hypex Corp's QIT constructed a Pareto Chart (before any improvement action was taken), shown in Figure 14.4. To remove the causes of the inefficiency problem, QIT developed and implemented a plan of countermeasures, the impact of which was tracked over a yearlong period. This was presented in a new Pareto Chart using the same categories of causes. The two charts were juxtaposed, producing a chart as illustrated in Figure 14.5. The value of the chart is in its ability to show the difference between the frequency of causes *before* and *after* the countermeasures, highlighting the improvement in NPD.

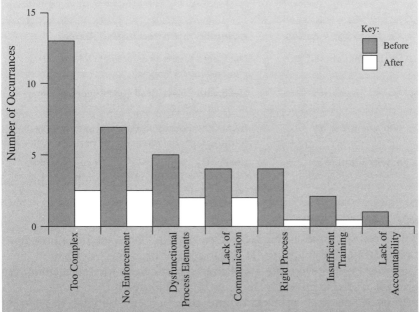

Figure 14.5 Before-and-after Pareto Chart: reduced problems in new product development process after implementing a quality improvement project.

Variations. The extensive use of the Pareto Chart in a whole variety of project environments is probably the reason for its many variations. In another variation—termed major cause breakdowns—the most frequent cause (highest bar) is further broken down into its components and sub-causes and is represented in another, linked Pareto Chart. The aim here is to get better insight into what makes up the most frequent cause.

Customize the Pareto Chart. The Pareto Chart described here may or may not be what your projects need. Customizing it to fit your specific situation is perhaps the best way to bring the expected benefits from the tool. The following examples listed give ideas about how to accomplish that.

Pareto Chart Check

Check to make sure you developed a proper Pareto Chart. The chart should show the following:

- Categories of causes or problems on the horizontal axis
- Bars representing categories' incremental frequencies in descending order from left to right
- The cumulative percentage curve.

Customization Action	Examples of Customization Actions
Define limits of use.	Use Pareto Charts when problem solving in larger mission-oriented projects.
	Use Pareto Charts for all quality improvement-oriented projects.
Add a new feature.	Add the number for incremental frequency next to the category of causes on the horizontal axis.
Modify a feature	Change the measurement units for incremental frequency to %. Each category's percentage is its portion of the total number of occurrences.

Summary

The topic discussed in this section was Pareto Chart, a type of histogram that indicates the frequency of occurrence of problems or causes in ranking order from left to right. The Pareto Chart can be used throughout the process of solving project quality problems. While early in the project it can help us decide which problem to study, later its contribution is in selecting which causes of the problem to tackle first. What makes a Pareto Chart valuable is its ability to help project teams concentrate on those causes of the problem that if removed will have the largest improvement impact. The following box summarizes the key points in developing the Pareto Chart.

Cause and Effect Diagram

What Is a Cause and Effect Diagram?

The Cause and Effect Diagram (CED) is a tool to identify, relate and graphically display causes of a problem (see Figure 14.6a). In a wider sense, since the problem is related to a process, CED visualizes the entire process. Very helpful in that regard is its hierarchical structure, indicating relationships of the effect to the main causes and their subsequent sub-causes. For example, main cause X has a direct relationship with the effect, while each of the subcauses is linked in terms of its level of impact on the main subcause. CED is also called the fishbone diagram because of its similarity to the skeleton of a fish or Ishikawa diagram for its inventor Kaoru Ishikawa.

(*a*) The Basic Cause and Effect Diagram

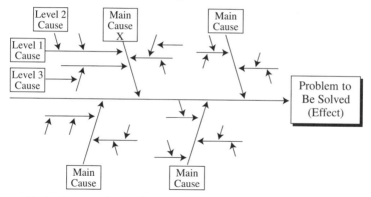

Figure 14.6 An example of a Cause and Effect Diagram of why projects get delayed.

Constructing a Cause and Effect Diagram

Collect Information Inputs. A CED is best used in the context of a quality improvement project that follows a Quality Improvement Map. Hence, the information about the project context, problem definition, goal, and targets are invaluable in efficiently building and applying the diagram.

Select the Format. There are two basic types of the diagram: dispersion analysis and production process classification. In this step, we need to determine which one is appropriate. Table 14.2, the construction process for the dispersion analysis, and details about the process classification diagram provided in section "Variations" may help with that decision. Our focus here is on the dispersion format.

Determine the Problem. So, what is the project quality problem we want to eliminate or at least improve on? If, for example, it is late projects, there must be a consensus when writing a problem statement—"Projects are late." Writing this statement on the right side of a sheet of paper (or computer screen) and drawing a box (an equivalent to the head of a fish) around it with an arrow (the "spine" of a fish) flowing to it will make up the effect side of the diagram. In a PM process, certain characteristics may be used as the effect. Examples may include budget overruns or dissatisfied customers on the level of project goals, or inefficient project kickoff meetings, poor milestone reviews, scope creep, or poor communication on a lower activity level.

Generate Ideas as to What Is Causing the Effect. Knowing the effect enables asking what the main causes of project delays are. As shown in Figure 14.6b, main causes (bones) are represented by branches running to the main branch. In this case, an organization experiencing late projects identified six areas as the main potential causes: PM process, management systems, metrics (performance measures), organizational culture, information systems, and organizational infrastructure. When there are difficulties in determining the main causes (branches), a usual strategy is to use traditional headings such as methods, machines, people, and materials [14], although they may not be very convenient for many types of projects.

Table 14.2: Types of Cause and Effect Diagram

Type of the Diagram	Basic Purpose	Major Advantage	Major Disadvantage
Dispersion analysis	Analyzes process dispersion (causes), relating causes and effect	Organizes, displays, and relates causes to effects Provides a framework for brainstorming or data analysis	May become complex Takes patience and commitment
Process classification	Identifies process steps, then lists factors affecting quality of each step	Visual image of factors impacting sequence of process steps Points to functional owners of steps that need improvement	Step-factor interrelation-ships may be difficult to identify

Identify All Possible Causes. Focusing on each main cause category and identifying all possible causes within the category is the essence of this step. Naturally, these will be shown on the diagram as subcauses. While practices vary, many a quality improvement team will resort to brainstorming to identify the subcauses, although the existing process data may also serve as a source of the subcauses' definition. Once the subcauses are known, there is a need to do some refinement work with them.

First, it is a legitimate practice to have the same subcause in several main cause categories, assuming that there is a direct, multiple relationship [15]. Next, having the subcauses of the main causes offers an opportunity to ask either of the following questions:

- ▲ What contributes to each of these sub-causes?
- ▲ Why does the main cause happen?

Continue asking the same questions until the lowest-level causes are uncovered. How many levels of subcauses should a CED have? The general rule is that you should keep digging for deeper causes but know when to stop. A rule of thumb is that when a cause is controlled by a manager that is removed from the team by more than one management level, there is no sense in further probing deeper causes [16]. Put simply, going deeper would not increase the effectiveness of the team actions.

With the lowest-level reached, the hierarchy of the CED is complete, revealing relationships that were not obvious. To clearly explain this, we can refer to Figure 14.6c, which illustrates the completed portion of the diagram for one main cause—management systems. Rewards, metrics, teamware, and planning and tracking methods are subcauses. In the case of teamware—also called collaborative technologies for teamwork—potential causes include the e-mail system, video conferencing, and intranet. More specifically, the lack of training and poor equipment may contribute to the substandard teamware. In summary, there are three levels of causes in this example, in addition to the main cause.

Review. Considering that a dominant approach is to build a diagram in a team setting, this is the time to have the whole group review the constructed diagram, making sure it is complete and meaningful (see the box that follows, "Tips to Construct a Cause and Effect Diagram"). Another way to review is to have others—not involved in the construction process but familiar with the process and problem—validate the diagram. Whether the former or latter method is preferred, it is highly recommendable to also confirm the diagram with data collection [16].

Utilizing the Cause and Effect Diagram

When to Use. The most valuable application of the diagram is in a quality project team environment, where all participants contribute, committing to make use of the diagram in the subsequent stages of the project. When that occurs, more value can be added to the diagram by quantifying the problem (effect) and as many of the causes as possible. With this information, the causes can be prioritized, targeted for improvement, and tracked.

Time to Construct. With proper facilitation, a five-member team can develop a diagram from half an hour to up to two hours. Larger teams and less skilled facilitation are bound to increase the necessary time.

Benefits. To improve project quality, you need to relentlessly gather more information about project processes. That is where CED provides value— it is the key to collecting information. Its design takes out the individual agendas, providing an invaluable focus on the content of the problem. Also, CED's benefit is its focus on issues at hand, rather than symptoms. Enabling the involvement of team members, CED helps build consensus, a collective view that inspires commitment to solving the problem.

Tips to Construct a Cause and Effect Diagram

- Think globally, act locally. Understand the impact of factors beyond issues in your control, but act on those you control.
- Listen to the ideas of participants. Capture their ideas about causes in only one or two words.
- Review. Have each team member review the CED the next day, or ask them to obtain an opinion of one or two additional people.
- You may state the desired results, instead of a problem. This helps to identify means (in place of causes) to achieve the results.

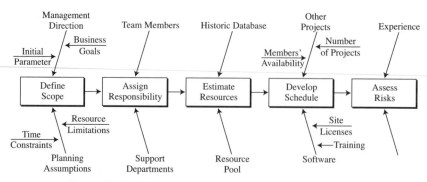

Figure 14.7 An example of the process classification diagram.

Advantages and Disadvantages. Aside from being a tool that can be easily learned by project participants at all levels of organization and applied immediately, the CED offers other advantages that have to be viewed in the context of its disadvantages. This is illustrated in Table 14.2, earlier in this tool's section.

Variations. Another format of the diagram is the process classification diagram (PCD). While the process of constructing it is similar to the one for dispersion analysis format, some differences are worth noting. The first task in building a PCD is to develop a basic Flowchart of the process targeted for improvement, laying out major process steps. The steps are in place of main cause categories used in the dispersion analysis. Then, all factors that might influence the quality of each major step, as well as the connecting steps between major steps, are added. Figure 14.7 offers an example of a completed PCD for project planning. The idea here is to fully understand all factors that may help improve the process. For major advantages and disadvantages, see Table 14.2.

Customize the Cause and Effect Diagram. Deriving the best value from a CED requires adapting it to your needs. Here are a few ideas to spur the inspiration to customize.

Customization Action	Examples of Customization Actions
Define limits of use.	Use dispersion analysis format for all but process redesign quality improvement projects.
Add a new feature.	Quantify the problem (effect) and causes to help cause priority setting.
Modify a feature.	Use *results* in place of the *effect* and *means to achieve results* instead of *causes*.

> **Cause and Effect Diagram Check**
> Check to make sure you developed a proper CED, including the following:
> - Effect (the problem)
> - Spine
> - Main causes with hierarchically related subcauses

Summary

This section described the Cause and Effect Diagram (CED), a tool to identify, relate, and graphically display causes of a problem. Since improving PM processes requires information, the value of the CED is in collecting such information. In addition, CED's structure helps eliminate the individual agendas while providing an invaluable focus on the content of the problem. By involving team members, CED helps build consensus, a collective view that inspires commitment to solving the problem. At the end, we focus our attention on the following box that highlights key points in creating a diagram.

Control Charts

What Are Control Charts?

The Control Chart is a tool that defines the voice of the process that we want to monitor, control, or improve over time (see Figure 14.8). It begins with a time-series graph, to which we add a central line as a visual location for discerning shifts or trends, and upper and lower control limits placed equidistant on both sides of the central line [17]. Lines on the Control Chart supply reference points that we use to decide whether the process behaves well or not. When a Control Chart shows time series that remains within the limits, with no apparent trend, nor any long sequences of points above or below the central line, the process is said to behave well and be in control. Such a process is predictable, consistent, and stable over time. Its opposite is the process that is not well-behaved, not in control—one that is unpredictable, inconsistent, and changes over time [17]. Thus, the essence of the Control Chart is its ability to reveal predictability, or lack of it. Key to it is the understanding of variation and its source (see the box that follows, "Understanding Variation"). While there are several different types of Control Charts, all of them are built on the same foundations (see the box "Pillars of Control Charts") and interpreted in the same way. Also, the same methodology is applied for their construction.

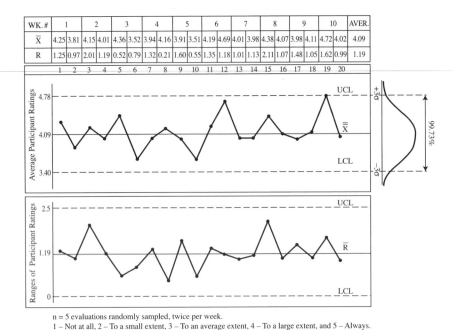

WK.#	1	2	3	4	5	6	7	8	9	10	AVER.
$\bar{X}$	4.25 3.81	4.15 4.01	4.36 3.52	3.94 4.16	3.91 3.51	4.19 4.69	4.01 3.98	4.38 4.07	3.98 4.11	4.72 4.02	4.09
R	1.25 0.97	2.01 1.19	0.52 0.79	1.32 0.21	1.60 0.55	1.35 1.18	1.01 1.13	2.11 1.07	1.48 1.05	1.62 0.99	1.19

n = 5 evaluations randomly sampled, twice per week.
1 – Not at all, 2 – To a small extent, 3 – To an average extent, 4 – To a large extent, and 5 – Always.

Figure 14.8 An example of $\bar{x}$- and R-chart for overall project management evaluation.

Understanding Variation

For each process or system, you can identify and measure indicators of performance [2]. In quality management, these indicators are called *quality characteristics* or *variables*. The difference between a project cost budget and the actual project cost, or the difference between a project's as-planned and as-is schedule are examples of quality characteristics for cost and time management processes, respectively. When the project is viewed as a system, quality characteristics can be related to its purpose. Some examples of the purpose might be project profits or customer satisfaction. All quality characteristics vary over time because of two types of causes [7]:

- *Common causes (also called chance, random, normal or natural causes, or background noise).* These are causes that are an inherent part of the process (or system), hour after hour, day after day, and influence each person involved in the process. As an example, the causes of variation in the time to develop a project schedule might include the workload of schedulers, their experience, and their familiarity with the scheduling software.

- *Special causes (also called assignable causes or signals).* These are not part of the process (or system) all the time, do not affect everyone in the process, and occur because of specific circumstances. For example, a new project member who is unfamiliar with the project risk response procedure may commit errors.

Understanding the difference—that is, variation—between the two types of causes is a prerequisite to effective use of the control charts and quality improvement.

Constructing Control Charts

Control Charts can help in three different areas [18]:

▲ Defining a process or to establish its state of statistical control
▲ Monitoring a process and signal when it goes out of control
▲ Estimating process capability

The following steps of planning a Control Chart, collecting data, determining control limits, and analyzing and interpreting the chart focus on the first application area. The second area is described in the step of using the chart as a problem-solving tool. Using the chart to determine process capability is beyond the scope of this book.

Plan a Control Chart. Crucial elements of the planning are

▲ Objective of the chart
▲ Selection of a process
▲ Measurement
▲ Sampling

Planning begins with the clear definition of the purpose of the chart [19]. If all of the involved do understand why the chart is being developed and what process is to be charted, their actions are more likely to be in line with the purpose and, thus, more effective. In our example in Figure 14.8, the project team FAB defines the objective of the chart—determine whether the PM process is stable for a selected quality characteristic (variable), so improvements can be pursued if necessary.

Based on the objective, one or more quality characteristics may be selected, and the types of data collected for the characteristic or variable will indicate the type of chart to use (for the procedure to choose the type of chart, see the "Variations" subsection). In our example, team FAB opts for one quality characteristic—the degree to which the project is managed per requirements in the project charter, leading to the use of $\bar{x}$ chart and R-chart (these charts are the focus of our discussion), which are charts for variables data. This characteristic is an aggregate of some 14 requirements, ranging from schedule to budget to communication to enforcement of the company's values.

The method of measurement is the degree of accomplishment of each of the 14 requirements, measured by means of a survey/evaluation of the Likert type on a 1 to 5 basis, with 1 being "Not at all per project charter" and 5 being "Always per project charter." All 15 team members complete two evaluations each week (frequency of sampling) after the regular Tuesday and Friday progress meetings (point of sampling), five of which will be randomly chosen for the charting (subgrouping strategy). For each of the five evaluations, a mean of ratings for the 14 requirements is calculated.

Pillars of Control Charts

According to Wheeler, these are the pillars of control charts [1]:

1st pillar. Control charts use three-sigma control limits. Whatever type of chart you use, the limits follow the same rule—they are set at three estimated standard deviations on both sides of the central line. There, their role is to separate probable noise (common cause variation) from potential signals (special cause variation).

2nd pillar. Use an average dispersion statistic to calculate three-sigma control limits. It doesn't matter whether you use ranges, standard deviation, or root mean square. When you use a proper approach, different statistics will produce similar results. In contrast, different statistics will produce similar incorrect results when a wrong approach is used.

3rd pillar. Control charts are rooted in rational sampling and rational subgrouping. This means that the way the data are gathered, placed in the subgroups, and charted is directed by several factors. These include the context of the data, sources of data variation, questions charts are addressing, and the intended use of the obtained knowledge.

4th pillar. Control charts are valuable only if you know to use them. Or, collecting data and building the charts without utilizing them is useless. Simply, if you do not utilize them, nothing will happen.

Collect Data. In this step, we will record the data, calculate relevant statistics, and plot the statistics on the chart. Usually, about 15 to 20 samples (subgroups of five) are collected and recorded on a Control Chart sheet, including any unusual events if they occur. More information about the necessary number of samples is in the box on page 481, "How Much Data Do You Need To Calculate Control Limits." For each sample, we calculate the mean $\bar{x}$ and range (R, the highest value minus the lowest value), and plot them on the their respective Control Charts. Next, the overall mean (also called grand average) and average range are computed, determining the central lines for:

$$\text{the } \bar{x}\text{-charts: } \bar{\bar{x}} = \frac{\sum_{i=1}^{k} \bar{x}i}{k} \text{ and R-charts: } \overline{R} = \frac{\sum_{i=1}^{k} Ri}{k}$$

Details about formulas can be found in Table 14.3.

Determine Control Limits. Use the average range and average mean to compute upper (UCL) and lower control limits (LCL) for R- and $\bar{x}$-bar charts with these formulas:

$$UCL_R = D_4 \overline{R} \qquad UCL_{\bar{x}} = \bar{\bar{x}} + A_2 \overline{R}$$
$$LCL_R = D_3 \overline{R} \qquad LCL_{\bar{x}} = \bar{\bar{x}} - A_2 \overline{R}$$

Here, D_3, D_4, and A_2 are the constants that depend on the sample size. Formulas to calculate the control limits and the constants are given in Table 14.3.

(a) Variable data formulas

Chart Type	Central Line	Control Limits
$\bar{x}$ and R (Average and Range)	$\bar{\bar{x}} = \dfrac{(\bar{x}_1+\bar{x}_2+\ldots\bar{x}_k)}{k}$	$UCL_{\bar{x}} = \bar{\bar{x}}+A_2\bar{R}$ $LCL_{\bar{x}} = \bar{\bar{x}}-A_2\bar{R}$
	$\bar{R} = \dfrac{(R_1+R_2+\ldots R_k)}{k}$	$UCL_R = D_4\bar{R}$ $LCL_R = D_3\bar{R}$
x and Rm (Individuals and Moving Range)	$\bar{x} = \dfrac{(x_1+x_2+\ldots x_k)}{k}$	$UCL_x = \bar{x}+E_2\bar{Rm}$ $LCL_x = \bar{x}-E_2\bar{Rm}$
	$Rm = \mid (x_{i+1} - x_i) \mid$ $\bar{Rm} = \dfrac{(R_1+R_2+\ldots R_{k-1})}{k-1}$	$UCL_{Rm} = D_4\bar{Rm}$ $LCL_{Rm} = D_3\bar{Rm}$
$\bar{x}$ and s (Average & Standard Deviation)	$\bar{\bar{x}} = \dfrac{(\bar{x}_1+\bar{x}_2+\ldots\bar{x}_k)}{k}$	$UCL_{\bar{x}} = \bar{\bar{x}}+A_3\bar{s}$ $LCL_{\bar{x}} = \bar{\bar{x}}-A_3\bar{s}$
	$\bar{s} = \dfrac{(s_1+s_2+\ldots s_k)}{k}$	$UCL_s = B_4\bar{s}$ $LCL_s = B_3\bar{s}$

k = Number of samples (subgroups)

(b) Attribute data formulas

Chart Type	Central Line	Control Limits
p-Chart (Fraction defective)	For each subgroup: p=np/n For all subgroups: $\bar{p}=\Sigma np/\Sigma n$	$UCL_p=\bar{p}+3\sqrt{\dfrac{\bar{p}(1-\bar{p})}{n}}$ $LCL_p=\bar{p}-3\sqrt{\dfrac{\bar{p}(1-\bar{p})}{n}}$
np-Chart (Number defective)	For each subgroup: np=Number of defective units For all subgroups: $n\bar{p}=\Sigma np/k$	$UCL_{np}=n\bar{p}+3\sqrt{n\bar{p}(1-\bar{p})}$ $LCL_{np}=n\bar{p}-3\sqrt{n\bar{p}(1-\bar{p})}$
c-Chart (Number of defects)	For each subgroup: c=Number of defects For all subgroups: $\bar{c}=\Sigma c/k$	$UCL_c=\bar{c}+3\sqrt{\bar{c}}$ $UCL_c=\bar{c}-3\sqrt{\bar{c}}$
u-Chart (Number of defects per unit)	For each subgroup: u=c/n For all subgroups: $\bar{u}=\Sigma c/\Sigma u$	$UCL_u=\bar{u}+3\sqrt{\bar{u}/n}$ $UCL_u=\bar{u}-3\sqrt{\bar{u}/n}$

np = Number of Defective Units   n = Sample Size
c = Number of Defects   k = Number of Samples

(c) Table of constants

	x- and Rm-chart			$\bar{x}$- and R-chart			$\bar{x}$- and s-chart		
	E_2	D_3	D_4	A_2	D_3	D_4	A_3	B_3	B_4
2	2.659	0	3.267	1.880	0	3.267	2.659	0	3.267
3	1.772	0	2.574	1.023	0	2.574	1.954	0	2.568
4	1.457	0	2.282	0.729	0	2.282	1.628	0	2.266
5	1.290	0	2.114	0.577	0	2.114	1.427	0	2.089
6	1.184	0	2.004	0.483	0	2.004	1.287	0.030	1.970
7	1.109	0.076	1.924	0.419	0.076	1.924	1.182	0.118	1.882
8	1.054	0.136	1.864	0.373	0.136	1.864	1.099	0.185	1.815
9	1.010	0.184	1.816	0.337	0.184	1.816	1.032	0.239	1.761
10	0.975	0.223	1.777	0.308	0.223	1.777	0.975	0.284	1.716

Table 14.3   Formulas and constants for Control Charts.

When Do You Recalculate the Control Chart Limits?

The issue of recalculating the limits can be operationalized by the following questions [3]:

Question 1: Does your chart exhibit a distinctly different type of behavior than in the past?

Question 2: Do you know why this change in behavior has happened?

Question 3: Is the new process behavior desirable?

Question 4: Do you expect and desire the new process behavior to continue?

If you answer "yes" to all four questions, basing them on data that you collected after the change in the process, you should recalculate the limits. If you answer "no" to question 1, keep the old limits. If you answer "no" to questions 2, 3, and 4, leave limits alone. Rather, look for the assignable cause.

> **Why Three-Sigma Limits?**
>
> Three-sigma limits are deliberately chosen by Shewhart because they make economic sense. They are sufficiently wide to eliminate the majority of the probable noise (common cause variation), enabling users to not waste time interpreting noise as signals (special cause variation). At the same time, the limits are sufficiently narrow to spot probable signals and allow users to not miss signals of economic relevance. Setting limits at three-sigma level provides a satisfactory balance of the two mistakes [5].

The central line and control limits for FAB's data from Figure 14.8 are as follows:

Central line for R-chart = $\overline{R}$ = 1.19

Central line for $\overline{x}$-chart = $\overline{\overline{x}}$ = 4.09;

UCL_R = 2.114 x 1.19 = 2.51 $UCL_{\overline{x}}$ = 4.09+ 0.577 x 1.19 = 4.77

LCL_R = 0 x 1.19 = 0 $LCL_{\overline{x}}$= 4.09– 0.577 x 1.19 = 3.40

This is the time to add the central line and control limits to the data already plotted on the respective charts and get the complete charts as shown in Figure 14.8. Again, when a Control Chart shows data points that remain within the control limits, with no apparent trend nor any long sequences of points above or below the central line, the process is said to behave well and be in control. If not so, the data falls outside the limits and shows unusual patterns, meaning that some special cause has probably affected the process. As a result, the central line and limits are biased. Essentially, this tells the team to locate the special cause and eliminate or control it in order to reach the state of statistical control. Take a new set of samples/subgroups and recalculate $\overline{\overline{x}}$, $\overline{R}$, and the control limits (for more details, see the box on page 475, "When Do You Recalculate the Control Chart Limits?").

When deciding whether a process is in control, look at the R-chart first. Because control limits in the $\overline{x}$-chart depend on the average range, special causes in the R-chart may create unusual patterns. When R-chart in is in control, we are ready to shift our attention and bring the $\overline{x}$-chart to an in-control state as well.

Analyze and Interpret the Chart. Only common causes are present in a process that is in statistical control. What does a Control Chart for a process that is in control really look like? Such a chart shows points that fluctuate randomly between the control limits without recognizable pattern. In particular, this means the following principles apply [18]:

1. There are no points outside the control limits.
2. The number of points above the central line is approximately the same as below the line.

3. The points seem distributed randomly above and below the central line.

4. The majority of the points are near the central line, and only a few are close to the control limits.

In other words, these principles assume that distribution of sample means is normal. The assumption is related to the central limit theorem of statistics that applies to subgroup means; that is, as the subgroup size increases, the distribution of the subgroup means will approach a normal distribution, regardless of how individual measurements are distributed (this is shown with a normal distribution curve in Figure 14.8). Recall that control limits are set at three sigma (standard deviations) from the overall mean (see the box on page 476 "Why Three-Sigma Limits?"), implying that the probability that any subgroup mean falls outside the control limits is as small as 0.27 percent, apparently very small. The remaining 99.73 percent should fall within the limits. This is the root of principle 1.

The symmetric nature of the normal distribution, with the same number of points falling above and below the central line, explains principles 2 and 3. Additionally, because the mean of the normal distribution is also median (the midpoint of the distribution), about half the points are on either side of the central line. Principle 4 is based on the knowledge that 68 percent—that is, the majority—of normally distributed points is near the central line, or more precisely, within one standard deviation (sigma) of it. Again, these principles hold as long as the process is stable and common causes reign, as is the case in Figure 14.8.

All too often, processes do not behave well. Rather, they go out of control when special causes come into play. What does a Control Chart for a process that is not in control look like? Figure 14.9 offers several such examples. When one point is outside of control limits (Figure 14.9a), the reason is either a special cause, or it is a common cause—a normal occurrence in the process, one of those points with the probability of 0.27 percent to fall outside the control limits. An uncharacteristic number of points on one side of the central line typically hints at the presence of special causes (e.g., bringing a new scheduler in an ongoing project), leading to a sudden shift in the process mean (Figure 14.9b).

Causes that come and go on a regular basis may produce cycles (Figure 14.9c). Alternating two cost estimators using their experiential (different) productivity standards is an example. A process may be influenced by special causes that gradually move points on the chart up or down from the central line, creating a gradual trend (Figure 14.9d). In one project, for example, when inexperienced programmers replaced the experienced ones on a team charting scope changes, the number of scope changes went up for some time before getting back to "normal."

Finally, when many points are near the central line (Figure 14.9e) or the control limits (Figure 14.9f), we experience hugging the central line and control limits, respectively. As an example, the former may be caused by systematically taking a sample from two work package managers, one of them with very high cost variances, the other with very low ones. Despite the high and low variances, the sample mean will not reflect this but rather average them out and hug the central line.

With this step, we conclude using the Control Chart to define a process or to establish its state of statistical control, and turn our attention to Control Charts as a problem-solving tool.

Apply the Control Chart as a Problem-Solving Tool. Control Charts allow us to characterize a process as being in control or out of control [20]. A process in control runs as consistently as possible, while one out of control does not. It is this difference that inspires people to venture into the continuous quality improvement. An out-of-control process means that variation is due to special causes that the team can search for, following the guidance of the Control Chart [21]. Uncovering and remedying the causes means improving the process, often with a little investment.

On the other hand, an in-control process has no special causes to look for, therefore, you should refrain from it. To improve it, we need to change something basic in its design. Before doing so, we should know that Control Charts define the voice of the process. But the voice of the process is very different from specifications, often called the voice of the customer. The two voices are different! That a process is in control does not mean it can meet the specs. Are we, then, getting the most of our process? If we think not, we should follow the guidance of the Control Chart and align the two voices. This would require taking action on the process—changing its basic design—to keep improving it until it is capable to meet customer specs. For example, in a PM process that consistently delivers project type X in 12 months, to meet a customer's requirement of 10 months, you may need to abandon a stage-by-stage approach in favor of overlapping stages. Also, you may reduce senior mangers' full control of slow milestone reviews in favor of an approach empowering teams to perform fast but equally good reviews on their own. These two are changes to the process design.

Utilizing Control Charts

When to Use. As already mentioned, in a process Control Charts can help you establish its state of statistical control, monitor it, and estimate its capability [22]. All three applications are useful in both across-projects processes and single-project processes. Simply because of the sheer amount of resources in larger projects, they are more frequent users of the charts.

Time to Construct. With the advent of computerized programs for Control Charts, setting up and maintaining them does not have to be time-consuming. In our example from Figure 14.8, a team of several skilled people took a few hours to plan the chart, develop the evaluation instrument, take first measurements, and set up the chart. With less complex measurements the time may decrease.

Benefits. The real value of Control Charts is in an organization's ability to use them to understand and improve their processes. By providing a common language to discuss process performance, the charts tell whether or not the process is predictable and in control, filtering out the noise (common causes) from the signal (special causes). With such knowledge, we know when to leave the process alone and when to act to improve it [23]. This encourages appropriate actions and discourages inappropriate ones [24].

Advantages and Disadvantages. Control Charts are considered to offer advantages of simple and effective tools that even less educated personnel can use with some training. At the same time, their statistical nature may challenge many a project manager.

Variations. There are many types of Control Charts. Figure 14.10 offers a tree diagram for selecting the appropriate chart. Begin by asking, "Do I have variables or attributes data?" Certain types of charts are only for variables data; others are for attributes data. Variables data are measured on a continuous scale, such as time (project schedule), cost (project budget), figures (number of scope changes), and so on. The $\bar{x}$-chart and the R-chart are most common charts for the variables data.

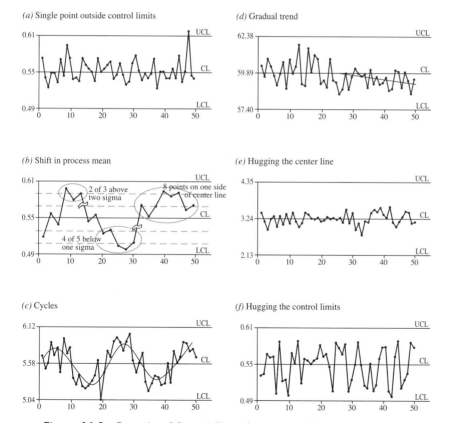

Figure 14.9 Examples of Control Charts for processes that are not in control.

From *The Management and Control of Quality*, 4th edition, by J. R. Evans and W. M. Lindsay © 1999. Reprinted with permission of South-Western College Publishing, a division of Thomson Learning.

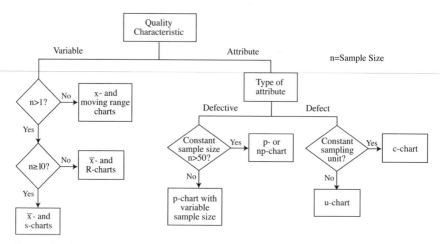

Figure 14.10 *Control Chart selection on the basis of the type of data and sample size.*

From *The Management and Control of Quality*, 4th edition, by J. R. Evans and W. M. Lindsay © 1999. Reprinted with permission of South-Western College Publishing, a division of Thomson Learning.

In attributes data, only two values are assumed: good or bad, yes or no. While they usually cannot be measured, it is possible to observe and count them. For example, in monitoring project milestones, a milestone is either accomplished (good) or is not (bad). Percent of milestones accomplished, an often-used performance metric, is then easy to track and chart. Some attributes data belong to the category of defect data, others to defective data.

A defect is a single nonconforming quality characteristic. For example, an organization surveys its 60+ customers on ten project quality characteristics—from cycle time to price/performance ratio—on a scale of 1 (not satisfied at all) to 6 (extremely satisfied). Every time the mean for a characteristic is rated 4 or below, it is considered a defect monitored on a u-chart. When a project (item) has two or more defects, it is considered defective. The organization uses p-chart to monitor defective projects. Once we clarified the type of data we have, we look for the sample size to determine which chart is needed.

Customize the Control Chart. Getting the best value out of Control Charts takes an effort to customize them for your own project needs. Several ideas for customization are offered in the box below.

Customization Action	Examples of Customization Actions
Define limits of use.	Use Control Charts in larger projects.
	Use Control Charts to monitor the stability of basic PM processes (e.g., cost estimating accuracy or schedule variance control).
	Use Control Charts to improve basic PM processes (e.g., customer satisfaction with the project delivery process).

> **Control Chart Check**
>
> Check to make sure you developed a proper Control Chart that includes the following:
> - The right type of chart
> - The central line
> - The time series
> - Properly calculated control limits

Summary

This section focused on the Control Chart, a tool that helps a PM process establish its state of statistical control, monitor it, and estimate its capability. All three applications are useful in both across-projects processes and single-project processes. Because they have more resources, larger projects are more frequent users of the charts. The box above highlights the key points in structuring the chart.

Concluding Remarks

This chapter presented four tools—Quality Improvement Map, Pareto Chart, Cause and Effect Diagram, and Control Charts—each one with a different purpose. Therefore, these tools do not compete with each other for the project manager's attention. Rather, they can be viewed as a suite of tools for quality control in projects. Quality Improvement Map leads a project quality team through a logical sequence of stages and steps, and enables a meticulous analysis of the project problem, its potential causes, and possible solutions. A significant help in this effort is the Cause and Effect Diagram, designed to identify, relate, and graphically display causes to a problem. Concentrating on those causes of the problem, which if removed will have the largest improvement impact, is what the Pareto Chart enables. Some of the problems can be eliminated by the Control Chart, a tool that defines the voice of the process that we want to monitor, control, or improve over time.

In the table that follows we provide a summary comparison, indicating situations that favor certain tools. If these situations are not sufficient to characterize the project, add more of them and mark how they favor the use of the tools. If a tool has many marks, chances are it is needed in the project. A careful study of the material covered in this chapter may help you better understand the need.

> **How Much Data Do You Need to Calculate Control Limits?**
>
> You may start calculating useful but soft limits with two subgroups of size four. These and other limits computed with less than 15–20 data points are considered "temporary limits," and should be revised as more data are generated. Once you hit 15–20 individual values, the limits start solidifying. Of course, you can use more data for the computation of limits, but exceeding 50 data points will provide little additional value [4].

A Summary Comparison of Quality Control Tools

Situation	Favoring Quality Improvement Map	Favoring Pareto Chart	Favoring Cause and Effect Diagram	Favoring Control Charts
Provide quality improvement methodology	✓			
Prioritize importance of problems and causes		✓		
Show causes of the problem			✓	
Control, monitor, or improve a process				✓
Individual small and simple projects		✓	✓	
Individual large and complex projects		✓	✓	✓
Organizations with a stream of smaller projects	✓	✓	✓	✓
Simple and easy to learn/use the tool		✓		
Not so simple and easy to learn/use the tool	✓			✓
Support a proactive approach to quality	✓			✓

References

1. Wheeler, D. J. 1996. "Foundations of Shewhart's Charts." *Quality Digest* 16(10): 59.

2. Nolan, T. W. and L. P. Provost. 1990. "Understanding Variation." *Quality Progress* 23(5): 70–78.

3. Wheeler, D. J.,"When Do I Recalculate My Limits?" *Quality Digest* 1996. 16(5): 79–80.

4. Wheeler, D. J. 1996. "How Much Data Do I Need?" *Quality Digest* 16(6): 59–60.

5. Wheeler, D. J. 1996. "Why Three-Sigma Limits." *Quality Digest* 16(8): 63–64.

6. Scholtes, P. R., B. L. Joiner, and B. J. Streibel. 1996. *The Team Handbook*. 2d ed. Madison, Wis.: Joiner Associates Inc.

7. Shewhart, W. A. 1931. *The Economic Control of Quality of Manufacturing Product*. New York: D. Van Nostrand Company.

8. Juran, J. M. 1974. *Quality Control Handbook*. 3d ed. New York: McGraw-Hill.

9. Gaudard, M., R. Coates, and L. Freeman. 1991. "Accelerating Improvement." *Quality Progress* 24(10): 81–88.

10. Kepner, C. H. and B. B. Tregoe. 1965. *The Rational Manager*. New York: McGraw-Hill.

11. Anderson, S. D. and E. L. Cook. 1995. "TQM Implementation Strategy for Capital Projects." *Journal of Management in Engineering* 11(4): 39–47.

12. Stevens, J. D. "1997. Blueprint for Measuring Project Quality." *Journal of Management in Engineering*. 12(2): 34–39.

13. Burr, J. 1990. "The Tools of Quality, Part IV: Pareto Charts." *Quality Progress*. 23(11): 59–61.

14. Mitra, A. 1998. *Fundamentals of Quality Control and Improvement*. Upper Saddle River, N.J.: Prentice Hall.

15. Sarazen, J. S. 1990. "The Tools of Quality, Part II: Cause-and-Effect Diagrams." *Quality Progress*. 23(7): 59–62.

16. Brassard, M. and D. Ritter. 1994. The Memory Jogger II. Salem, N.H.: GOAL/QPC.

17. Wheeler, D. J. 1996. "What Are Shewhart's Charts?" *Quality Digest*. 16(1): 72.

18. Evans, J. R. and W. M. Lindsay. 2001. *The Management and Control of Quality*. 5th ed. Cincinnati: South-Western Publishing Company.

19. Nolan, K. M. 1990. "Planning a Control Chart." *Quality Progress* 23(12): 51–55.

20. Wheeler, D. J. 1997. "Three Types of Actions." *Quality Digest* 17(8): 23.

21. Pyzdek, T. 1998. "I Hate SPC." *Quality Digest* 18(9): 26, 54.

22. Goetsch, D. L. and S. B. Davis. 2000. *Introduction to Total Quality.* 3rd ed. Upper Saddle River, N.J.: Prentice Hall.

23. Shainin, P. D. 1990. "The Tools of Quality, Part III. Control Charts." *Quality Progress.* 23(8): 79–82.

24. Godfrey, A. B. 1999. "Statistical Quality Control." *Quality Digest.* 18(3): 18.

Performance Reporting and Closure

M ajor topics in this chapter are performance reporting and closure tools:

- ▲ Risk Log
- ▲ Summary Progress Report
- ▲ Postmortem Review

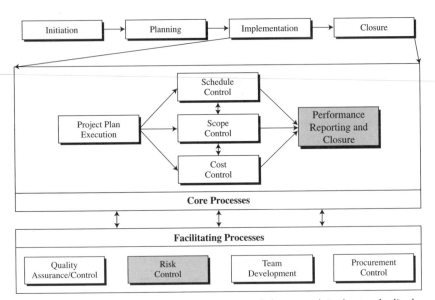

Figure 15.1 The role of performance reporting and closure tools in the standardized project management process.

These tools are designed to aid in monitoring and reporting progress, and closing the project (see Figure 15.1). In particular, they first enable us to monitor, report, and control risks. Further, with their help we can collect and disseminate performance information to familiarize stakeholders with project progress toward its objectives. Lastly, they can assist us in learning from our project mistakes and successes in order to improve future performance. The value of what they help produce is strongly dependent on the information they receive from the scope, schedule, cost control, and other tools from facilitation processes. The target of this chapter is to offer practicing and prospective project managers the following learning opportunities:

▲ Become familiar with performance reporting and closure tools

▲ Choose performance reporting and closure tools that match their project situation

▲ Customize the tool of their choice

These skills are vitally important in implementing a project and building a standardized PM process.

Risk Log
What Is a Risk Log?

Simply said, a Risk Log is an extension of the Risk Response Plan. When the latter lays down actions to mitigate risks, the former helps control that they happen. Such control boils down to three major activities: establishing the risk baseline, evaluating the actual risk status, and defining further actions (see Figure 15.2). It is these three activities that fully define the purpose and content of the Risk Log.

RISK LOG

Project Name: Genesis

Page #: 1 of 1
Date: May 23, 02

Work Package/ Task	Risk Event # and Description	Owner	Criticality*	Plan			Actual			
				Preventive	Trigger Point	Contingent	Status	Date of Impact	Risk Impact	Recovery Action
Code and Unit tests	9. Internal S/W testing slow	Marsha M.	C	Outsource	Vendor not chosen by 6/1/02	Use old vendor	Preventative action two weeks behind schedule	6/7/02	Two-week delay	Activate contingency action

* C–Critical
  NC–Near-critical
  NNC–Noncritical

Figure 15.2 An example of the Risk Log.

Keeping the Risk Log

Despite its deceivingly simplistic appearance as a spreadsheet, the Risk Log reflects the process of controlling risks. That calls for a certain level of discipline and mental concentration, which in turn mandates the existence of an arranged sequence of steps that we describe in continuation.

Prepare Information Inputs. Disciplined and meaningful process of keeping the Risk Log would not be possible without these three inputs:

- ▲ Risk management plan. A document that provides a roadmap for dealing with risk throughout the project's life, including details on establishing and running the Risk Log
- ▲ Risk Response Plan. The baseline to compare against the actual risk status
- ▲ Performance reports or verbal information about actual performance indicating the actual risk status.

Refer to the Baseline. It is the baseline risks that we target to mitigate when executing the project. Therefore, we go back to where the baseline is—in the Risk Response Plan—to refer to our target. The target definition that includes the first seven columns in Figure 15.2 begins with the description of the individual risk and its number related to the Work Breakdown Structure. The name of the person who is responsible to mitigate the risk (i.e., the owner), the risk's level of criticality, preventive action, trigger point, and contingent action are all parts of the baseline. When this information is provided, the target is discerned, and it is time to move on and acquire information about the actual risk status.

Establish the Actual Status of Risk. Regular progress meetings are a perfect place to establish the actual status of the project. Most project managers will cleverly seize such an opportunity to do this by comparing the baseline schedule and budget to their baselines. While that is a good approach, it would be even more meaningful if they included the progress statusing of risk in their analysis. Risks, like schedule and budget, need to be reviewed regularly in progress meetings (this does not preclude having unscheduled risk reviews), especially in risk-abundant projects.

In line with the Risk Response Plan, taking the preventive action is the first step in mitigating a risk. Typically, the action is broken down in lower-level tasks with identified schedule and responsible persons (typically not shown in the Risk Log). In Figure 15.2, for example, the preventive action for risk # 9 (second row) is to outsource the software testing. To make it happen, the owner divided the preventive action into tasks of defining the scope of testing, identifying vendors, issuing a request for a proposal, gathering and evaluating proposals, shortlisting and negotiating, and selecting the vendor to be given the job. The schedule and responsibilities for the tasks are also defined. In the progress meeting of May 23, 2002, she reports the preventive action status. It is two weeks behind schedule; in other words, it is failing. She also informs the audience that the date of risk impact (when the software testing is planned to start) is just around the corner. If the preventive action is not successfully completed by July 1, 2002, the project will be hit with a two-week delay. This is, of course, accompanied by a detailed discussion about the cause of the failure and possible course of recovery.

Define Further Actions. Determining what is the further action to recover and mitigate the risk is the next step. The consensual opinion in the progress meeting discussion is that continuing with the preventive action cannot mitigate the risk. The meeting culminates in the decision to suspend the preventive action and activate the contingent action of using the old vendor. The risk owner is asked to develop an action plan to be reviewed and approved in an unscheduled risk review the next day.

Overall, the sequence of

- ▲ Referring to the risk baseline
- ▲ Establishing the actual risk mitigation status
- ▲ Identifying further actions via the Risk Log

is as logical as any schedule or budget control action. What is also logical is that it takes the step of

- ▲ Enforcing the log keeping and the related process as a regular practice and behavior

if the risk response plan is to be executed successfully (for a good example, see the box that follows, "The Discipline of Risk Leaders").

The Discipline of Risk Leaders

This is a story of a project manager from a leading company whose server development group sees regular and disciplined risk response as a high priority. "I always put in place a mitigation plan for absolutely everything that we risk. Sometimes I may have a list of 100 risks. It doesn't matter, since each one has to be mitigated. I analyze each one per our scale of showstopper, high, medium, low, and no impact. I've got a spreadsheet set up that automatically assigns value to each risk and come up with a score. Based on the length of the project and assigned values, I put together a smooth curve that shows me over time how to take mitigation actions to get to zero risk. So, I have a target, 'cause I know I am not going to clear all my risks up front. And, then, I attack. I go to my team and define a mitigation plan for each risk. It can be as simple as 'Monitor weekly meetings for this risk.' Or, it may be more complex. Take, for example, a showstopper risk—the kit group doesn't provide us with rules for routing particular bus on the board. If that happens, the board can't be laid off, and the server project can't be completed timely. So we ask in the team meeting, 'If they don't give it to us, what can we do? Our mitigation plan may be branch-predict or go ahead, get it all done, making a provision for modifying the routing rule. Or, work on a daily basis with the kit group, get them more resources, elevate the priority of their task, whatever. Get onto it. Monitor, until you close the risk up."

Utilizing the Risk Log

When to Use. If a project is a reasonable ground for risk events to occur, it deserves the Risk Log. In such a situation the extent of formality in keeping the log will hinge on the size and complexity of the project. For large and complex projects faced with an abundance of risks, insistence on the disciplined monitoring of risks and consequently keeping a track of them is a must and also something they have time and resources to do. As a high technology project manager put it, "My Risk Log is one of the top three things in my project (the other two are the schedule and communication with team members)." Conversely, the reality of limited resources will force smaller projects to stick with informal treatment of risks and the Risk Log.

Time to Develop. It is good to know that keeping the log is not time-consuming, requiring no more than a minute or two to make an entry. Even the long list of risks in larger projects is not a time killer. Where time may be consumed, and deservedly so, is in what precedes making an entry—monitoring risks and discussing actions deployed to mitigate them. Such management activities, however, are value adding and in many ways destined to determine the project future.

Benefits. As a repository of risk management actions, the Risk Log provides an important overview of major risk events that are identified, assessed, quantified, responded to, and mitigated or not. Its value can be comprehended by asking what if we did not have it? In a project fraught with risks, without the Risk Log, the project might feel the heat that Damocles felt when sitting under a sword suspended by a single hair. In contrast, having a log may be viewed as an equivalent of eliminating or reducing the heat. This is done by providing clarity of threats posed to the project that might lead to both cost losses and project delays. It also offers transparency of actions taken to avert such undesired outcomes. Most importantly, the Risk Log creates a potential to make decisions that impact the undesired project events.

Advantages and Disadvantages. The Risk Log has an appeal for its

- ▲ *Clarity.* You can find succinct information about major risks status clearly displayed in one place. That convenient one-stop risk overview, rid of excessive details, is an efficient time-saver.
- ▲ *Simplicity.* A quick look at the Risk Log is perhaps sufficient for someone to construe its message and even start using this tool. To those with tight time budgets, this is a blessing.

With its appearance of a spreadsheet, the Risk Log may get a label of one more bureaucratic thing to deal with, increasing the level of documentation. Additionally, its format of presenting individual risks as line items isolated from each other may appear as lacking the reality of risk interactions.

Variations. Perhaps the most widespread variation is called the issue log. As discussed earlier in Chapter 12, in the box "Issues and Risks," many project managers use the terms issues and risks interchangeably [2]. Consequently, the issue log tracks issues in the project, including both issues that have already occurred and future issues (essentially, risks). Such a tool is very similar to the Risk Log and may include, as an example, these columns: issue number, issue description, date of issue, activity affected, issue criticality, trigger point date, resolution responsibility, required action, next status date, and whether or not the issue is resolved.

A natural extension of the issue log is the issues aging chart (see Figure 15.3). Using curves, the chart tracks the cumulative number of unresolved issues of low priority, high priority, and total issues over time. The purpose is to identify the trend and act if it is not favorable. For example, you would not want to have the unfavorable trend of the high-priority curve going upward all the time. Rather, you want to discern such a trend in a timely manner and take actions to reverse it by resolving the majority of issues, if not all.

Another outgrowth of the issue log is the issue database (see the box that follows, "Issue Database"). Just as some companies call these tools issues aging charts and issue database, others name them risk aging charts and risk databases.

Customize the Risk Log. Although the Risk Log's simplicity does not offer unlimited opportunities for customization, it is worth considering some changes to its generic format that we described. That should help make it better match your own needs and culture of monitoring risks. The purpose of the few ideas listed is no more than to stimulate thoughts about customizing the tool's scope.

Issue Database

Learning from past experience is a way for many organizations to continuously improve. A tool that helps classify such learning and offers improvement strategies to future generations of projects is the issue database. Simply, it records three types of information: (1) issues of significant impact that occurred in past projects, (2) the nature of the impact that happened or was prevented from happening, and (3) what actions were or could be used to successfully resolve such issues.

How is the database developed? Searching through risk logs of past or current projects and postmortem reviews helps identify the preceding types of information. Issues of a similar nature are then grouped. For example, groups may include team issues, technology issues, vendor issues, scheduling issues, risk management issues, and so on. [1]. Computerized databases of this sort that are searchable are of special value.

What can the database do for a project? It can serve as a checklist in the project planning. Also, it can serve as a checklist in the proactive cycle of project control (PCPC) to identify issues and risks in monitoring schedule, cost, and scope, such as:

- What are the issues causing the variance?
- What new risks may pop up in the future and how could they change the preliminary predicted project completion?

Additionally, the database offers "premade" impact assessment and actions to mitigate the impact

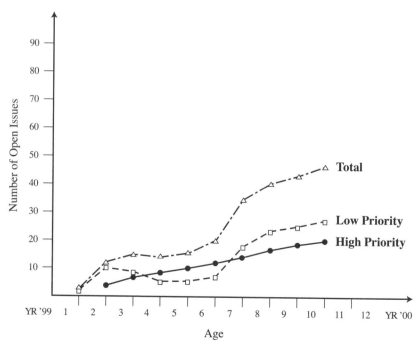

Figure 15.3 An issues aging chart.

> **Risk Log Check**
> Check to make sure your Risk Log is properly configured. The log should
> - Be based on necessary information inputs: risk management plan, risk response plan, and information about actual performance.
> - Indicate the risk baseline: owner, criticality, preventive action, trigger point, and contingent action.
> - Describe the actual status: risk is mitigated or not; if not, list its status and the date of impact.
> - Define details of the further action, if not already mitigated.

Customization Action	Examples of Customization Actions
Define limits of use.	Use Risk Log in all projects facing abundant risks.
Add a new feature.	Add a column indicating project activities affected by the risk (useful for cross-functional projects).
Modify a feature.	Use "time to impact" measured in time units (e.g., days) instead of "date of impact."

Summary

The Risk Log is the tool presented in this section. It helps evaluate the actual risk status and define further actions. If a project is a reasonable ground for risk events to occur, it deserves a Risk Log, which creates a potential to make decisions that impact the undesired project events in the course of project implementation. Tailoring the Risk Log to match specific project needs can make it even more beneficial. The following box recaps the key points in organizing the Risk Log.

Summary Progress Report
What Is the Summary Progress Report?

The Summary Progress Report is a document typically a page long that highlights and briefly describes the status of the project, reporting on the scope, cost, time, and quality variance, showing significant accomplishments, identifying issues, predicting trend, and stating actions required to overcome issues and reverse the negative trend. Contrary to the beliefs of many that the report is about the history of the project, the report should be about the future of the project that is based on its past. An example of such a proactive Summary Progress Report is illustrated in Figure 15.4.

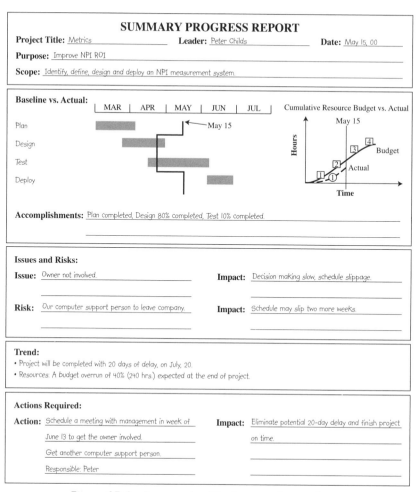

Figure 15.4 An example of the Summary Progress Report.

Developing a Summary Progress Report

Prepare Information Inputs. Producing a meaningful Summary Progress Report starts off quality information inputs such as:

- ▲ Project baseline plan
- ▲ Work results and project records
- ▲ Proactive cycle of project control (PCPC).

Lacking a solid project baseline plan makes it practically impossible to establish where the project is relative to its goals. When the project baseline plan with various baselines—scope, cost, and schedule, for example—is available, then they are compared to the actual state of the project to assess its performance. The actual state is derived from work results and other project records. Through work results, for example, we report which tasks or deliverables are completed and resources expended, presenting them by means of schedule and cost control tools such as the Jogging Line and Earned Value Analysis [2]. Other information describing the project execution may be included in project records such as correspondence, minutes of meetings, and progress statements. While the project baseline plan, work results, and project records supply information, PCPC provides the algorithm for processing information in the report.

Literally, the Summary Progress Report's design is no more than a refined application of the PCPC, following its five questions in a somewhat modified form:

- ▲ What is the variance between the baseline and the actual project status?
- ▲ What are current issues causing the variance?
- ▲ What risks may pop up in the future and how could they further derail the project?
- ▲ What is the trend—the predicted completion date, budget, scope, and quality—if current issues and risks persist?
- ▲ What actions should we take to prevent the predicted completion date, budget, scope, and quality from happening and deliver on the baseline?

Obviously, the majority of the questions are about the project future, attesting to our proclaimed goal of giving the report a proactive nature.

Design the Reporting System. Defining the purpose, hierarchy, frequency, responsibilities, and distribution of the Summary Progress Report is what we call the *reporting system*. This provides consistency in style and format, enabling comparison with previous and future reports in the project and other projects. "Is this report for external or internal customers?" is the essence of the purpose. In most cases, the detail and amount of information for internal reports will be different from external requirements. For example, external reports generally focus on helping the customer determine the status of work being funded and need to be structured accordingly. Each of the two types of reports may need a hierarchical structure, including the Summary Progress Report, the detailed progress report, and backup data.

Covering the highlights in one page, the Summary Progress Report provides the crucial points of the overall report while enabling management to review performance progress and trend at a glance. Parts of the detailed report are the general status of the project and its main parts, major developments, significant variances, major problems, predictions of final schedule and cost, and specifications of corrective actions. Essentially, both the summary and detailed report address the same types of information but in a very different level of detail. In addition, both are built around PCPC. Because of these similarities, we focus only on the methodology of the

Summary Progress Report, which when the level of detail is accounted for is applicable to the detailed report as well. Various tools of schedule, cost, scope, and quality control provide backup data. Examples are the B-C-F and Milestone Analyses.

The purpose and hierarchy set the stage for the frequency of the report. For example, in a firm, internal Summary Progress Reports are issued as follows:

- ▲ Small projects, typically six months long, weekly
- ▲ Medium projects, typically 12 months long, biweekly
- ▲ Large projects, typically 24 months long, monthly.

These experientially developed benchmarks are based on the firm's belief that each of the projects should have a similar number of summary reports— 26, 26, 24, respectively—that is, control cycles, if it is to be controlled properly. The point is not the sheer number of reports. Rather, it is that the firm's situation, defined here in terms of its control cycle policy, mandates the frequency and schedule of production, making the report a regular scheduled event. Added to those should be the so-called unscheduled reports, prepared in response to unexpected events of a critical nature that F.L. Harrison calls "Red Bandit" to aptly describe their potential to act as showstoppers [3]. Finally, defining responsibilities in preparing the report and who needs to see it completes the design of the reporting system.

Determine the Variance. This is an application of the first step in PCPC: What is the variance between the baseline and the actual project status? A comparison of the project baseline plan and the actual work results should easily yield the variance—that is, the difference between the two. Take, for example, Figure 15.4. If the project the figure reports on was a small departmental project, it might use the Jogging Line to indicate the schedule variance. Such a non-project-driven department might easily opt to show only this type of variance. In a different situation, where this would be a large project to design and deploy performance metrics throughout a project-driven firm, the Milestone Analysis might be used to identify the schedule and cost variance, or even the full-scale Earned Value System might be employed and supported by verbal descriptions about the quality and scope variances. Additionally, stating major accomplishments such as milestone completions of the report helps better acknowledge and visualize the progress.

Clearly, this is the report on the project level. To be able to develop it, you need first to have similar reports for project activities. A good strategy is to use the WBS as a framework, as described in the section on WBS in Chapter 5. The process of reporting starts on the work package level, identifying their variances and aggregating them up the WBS hierarchy to establish the variances for the whole project. Subsequent steps of identifying issues, predicting trends, and specifying corrective actions should follow the same approach of using the WBS as the framework. One way to do that is to use the system of rehearsals and progress meetings described in the section on the Jogging Line (Chapter 12).

The First Extreme—The Case of Misreporting

This is a short, no-thrills progress-reporting story of a project manager in a software development company. "Project managers do project status report every month for every project they manage. If something went wrong or certain deliverables in a project phase were not accomplished, we have to have a full explanation on it. Also, I have to show how much time I spent on each project and administrative work. Frankly, every project manager kind of reports 100 percent of her regular work hours being used, no matter what. We may have been working 120 percent or 130 percent of our regular work hours, but we don't report that; we just say 100 percent. Why? Because it doesn't make any difference what you report... Those are really very long reports, almost always 6 to 7 pages. I usually manage four to five projects at a time and do a report for each of them. It takes a lot of time to write them but I have to write them because all vice presidents read them... No, we don't do summary reports. And, we don't forecast the completion date, nor do we suggest what to do to make up for the delay. Sometimes, you may get a call from your boss asking you what can be done to catch up with the schedule." The lack of purpose, lack of focus, lack of proactive orientation are terms that come to mind after reading this story.

Identify Issues and Risks. If there is a variance, especially an unfavorable one, what are the issues causing the variance? Also, what risks may pop up in the future, and how could they further derail the project? The first of the two questions—both are from PCPC—probes to learn what present problems are at the root of the variance and what their impact on the project is. The second one forces you to look into the future, detect possible troubles and assess their future impact (see the section on issues and risks in Figure 15.4 for an example). The point, of course, is to figure out how the project can deal with them now. For instance, a project that asked this question learned that one of their major vendors might go bankrupt. The team immediately developed a contingency strategy instead of waiting to hear a few months later about the bankruptcy, at which point they might be helpless to correct the course of the project.

Generally, issues impacting the project progress may come from any area of work. This is why it may be useful when asking this question about the future issues to use a checklist for progress reporting. One such checklist is the issue database with potential solutions (discussed earlier in the "Issue Database" box).

Predict Trend. What is the trend—the predicted completion date, budget, scope, and quality—if current issues persist? Although forecasts of this type are not easy and are notoriously vulnerable, their essence is less in their accuracy and more in their creating of early-warning signals, signaling where the project is going. For example, in Figure 15.4, "project will be completed with 20 days of delay, on July 20" is a stern warning, one that mandates action to reverse the trend. The ability to forecast the trend, week after week, or whatever the report frequency is, is paramount in building an anticipatory climate where project teams are alert about the project's past progress but even more about what the future bears. Most of schedule, cost, scope, and quality control tools in this book provide a sound repository of prediction methodologies.

Specify Actions. If trends are unfavorable, what actions should we take to prevent them and deliver on the baseline? With a look into the future, which is our trend, we need to specify corrective actions, assess their impact, and assign their owner in the report (see Figure 15.4). Along with the trend, the specification of corrective actions is perhaps the most valuable part of PCPC and the report because it enables project teams to be proactive (see the box on page 496, "The First Extreme—The Case of Misreporting"). While the performance progress is important in telling where we are, it is no more than the project history—there is really nothing that we can do to correct or change it. Our only opportunity to change the project is in the future, and that is what the trend and corrective actions offer: an opportunity to anticipate and shape the future by acting—now.

Utilizing the Summary Progress Report
When to Use the Summary Progress Report. Small or large, projects need the Summary Progress Report. Pressed for resources, small projects—especially in a multiproject environment—will likely issue the summary report as their only report, doing away with a more detailed main body and backup data of the report.

Although many will prepare the report in a formal, written format per predetermined frequency, it is not unusual for small projects to use its proactive cycle informally for verbal reporting (see the box that follows, "The Second Extreme—The Case of Underreporting"). Quite to the contrary, formality, regularity, and detail necessary to cover all hierarchical levels characterize large projects' Summary Progress Reports, almost certainly supported by the main body of the report and back-up data.

The Second Extreme—The Case of Underreporting
We heard this story during a ten-minute lunch with a project manager in a technology firm. "We develop components for our internal customers who build them into their new products for external customers. With seven projects that I am managing right now, I don't really have time to write any reports. It is not just I; it is with all project managers. All of us run multiple projects at a time, too many we believe, and no one really has time for reporting. How could we have time for that if we work 70-hour weeks...No, my internal customer doesn't require any reports. Since they never told me what the deadline for their project is, I don't see why I would give them any report anyway...Well, my boss would like to have our reports but knowing how busy we are, he doesn't require them. He was in our shoes before he was promoted to this position, so I guess he understands what kind of situation we are in. He does ask us in our weekly staff meetings if we have any problems he can help with. But he can't really help much because he has no one else to give us to help out. And, when our schedules slip, nothing happens...Yes, I developed a Gantt Chart for each project initially, but with this pace of work, I just haven't had time to update it." The lack of purpose, lack of focus, lack of proactive orientation are terms again that come to mind after reading this story.

Time to Prepare the Report. An hour may be enough for a simple and small project team to run a progress meeting and prepare a typical Summary Progress Report. Even as time requirements go higher with the size and complexity of projects, it is clear that a few hours of a large project team's time should suffice for the summary report production. This assumes that extra time—perhaps running in tens of hours—was spent to develop backup data and write the main body of the report.

Benefits of the Report. If we picture time spent on developing the report as an investment, then return on this investment can be very lucrative in multiple ways. First, the process resulting in the report ensures a proactive cycle of project control, communicating information about project problems and status to all concerned, including higher management, and taking actions to put the project ship on track. Second, the Summary Progress Report is a vehicle to preserve stakeholders' involvement in the project. By feeding them with information about the past and future of the project, we help them see the big picture of the project and understand the impact of their contribution. This, in turn, helps maintain their motivation and coordination with others, further strengthening the team cohesion [3]. Third, the cycle of reporting instills discipline. To busy managers often carried away by daily pressures, the regularity of reports is a force that makes them sit down, collect data on project health, look into the project future, and form opinions dictating actions. Such work, unlike their daily firefighting of project problems, is the clear essence of PM—think, predict, act. If a project manager functions in an influence without authority environment, the report offers the fourth benefit. The time of reporting is when higher management pays attention to the project manager and his announcements about schedule variances, resources, and actions. Having higher management's ear is what helps the project manager increase the influence. Finally, the last benefit is one of using the report to create a historical trail and record that can be analyzed in the Postmortem Reviews and leveraged for the general purpose of project improvement.

Advantages and Disadvantages. Major advantages of the Summary Progress Report are its simplicity and visual power. Its simple proactive cycle and brevity, combined with visual representation of information, enable readers to easily glance and grasp where the project is and where it is going.

On the other hand, preparation of such a report may be a disadvantage, since it can be time-consuming in larger projects. This is a trade-off worth taking when you are aware of the report's benefits.

Variations. The summary report is one of those tools that has a plethora of forms. While they vary in contents, detail, and length, most of them tend to emphasize the history part of the report. This significantly reduces their usability, making one sympathize with those who complain about the amount of time involved in preparing such reports with little value.

Customization Action	Examples of Customization Actions
Define limits of use.	Use the Summary Progress Report in each project on a regular periodic basis.
	Must be in a written format, except for very small projects.
	Stick with one-pager format.
Adapt a feature.	Focus on project schedule (for large, time-to-market projects).

Use an Earned Value Analysis version for establishing the variance and predicting trend.

Customize the Report. The generic characteristics of the Summary Progress Report described here may not reflect the specific situation in your projects. To avoid this problem and use the tool in the best possible way, customize the report for specific project needs. In the table above, we list some ideas about the customization.

Summary

This section dealt with the Summary Progress Report, a page-long tool that highlights the status of the project, predicts trend, and states actions to overcome the negative trend. Small or large, projects need the report. While the former can use it in an oral form, the latter require formality and regularity. In any case, several benefits may ensue. First, the process resulting in the report ensures a proactive cycle of project control. Also, the report helps preserve stakeholders' involvement in the project and creates a historical trail and record. In addition, the cycle of reporting instills discipline. The following box summarizes the key points in structuring the Summary Progress Report.

Summary Progress Report Check

Check to make sure that the report is appropriately structured. It should include the following:

- Variance between the baseline and actual project/activity status
- Issues (or problems) currently causing the variance
- Risks (Issues) that may emerge in the future and further impact the variance
- Trend resulting from the variance and risks
- Actions required to reverse the negative or maintain the positive trend
- Backup data.

Postmortem Review

What Is the Postmortem Review?

A Postmortem Review is both a process and a document capturing crucial learning about what was done well and what was done poorly over the course of the project. The idea is to find ways to use such learning to avoid past conscious or subconscious mistakes and repeat successes with the goal of improving projects in the future. Since humans are fallible and mistake-prone, projects abound with flaws, and a potential for learning from postmortems is significant. As a result, many view postmortems as being a formal part of knowledge management [1]. Figure 15.5 offers an example of the Postmortem Review document.

Implementing the Postmortem Review

Prepare Information Inputs. The following inputs may help prepare and run a successful postmortem:

 ▲ Postmortem guidelines
 ▲ Examples of past postmortems
 ▲ Project records

Major Ground Rules

- *Don't get back at people.* Make it clear from the first second of the Postmortem Review that this is not a place for getting back at people, finger-pointing, blaming others, venting anger, or developing intelligent solutions for too many project problems. Failing to explain and enforce this led many companies to abandon the practice of postmortems. Also, clarify that no comment will be used in performance reviews of attendees.

- *Don't be oversensitive.* Check your ego at the door and be humble. Postmortems exist in order to identify mistakes with the process and participants' work. If mistakes are found in your work, be self-critical and look at them as an opportunity to improve and grow.

- *Don't attack anyone.* The focus is on problems, not people. Concentrate on any issues from process to product to team dynamics, and keep comments constructive. Placing blame on a person or finger-pointing killed too many postmortems, taking away an invaluable opportunity to learn and be better.

- *Don't forget facts.* Think again: what gets measured, gets improved. Gathering data and facts and making them the heart of the postmortem discussions and report provides metrics for learning and improvement in future projects.

- *Don't write a postmortem book.* People in future projects won't read long reports. Although many important issues will come up in the discussion, be parsimonious in writing the report. Do so by being brief, focusing on a vital few specific recommendations that have the highest potential for improvement. See details that follow in the "Document the Review" subsection.

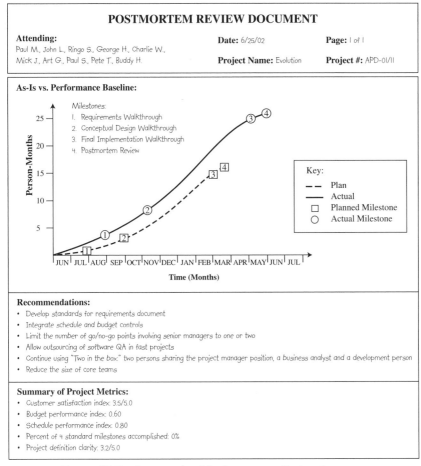

Figure 15.5 An example of the Postmortem Review document.

Organizations that have reached the point of having a standard methodology for projects usually possess guidelines that clearly chart the process for postmortems, reducing uncertainties and saving the time it takes to execute them professionally. Perhaps the most important part of the guidelines are behavioral or ground rules. Discussing the flaws of the project is an extremely sensitive issue that can easily derail the postmortem by creating an atmosphere of personal attacks and animosity, and resulting in people feeling singled out as the culprits of the project troubles. Such an atmosphere is certain to disable learning, defeating the very purpose of the Postmortem Review. Every care must be taken in the guidelines to create ground rules that will set the tone of constructive and cooperative spirit, free of placing the blame for project lows (see the box on page 500, "Major Ground Rules").

Whether or not the guidelines are available, having participants review past postmortems of similar projects and take care of the necessary logistics ahead of time (see the box that follows, "Logistics of the Postmortem Meeting") are very useful practices. Another important input is project records that document and help reconstruct the project history. Among them a special place belongs to the Risk Log and Summary Progress Reports.

Prepare the Review. Well-run reviews feature focus and efficiency. A prerequisite to this is a good agenda that plays the role of blueprint for the postmortem meeting, defining steps of the discussion and allotting time to the steps, although in the case of strong interest in a topic, the time can be extended on the spot. For example, the steps may include the following:

▲ Reconstructing an "as-is" time line and comparing it with the baseline

▲ Project low points ("lows"), or which parts of the project went wrong

▲ Project high points ("highs"), or which part of the project went well

▲ Recommending what should be done differently and what should be continued in future projects

---

Logistics of the Postmortem Meeting

● *Who Should Attend?* Invite key participants, functional groups, and stakeholders whether they are from areas related to project "lows" or "highs, " because to learn and improve, the review needs to look at both successes and failures. If there are too many attendees, they can be divided into a core group and functional groups. After functional groups hold their mini-postmortems about their portions of the project, representatives from each group form the core group tasked with conducting the final postmortem. Also, make sure every attendee is equally involved in the discussion.

● *Room.* When a postmortem features a heated debate in a small room, people may feel cornered, possibly reacting with a "fight or flight" attitude. To prevent these behaviors, get a large, spacious room, comfortably seating all attendees. Arrange for seating that promotes equality of participants, for example, around the round table where all participants face each other.

● *Facilitator.* Getting an impartial person who neither participated nor has a stake in the project is key to creating a conducive environment to an effective facilitation. Such a person will provide and direct the process in the meeting without being involved in the discussion. Her main job is to ensure that the agenda is followed and discussion is focused on issues. Also, she should make sure personal attacks are avoided, constructive comments are given, all participants are evenly involved, and time is carefully looked after. For this to happen, she needs to establish firm and fair ground rules.

● *Recorder.* Visual communication that supports focus on issues, not people is what recording needs to provide. Capturing and organizing crucial information from the discussion on sheets of flip chart papers taped on the room walls or white boards, the recorder enables effective flow of this communication. Subtleties such as using color markers to denote the nature of the comment (e.g., "red" means showstopper) or symbols to express the group feelings about a comment (e.g., "?" means conflicting views about the comment) further refine the quality of recorded information.

Logically, the agenda will vary across industries and strategic foci of their projects. While the preceding example above is related to a high-tech company with rapid time-to-market interests, a conventional manufacturer competing on cost may concentrate on an as-is budget instead of the time line. Whatever the focus is, once the agenda is ready, it should be distributed as the preliminary one well ahead of time, and it should ask participants to contribute ideas on project lows and highs, as well as other items in the agenda. Some of these ideas should be incorporated in the final agenda, while others may be used for the discussion in the meeting. Generally, the project team should do as much of the steps in the agenda as possible before the meeting, allowing the meeting to focus on the essential issues. If work schedules of the participants are too tight to allow them to generate any ideas before the meeting, making the preliminary agenda the final one is a valid option.

Run the Review. After making the opening remarks about the purpose and establishing ground rules of the Postmortem Review, the project team is ready to take the following chronological steps:

- Review and rank issues.
- Ask what went wrong.
- Ask what should be done differently in the future.
- Ask what went well.
- Prioritize recommendations.

In the first step, participants should review and rank issues that dominated the project execution. A very helpful strategy to direct the review is to get to the participants a checklist of issues before the meeting, or at least in the meeting, putting them on the same page. As shown in the box that follows, "Examples of Questions from the Postmortem Meeting Checklist," the issues may range from project planning to scheduling/budgeting to team organization to product design, essentially making any area of the project open to review. Such a wide-open approach begs for funneling of the participants' energy. A technology company secures the funneling by asking each participant to go through the checklist and identify the top five issues or project lows. These are posted on the flip chart sheet, clarified by those who identified them, and ranked by a voting method to get the top five issues for the whole group that will be taken to the next step.

"What went wrong with the top issue?" is the leading question for the second step. For example, in a company focused on time-to-market, the top issue in a project postmortem was schedule slippage, so the question was "What went wrong with the schedule?" To enable the discussion, the project team prepared before the meeting an as-is versus baseline schedule supported by the time line and resource hours each participant spent on the project. Whether the chart is prepared ahead of time or on the spot, it should be visibly displayed on the wall, like in our case, to channel the "wrong" discussion. There are several methods to generate the list of what went wrong. One is to simply brainstorm, letting participants state whatever occurs to them as they ponder the issue. When the nominal group technique is used, participants do silent, written brainstorming, and the facilitator asks each one of them to provide one comment at a time, until their comments are exhausted. Another possibility is to have participants volunteer their comments on Post-it notes, one comment per Post-it note, and combine and sort them by topic to generate an Affinity Diagram.

Examples of Questions from the Postmortem Meeting Checklist

Project Planning

- Were the business goals of the project clear to you?
- Were other functional goals (development, marketing, manufacturing) clear to you?
- Were scope, time, cost, and quality goals clear to you?
- How adequate and complete was the project plan when actual work started?

Customer Voice

- Was customer voice identified and built into the project planning, design, and implementation?
- Were the project process, product and deliverables in line with customers' (clients, stakeholders, sponsor) expectations?
- Was communication with customers effective?

Product and Deliverables

- Are you satisfied with the project product? Other project deliverables?
- What are the variances between the actual and planned product and other deliverables?
- How effective were actions to control the variances?

Design and Specs

- To what extent was information in specs adequate to proceed with development work?
- Were they done on time? Were they ever "frozen"?
- Did each functional group know which piece of functional design they owned?
- Were cross-functional interfaces in design/specs development clearly defined?

Scheduling and Budgeting

- Were the schedule and budget baseline realistic? Detailed enough? Sufficient resources?
- How did milestones help with tracking the schedule/budget?
- Did significant variance between the baseline and actual schedule/budget occur?
- Did project progress metrics work?

Organization, Team and Resources

- Was project organization adequate? Functional versus matrix versus dedicated team?
- Did the team work well?
- Did the team have necessary skills? Resources?
- Were roles of team members and functional groups clear and fulfilled?

Risk Management

- Were risks associated with the process, product and deliverables identified?
- How correct were assumptions about the risks?
- How effective were contingency actions against the risks?

Communication

- How effective was communication with management? Functional groups providing resources? Other project teams?
- How effective was communication within the team?
- Was reporting timely? Proactive? Helpful? Too time-consuming?

Higher Management

- Were managers involved in major project reviews? Were they effective?
- Were management decisions communicated to you?
- Were the decisions clear? Fast enough? Did you understand how they were made?
- How much did they help you get your job done?

Project Management Systems and Software

- How did the company's PM methodology work—project planning, change controls, sponsor support, and so on?
- How effective were PM software programs used in the project?

Several factors will help determine which of the methods to apply. Brainstorming is good for situations where there are no "squeaky wheels"—that is, people dominating the discussion. You can eliminate their unwanted influence by going to the nominal group technique, unless you don't want the person who originated a certain comment to be known. If the anonymity of comments' owners is the goal, or you want to see how many times a certain comment is made, sticking with the Affinity Diagram is an option. If these wrongs were collected before the meeting, this is the time to ask for additional ones. Once the list of the wrongs is finalized, they should be prioritized, for example, using again the voting system. Selecting the top three wrongs allows the team to concentrate on the vital few.

While this step—what went wrong with the top issue—appears slow, well-facilitated postmortems will get it done in five to ten minutes. Then they will shift their focus to the third step. As important as learning what went wrong is, it is at least equally if not more important to develop ways to correct the lows in future projects. Asking what should be done differently in the future with relation to the first issue—note that we are still dealing with the first issue—is that third step. For each of the wrongs, participants need to ask [4] the following questions:

▲ Were there any early warning signals of the wrong?
▲ What should we have done differently?

Generating and selecting a few potential corrective strategies can be done with the method of collecting and selecting top wrongs. At this time, note again that only the first issue is analyzed, meaning that steps 2 and 3 for the remaining top issues need to be repeated.

What follows is an emotionally more pleasant experience of the fourth step that looks at what went well, or as the company in our case calls it, project highs. Using the same method of collecting and selecting the top wrongs, our company's participants in the postmortem will identify the top five highs, followed by the discussion around these questions:

▲ Which practices caused the highs to happen?
▲ Which of these practices were not used before in projects and are recommended to be introduced to future projects?
▲ Which of these practices were used before in projects and are recommended to be continued in future projects?

The completion of the fourth step leaves participants with

▲ Potential top corrective strategies that emanated from project lows and related wrongs
▲ Practices proposed for use in future projects that stemmed from project highs

Since not all of them are equally important and powerful, participants take on the fifth step of prioritizing these recommendations. Although this is an effective action to wrap up the postmortem, some companies choose to prioritize after the meeting, in the process of writing the Postmortem document.

Document the Review. Ending a postmortem without a Postmortem Review document is a frequent practice, especially in smaller projects. Although a postmortem without the document is a better option than not having a postmortem at all, having the document offers significant advantages. Specifically, the document makes it possible to disseminate the messages from the postmortem as described in the next subsection. Therefore, to begin documenting, appoint a person to write the document, possibly as early as in the preparation stage of the Postmortem Review. It is also at this time that the document content should be defined, in case the organization did not standardize it to save time to the document writers, while providing consistency and comparability across projects. For example, the

Postmortem Review document may consist of the body and the attachment. Included into the one-page body may be the time line and time-phased budget (as-is versus baseline), attendee list, a set of recommendations, and major project metrics (e.g., coding productivity data or earned value measures). Such an approach is applied in the example from Figure 15.5. The purpose of the multiple-pages-long attachment is to preserve as much as possible the authentic comments of participants recorded before, during, and after the postmortem meeting. Vital to writing the document is also having all participants review it and suggest changes as they see fit.

Use Postmortem's Lessons. What comes to mind when the results of the postmortem are not used are George Santayana's words—those who cannot remember their past are condemned to repeat it. Still, this is what may easily happen to many project teams who perform the postmortem, document it, and then forget about it. Instead, the teams and their organizations have several great options to capitalize on the knowledge captured in the postmortem. They can apply the lessons learned in the review to avoid similar mistakes in the next project, mitigate risks in other projects, or improve the overall PM process of the organization to mention a few [1]. For example, an organization in the information technology business requires that the planning for the new project start with reviewing the Postmortem Review document from the previous project and using it as a self-check. Another organization takes major issues and their solutions from the postmortem and puts them in a Web-based issues database. All projects are required to periodically assess how they do on the issues and, if they don't do well, consider mitigating strategies similar to the issues' solutions from the database. In a high-tech manufacturing company, each lesson learned is associated with a work package in the standardized WBS that has to be used in planning for each new project. Possibly the highest value of the Postmortem Review comes from evaluating its recommendations to determine which ones have an across-projects impact and making changes to the overall PM process accordingly [5].

Utilizing Postmortem Review
When to Use. Nominally defined, the Postmortem Review is a review performed after the project is complete. In larger projects, that may be a few weeks after when team members' emotions related to the pressures of the last days of the project are toned down and minds are clear for an objective analysis. It is not unusual that some product development teams do another review—for example, six months after the completion—to assess the market impact of their project product and develop strategies for new product releases. These same larger projects also tend to conduct postmortems at the end of a major phase or major milestone accomplishment. Although postmortems are a common practice in larger projects with significant effort and commitments, there is a growing tendency to make them a routine part of smaller projects as well, performing them informally and quickly.

Time to Complete. Policies of companies differ with regard to the duration of Postmortem Review meetings. While some of them are in favor of one- or two-day-long review meetings, others attempt to go no longer than four hours, even if it means breaking it into two separate meetings [4]. These examples for larger projects are made possible by doing lots of work outside of the review meeting. Unlike them, smaller projects would typically opt for a very short review meeting, ranging from half an hour to an hour.

Benefits. The logic of the postmortem's value is relatively simple: Even if informal, the postmortems accelerate learning, improving performance in projects. Some research studies undoubtedly confirmed such value and qualified postmortems as unique opportunities to learn the lessons for future projects [6-8]. Identifying highs and lows of the project, such learning captures what needs to be enhanced, preferably in the form of an action plan. Another value is in the fact that a Postmortem Review brings closure to a project. For those who work hard for many months and attain project goals, the significance of the closure is the indisputable recognition of success. Despite its benefits, too many projects miss this invaluable opportunity to learn faster. To not have a postmortem is a pennywise, pound foolish approach that perhaps only ignorant companies can afford, often excusing it by the pressure to move people to new projects that are likely destined to repeat flaws of their predecessors without a Postmortem Review.

Advantages and Disadvantages. The postmortem is often viewed as low-hanging fruit. It is a relatively simple project tool with low resource requirements that offers a significant potential for high return on investment in the form of benefits explained in the preceding text.

However, this major advantage must be considered in relation to its major disadvantage that when poorly conducted the postmortem may easily create dysfunctional conflicts stemming from personal attacks and finger-pointing for the project's lows and wrongs.

Variations. Some of the synonymous terms for the postmortem that circulate in the PM community are post-project, lessons learned, and post-implementation review. Their rationale, methodology, and effects are identical to those of the postmortem. Traditional tools for post-project analysis such as final project reports, "do's and don'ts" reports, and edification reports have both similarities and dissimilarities with the postmortem. For example, all of these tools tend to identify project lows and highs. On the other hand, they are typically the work of project managers, meaning they are not as participative as are team-based review meetings in Postmortem Reviews. Also, only some of them offer recommendations for future improvements of other projects.

Customize the Postmortem Review. Generic in its nature, the postmortem we described may not fit some projects' needs. Therefore, in order to get the most out of this tool, it is appropriate to tailor it for specific project situations. A few ideas that follow may give clues on how to customize the postmortem.

Customization Action	Examples of Customization Actions
Define limits of use.	Use the Postmortem Review in all projects.
	Prepare a Postmortem Review document for each review.
Add a new feature.	Submit each Postmortem Review document to the PM process owner (e.g. body or individual responsible for continuous improvement of PM process).
Adapt a feature.	Limit Postmortem Review meetings for large projects to four hours. Limit Postmortem Review meetings for small projects to one hour.

Summary

The Postmortem Review is performed after the project is complete, or at the end of a major phase or major milestone accomplishment. Although postmortems are a common practice in larger projects, there is a tendency to do the same in smaller projects—informally and quickly. The benefits of the review are in accelerating learning. In particular, by identifying highs and lows of the project, such learning captures what needs to be enhanced in future project performance, preferably in the form of an action plan. Another value is in the fact that a Postmortem Review brings closure to a project. The following box summarizes highlights in structuring a Postmortem document.

Postmortem Review Check

Check to make sure that the Postmortem Review document is appropriately structured. It should include the following:

- Up to a page-long body of the report with attendee list, "as-is" versus baseline time line and time-phased budget, a set of recommendations, and major project metrics
- An appendix with the information recorded in the meeting or from correspondence collected by the postmortem team before and after the meeting.

Concluding Remarks

In this chapter we presented three tools—the Risk Log, Summary Progress Report, and Postmortem Review. As a set of complementary tools, they first enable us to monitor, report, and control risk and progress of the project in the course of its implementation. Second, they help us study the implementation and learn the lessons for future improvements.

Each of the three tools is designed with a distinct purpose. The Risk Log helps monitor and control risks in the course of project implementation. With its page-long format, the Summary Progress Report highlights the overall status of the project, predicts trend, and states actions to overcome the negative trend. The Postmortem Review allows capturing crucial learning about what was done well and what was done poorly over the course of the project, with the intent of avoiding past mistakes and repeating past successes. While using them as a toolset is preferred, they may be employed individually.

More details on situations in which to use each of the tools are described the summary comparison that follows. Specifically, we identified several project situations and marked how each one favors the use of the tools. If they do not provide enough details to decide which tool to deploy, brainstorm to identify more such situations. Mark how they favor the tools. A tool with many marks is probably good to use.

A Summary Comparison of Performance Reporting and Closure Tools

Situation	Favoring Risk Log	Favoring Summary Progress Report	Favoring Postmortem Review
Small and simple projects		√ (informal)	√ (informal)
Large and complex projects	√	√	√
Evaluate the actual risk status	√		
Define actions to control risks during project implementation	√		
Need formal or informal progress reporting		√	
Summary detail needed		√	
Pursue proactive cycle of project control	√	√	
Provide early-warning signal		√	
Need to learn lessons from accomplished major milestones			√
Need to learn lessons from project implementation			√
Want to accelerate PM learning			√

References

1. Lientz, B.P. and P.R. Kathryn. 1999. *Breakthrough Project Management*. San Diego: Academic Press.

2. Project Management Institute 2000. *A Guide to The Project Management Body of Knowledge*. Drexell Hill, Pa.: Project Management Institute.

3. Harrison, F.L. 1992. *Advanced Project Management: A Structured Approach*. 3d ed. New York: Halsted Press.

4. Thomke, S. and S. Sinofsky. 1999. "Learning from Projects: Note on Conducting a Postmortem Analysis" Harvard Business School note #9-600-021 Cambridge, MA: Harvard Business School Press.

5. Whitten, N. 1995. *Managing Software Development Projects*. New York: John Wiley & Sons.

6. Wheelwright, S. and K. Clark. 1992. *Revolutionizing Product Development*. New York: The Free Press.

7. Adler, P. 1995. "Interdepartmental Interdependence ad Coordination: The Case of the Design/Manufacturing Interface." *Organizational Science* 6(2).

8. Cusumano, M. and K. Nobeoka. 1998, *Thinking Beyond Lean: How Multi-Project Management Is Transforming Product Development at Toyota and Other Companies*. New York: The Free Press.

Industry
Applications

16

Selecting and Customizing Project Management Toolbox

You are unique, and if that is not fulfilled then something has been lost.

Martha Graham

Which of All These Tools Do You Really Need?

Now that we finished our review of 50+ project management tools in the previous chapters, a natural question comes to mind: "Which of all these tools and in which form do I really need?" The goal of this chapter is to answer this question by developing a framework for selecting and adapting a project management toolbox that supports the SPM process that we discussed in Chapter 1. Specifically, the framework should help customize the use of the toolbox while accounting for the specific situation of various companies and projects.

Process for Selecting and Adapting Project Management Toolbox

There are three major steps, each including several tasks, in selecting and adapting a project management toolbox for specific projects (see Figure 16.1):

- ▲ Secure strategic alignment
- ▲ Customize the project management toolbox
- ▲ Improve continuously

Aligning the project management toolbox with the organization's competitive strategy tells us in broad terms what categories of project management tools to select and adapt. This alignment drives the next step—customization of the project management toolbox by selecting specific project management tools. The deployment of the project management toolbox in real-world projects will reveal its glitches and generate new learning, resulting in a need for the toolbox's continuous improvement, the third step. Details about each step follow.

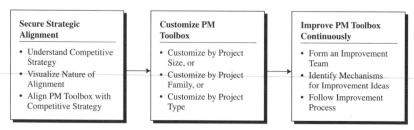

Figure 16.1 Process for selecting and adapting a project management toolbox.

Secure Strategic Alignment

The purpose of the project management toolbox is to enable the implementation of projects that will effectively support an organization in pursuing its competitive strategy and goals. To make the purpose happen, you need to fully align the project management toolbox with the competitive strategy. This can be achieved through the following steps:

- ▲ Understand the organization's competitive strategy.
- ▲ Visualize the nature of alignment.
- ▲ Align the project management toolbox with the organization's competitive strategy.

Understand the Organization's Competitive Strategy. In Chapter 1 we showed that an organization's project management toolbox has to be aligned with its competitive strategy. To be successful in designing the toolbox, therefore, project managers must understand the strategy. However, many of them do not. Why? Among many reasons for this, two matter here. First, in many organizations, strategy formulation and implementation is viewed as the executives' domain of work—they are tasked to chart a competitive strategy course in an organization—to which mere mortals such as project managers have no access. Second, because of such a view many project managers do not show or show a very little interest in learning about the competitive strategy. As a result, neither is the strategy communicated to project managers nor do they work to understand it. Project managers should be tenacious, probing and digging to comprehend the strategy, even if the strategy is not communicated.

These two reasons create substantial obstacles for project managers. To remove the obstacles, project managers need to talk to senior managers and convince them that competitive strategy is the guide to planning and implementing projects, the SPM process, and the project management toolbox, and that project managers need to know it in order to secure expected returns on their projects Therefore, our mandate is clear: Understand the strategy or designing the toolbox will be like shooting in the fog—we neither know where the target is nor whether we hit it.

Visualize the Nature of Alignment. Part of convincing the senior managers is the ability to clearly visualize the nature of the alignment between the toolbox and the competitive strategy. In Chapter 1 we laid the foundation for the alignment. In particular, we used examples of three companies—Intel, AWI, and OAG—to illustrate how the project management toolbox can be focused to support competitive strategies. To support its differentiation strategy, Intel pursues a schedule-driven toolbox. A cost-focus toolbox is what AWI does to back its cost leadership competitive strategy. OAG uses cost-quality focus in its toolbox to sustain its competitive strategy of best cost. Obviously, the focus in each company's toolbox is aligned with their competitive strategy. This is summarized in Figure 16.2.

To keep our focus pragmatic, we will illustrate how these examples of the project management toolbox work in real-world projects. For that purpose, in Figure 16.3 we show what we conveniently call investment curves—a more precise term is the Net Present Value curves—for three comparable projects performed in tune with the competitive strategies of Intel, AWI, and OAG. Each curve shows four important points: project start, time-to-deployment (TTD), time-to-breakeven (TTB), and salvage point. Project start is the time when the project is launched and begins to spend resources; therefore, the cash flow becomes negative. Investment and negative cash flow continue to increase until the project is completed. At that time the project product can be deployed, or in other words used by customers. Instead of TTD, project managers in product and software development use the term time to market. Most of other project managers prefer the term project cycle time or, simply, project completion time. Note that negative cash flow usually reaches its peak at the TTD point. After that, the use of the project product begins to generate the returns, and as a result, the curve turns up until it reaches TTB. This is the point where all investments in the project are equal to returns generated by the use of the project product. Beyond that point, the cash flow turns positive and typically continues to do so until the project product is salvaged.

		DIFFERENTIATION	
		Low	**High**
COST	**High**		**Competitive Strategy:** Differentiation **PM Toolbox:** Schedule-Driven
	Low	**Competitive Strategy:** Low Cost **PM Toolbox:** Cost-Driven	**Competitive Strategy:** Best-Cost **PM Toolbox:** Cost-Quality-Driven

Figure 16.2 Examples of the alignment of project management toolbox with competitive strategies.

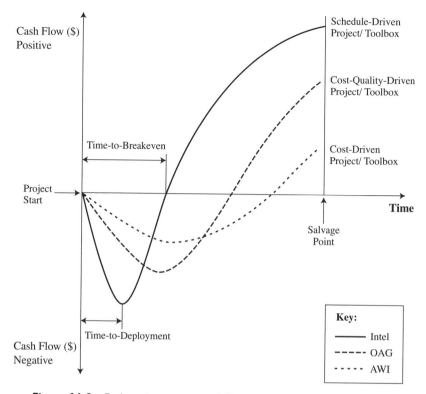

Figure 16.3 Each project curve uses different project management toolbox.

We will use the curves to explain the nature of the project management toolbox's alignment with the competitive strategy for each of the three companies. Consider, for example, Intel's case. One element of Intel's differentiation strategy is speed in projects. Figure 16.3 illustrates that point; TTD and TTB in Intel's curve are reached much sooner than for other two companies. For this to be possible, Intel needs to use a schedule-driven toolbox in which the central role and priority belong to tools that can help accomplish as fast TTD and TTB as possible. These are schedule planning and control tools that set the tone for company's projects—Gantt Chart, Time-scaled Arrow Diagrams, CPM Diagrams, Milestone Chart, and so on. Simply, most of the management time and attention goes to these tools, and they also serve as the primary basis in crunch times of making key decisions. This does not mean that other components of the typical project management toolbox such as cost and quality tools are ignored. Quite to the contrary, they are important and have their role in the toolbox. That importance and role are dictated by the importance and role of cost and quality goals in Intel's competitive strategy. Obviously, Intel cares about building quality products to keep customers happy and uses Control Charts, Flowcharts, and Affinity Diagrams for that purpose. In efforts to keep cost goals as low as possible, Intel deploys Cost Estimates and Baselines.

But in Intel's hierarchy of goals, quality and cost goals are lower than schedule goals. That is why Intel's curve has more negative cash flow at TTD than the other two curves; speed, apparently, has its cost [1]. As a result, the driving force of project management toolbox are schedule tools. Quality, cost, and other tools are subjugated to schedule tools.

The case is different in the toolbox of AWI, a company bent on cost leadership. Logically, then, its projects are cost-driven, searching to minimize project cost whenever possible. This logic is apparent in Figure 16.3. The AWI curve has a lower point of the most negative cash flow than those of Intel or OAG. Not by accident! It is the intended goal and realized outcome of project actions. To accomplish such project goals, AWI is willing to take the longest time to reach TTD and TTB. Crucial in this effort is a cost-driven toolbox. Its emphasis is on cost, cost, and cost. Correspondingly, cost estimates and cost baselines are carefully prepared, as is the assessment of return on investment, even for small cost-cutting projects. Obviously, the primary attention and management time are devoted to cost planning and control tools in order to lower the cost and sustain AWI's strategy of cost leadership. Even schedule and quality tools are adapted to support the low cost goals. Gantt Charts, known for their simplicity and low cost of development, are the schedule tools of choice, shooting for project durations that offer the lowest cost of project execution. CPM Diagrams, which take time and, hence, cost to train people and prepare are used infrequently, primarily in large projects. Control Charts, Flowcharts, and other quality tools are focused on making the project management process fully standardized and built on templates for each tool from WBS to Cost Estimate to Risk Response Plan. Other tools are applied with the same low cost reasoning. Risk Response Plan is focused on cost risks. Meeting agendas are standardized to lessen the cost of meetings. A similar approach is with the WBS development—stick with the minimum number of levels that enable cost-effective management.

The intent of aligning the project management toolbox with the competitive strategy is aggressively pursued in OAG as well. The driving force is the best-cost strategy that is also translated to the project level. As can be seen from Figure 16.3, TTD and TTB are shorter than AWI's but longer than Intel's. This means that cost focus is lower than AWI's but higher than Intel's. Such cost philosophy is closely intertwined with the need for the project to emphasize quality goals more than the other two companies. Given this situation, how does one shape a quality-cost driven project management toolbox?

A combination of well-balanced quality and cost tools has the priority. It is so much so that management pays the highest attention to them, giving them most of their time and relying on them for major decision making. Formal or informal customer voice tools and quality control programs are crucial for hitting the customer-set quality level, as are Cost Estimates and Baselines. Schedule tools, generally simpler and less costly such as Gantt

Charts, are part of the game. To OAG and its customers, delivery on schedules matters, because keeping customers is not possible without being punctual. Nevertheless, schedule goals are not overly ambitious, resulting in not so fast schedules. Clearly, in OAG's food chain of project goals, schedule tools are on the lower level than the quality-cost combination. Other tools such as scope or risk tools are also adapted to support the combination. For example, it is usual to prepare the Risk Response Plan focused on cost, rather than on schedule (as a schedule-driven toolbox would stress).

As you can see from our discussion, the nature of alignment of the toolbox is reflected in the balance of two issues. First, most of the tool types show up in all three toolboxes. Obviously, this is where the one-size-fits-all approach makes sense. The second issue concerns the situational approach, adapting tools to account for the character of the three toolboxes. With this knowledge in mind, we are now ready to offer a few guidelines for the alignment of the toolbox.

Align the Project Management Toolbox with the Organization's Competitive Strategy. Four tasks are important in achieving the alignment:

- ▲ Set the desired level of the alignment.
- ▲ Establish the current level of the alignment.
- ▲ Identify gaps between the current and desired alignment.
- ▲ Act to provide the desired alignment.

In many companies, the process team responsible to manage the project management process (or specially appointed teams) sets the desired level of the toolbox alignment. For example, refer to Table 16.1. This table identifies the characteristics that each of the three types of the project management toolbox—schedule driven, cost driven, and quality-cost driven—should have. Earlier in this section, we described the nature of these characteristics in examples of Intel, AWI, and OAG.

In setting the desired level, the team may decide to go for an incremental implementation of the schedule-driven toolbox. Accordingly, they can choose to give simple schedule tools the central role and priority, while leaving the adaptation of other tools (e.g., cost or risk planning tools) to support schedule tools for some later time.

If the process team understands the issue of the alignment, the path is clear. Lots of discussion may be needed to build a consensus about the level, especially if many of the involved people who have a say are busy. Good planning, leadership, and communication will certainly help in accomplishing the task. Add to this good change management skills and endless patience, when there is a need to first educate the process team about the alignment substance.

Table 16.1: Characteristics of the Strategically Aligned PM Toolboxes

	Company's Competitive Strategy		
	Differentiation	Low Cost	Best-Cost
	Nature of Project Management Toolbox		
Characteristics of the Project Management Toolbox	Schedule-Driven	Cost-Driven	Quality-Cost-Driven
The central role and priority belong to schedule tools	✓		
Most of management attention goes to schedule tools	✓		
Most of management time goes to schedule tools	✓		
Schedule tools are primary basis for making decisions	✓		
Other tools are adapted to support schedule tools	✓		
The central role and priority belong to cost tools		✓	
Most of management attention goes to cost tools		✓	
Most of management time goes to cost tools		✓	
Cost tools are primary basis for making decisions		✓	
Other tools are adapted to support cost tools		✓	
The central role and priority belong to quality–cost tools			✓
Most of management attention goes to quality–cost tools			✓
Most of management time goes to quality–cost tools			✓
Quality–cost tools are primary basis for making decisions			✓
Other tools are adapted to support quality–cost tools			✓

Interviewing project customers, managers, and sponsors is very helpful in obtaining information about the current level of the alignment. Observing and auditing key projects will further improve the knowledge of the actual alignment. Skipping these tasks and assuming that we know where the current notch for the alignment is may result in serious problems.

Once the desired and current levels of alignment have been known, the time has come to clearly identify the gaps between them. The significance of the gaps—whether they are major or minor—will be a basis for the decision to what extent the gaps will be acted on and closed.

In this section we explained how to go about securing strategic alignment of the toolbox. Necessary for this are understanding of the organization's competitive strategy, visualizing the nature of the alignment, and performing the alignment. Note that the alignment act is truly strategic. It provides a broad strategic direction for the toolbox customization of a coarse granularity. To use the toolbox in real-world projects, you need to customize the toolbox with fine granularity. That is the focus of our next step.

Customizing the Project Management Toolbox

There are multiple options for customizing a strategically aligned project management toolbox. Three of them are perhaps the most viable:

- ▲ Customization by project size
- ▲ Customization by project family
- ▲ Customization by project type

The three options are three different ways to select and adapt the toolbox. Each option has the purpose of showing which specific project management tools to select and adapt for the project management toolbox. For this to be possible, each option is based on the SPM process, which actually dictates the choice of the tools. As discussed in Chapter 1, each tool in the toolbox is chosen to support specific managerial deliverables in the SPM process. Also, the PM toolbox is constructed to include all tools necessary to complete the whole set of the SPM process's managerial deliverables. Therefore, in any of the options the first step is to lay down the SPM process—that is, its phases, project management activities, and managerial deliverables (technical activities and deliverables are beyond the scope of the book). Then comes the selection of individual tools to support the deliverables.

An in-depth knowledge of individual tools is a prerequisite to each of the options, because you need to understand how each tool can support a managerial deliverable. We will describe the options in turn and offer guidelines for selecting one of them for the implementation. Whichever option has been chosen, for those who select individual project management tools or design toolboxes relying on *A Guide for Project Management Body of Knowledge* (popularly called PMBOK), Appendix A provides linking of the tools covered in this book to PMBOK.

Customization by Project Size. Some organizations use project size as the key variable when customizing a project management toolbox that is already strategically aligned. Their view is that larger projects are more complex than smaller ones. Or, the size is the measure of the SPM process complexity. The reasoning here is that as the project size increases, so does the number of project management activities and resulting managerial deliverables in a project, and so does the number of interactions among them. Worst of all, this number of interactions grows by compounding, rather than linearly [3]. Such increased complexity, then, has its penalty—larger projects require more managerial work and deliverables to coordinate the increased number of interactions. This translates into SPM processes and project management toolboxes for larger projects.

Since different project sizes require different SPM process and project management toolboxes, we first need a way to classify projects by size and then customize their toolboxes. For size classification we will draw on the experience of some companies. In Table 16.2 we present three examples. All companies created three classes of project size: small, medium, and large. The units to measure the size are dollars or person-hour budgets. On the basis of the size, the companies determined the managerial complexity of its project classes and SPM processes. The complexity, further, dictated the project management toolbox, a simplified example of which is illustrated in Table 16.3 (for more examples, see Appendix B). For the sake of simplicity, only the toolbox is shown, leaving out technical deliverables.

As Table 16.3 indicates, some of the tools in the toolboxes for projects of different size are the same, others different. For example, all use the Progress Report because all projects need to report on their performance. Since managerial complexity of the three project classes and their SPM processes call for different tools, some of the tools differ. A Risk Log, for example, is needed only in large projects. To be successful, the process team designing the toolbox should carefully balance the one-size-fits-all tools with those that account for the specific situation of the SPM process.

- -

Table 16.2: Examples of Project Classification per Size in Three Companies

| Company | Project Size | | |
	Small	Medium	Large
Product development projects in a $1B/year high tech manufacturer	$ 1–2M	$ 2–5M	> $ 5M
Design projects in a $150 M/year low tech manufacturer	<$50K	$50–150K	> $150K person-hours
Software development projects in a $40M/year company	300–400 person-hours	1000–3000 person-hours	> 3000 person-hours

Experience of these companies offers several guidelines for customizing the project management toolbox by project size:

- ▲ Identify few classes of projects and their SPM processes.
- ▲ Define each class by the size parameter.
- ▲ Match the class complexity with the proper toolbox, each tool supporting a specific managerial deliverable.

Note that while customization by project size offers advantages of simplicity, it also carries a risk of being generic, disregarding other situational (contingency) variables. To some, these other variables may be of vital importance, as will be pointed in the next section on customization by project family.

Customization by Project Family. When the project management toolbox is strategically aligned, you can opt to customize it by family types in their industry. Many companies choose such options in a belief that project families in their industry are sufficiently unique to merit a family-industry-specific SPM process and toolbox [4]. Such an approach can be easily understood with the help from the situational (contingency) view [2].

Table 16.3: Example of Customization of PM Toolbox by Project Size

| Project Size | Definition | Project Phases | | |
		Planning	Execution	Closure
Small	Project Charter	Scope Statement WBS Responsibility Matrix Milestone Chart	Progress Report	Progress Report
Medium	Project Charter Skill Inventory	Scope Statement WBS Responsibility Matrix Cost Estimate Gantt Chart	Progress Report Updated Gantt Chart	Progress Report
Large	Project Charter Four-Stage Model Stakeholder Matrix Skill Inventory	Scope Statement WBS Responsibility Matrix Cost Estimate Time-scaled Arrow Diagram (cascade) Probability-Impact Matrix Commitment	Progress Report Updated Time-scaled Arrow Diagram Slip Chart Change Request Risk Log Commitment	Progress Report Postmortem

As a group of organizations that compete directly with each other to win in the marketplace [5], an industry is characterized by the nature of its environment and business task. For example, companies in a high-technology industry face an environment of dynamic technology change. Because of this, their business task abounds with fast time-to-market projects [6]. Combined, the described nature of environment and business task work to create similar challenges in families of projects. For example, a project family of new product development projects in high-tech industries faces similar challenges. So do facilities management projects, manufacturing projects, marketing projects, and information technology projects in the high-tech industry. These same project families also exhibit similarities in other industries.

A resulting phenomenon is, then, that SPM processes and related project management toolboxes to resolve the challenges of a project family in a certain industry tend to converge. This creates a situation in which several essentially similar models of the SPM process and toolbox are perceived as dominant family-industry standards. In response, some companies tend to select toolboxes per family-industry standards, typically exploiting the opportunity of benchmarking these dominant models. One of such models was illustrated in Figure 1.1 in Chapter 1.

Many organizations take this customization by project family-industry standards further. In particular, they recognize that each family has multiple project classes of different technical novelty, each calling for a customization of the toolbox. Consider, for example, the case of a low-tech manufacturer (Table 16.4).

This company classifies all projects of its new product introduction project family into three groups: simple, medium, and complex. To distinguish them, the company uses 11 characteristics, mostly related to the project technical novelty. Some of them include customer, product features, end use application, manufacturing process, and assembly process.

Table 16.4: Example of Classifying Projects in a Project Family by Technical Novelty

Characteristic	*Simple*	*Project Class* *Medium*	*Complex*
Customer	Existing	Existing or new	Existing or new
Product features	0-5 feature changes to the existing product	> 5 feature changes to the existing product	No similarity to any existing product
End use application	Same	Same	Same or new
Manufacturing process	Existing	Existing	New
Assembly process	Existing	Some changes to existing process	New

Generally, the more highly technical the novelty, the more complex the projects are [7]. This is because the increasing technical novelty in projects leads to more uncertainty, elevating the need for more flexibility in the SPM process and supporting toolbox. In particular, as the technical novelty of projects increases, the SPM process

▲ Takes more iterations and time to define the project scope.
▲ Requires higher technical and managerial skills.
▲ Deepens communication.
▲ Calls for more effective management of change.

Adapting to such SPM process, you also need to adapt the project management toolbox. As technical novelty grows

▲ The more evolving the Scope Statement and WBS are.
▲ The Gantt Chart or CPM Diagram become more fluid, often updated to account for scope changes.
▲ The Cost Estimates are as fluid as the schedules are.
▲ The risks increase.

A simple example reflecting these trends in adapting the toolbox for the three classes of new product introduction project family from Table 16.4 is illustrated in Table 16.5 (for more examples, see Appendix C).

As the table shows, the toolboxes of the three classes of projects are similar in some and different in other aspects. For example, all use schedules and Progress Report, clearly an offshoot of the one-size-fits all approach. Still, the schedules differ in that simple projects rely on a simple Milestone Chart, while complex projects use a rolling wave type of the Time-scaled Arrow Diagram. Obviously, the variation in the technical novelty of the project is the source of the differences.

In customizing by family type, you need to be aware of certain advantages and risks. The advantages include the following:

▲ Simplicity. Because it is built on one variable (technical novelty), this customization type is simple.
▲ Easy to understand. This customization relies on the technical aspects of projects, the center of the professional background of project managers. Obviously, this makes the customization easy for project managers.

This customization also creates risks in the following:

▲ Projects light on technology. In these projects, technical novelty is not a relevant issue, rendering the customization irrelevant.

▲ Introducing too many models. The application of the customization in a company with multiple project families generates as many project management toolboxes, reducing the ability to integrate the families into a coherent company system.

In summary, here are several deployment guidelines when customizing a project management toolbox by project family:

▲ Classify your projects and their SPM processes into few classes.

▲ Define each class with a few technical parameters.

▲ Support each class with the proper toolbox, each tool backing a specific managerial deliverable.

Table 16.5: Example of Customization of PM Toolbox in Project Family Type by Technical Novelty

Project Class	Definition	Project Phases		
		Planning	Execution	Closure
Simple	Project Charter Scoring Model, emphasis on NPV Bubble Diagram Pie Chart for Portfolio	Milestone Chart	Progress Report	Progress Report
Medium	Project Charter Scoring Model, emphasis on NPV Bubble Diagram Pie Chart for Portfolio	Discussion Guide, Customer Interview Report Scope Statement Gantt Chart with Milestones	Progress Report Updated Gantt Chart	Progress Report
Complex	Project Charter Four-Stage Model Skill Inventory Scoring Model, emphasis on NPV Bubble Diagram Pie Chart for Portfolio	Various Voice of the Customer tools Scope Statement WBS Responsibility Matrix Milestone-Based Cost Estimate Time-scaled Arrow Diagram (cascade, rolling wave) Milestone Chart Risk Response Plan (qualitative) Commitment	Progress Report Updated Time-scaled Arrow Diagram, Milestone Chart Change Request Project Change Log Risk Log Commitment	Progress Report Postmortem

Customization by Project Type. While the previous two options of customization rely on one dimension (variable) each—project complexity (measured by size) and technical novelty, respectively—customization by the project type uses both of the variables, once the toolbox is strategically aligned. We call such a two-dimensional model by the name of its creator—the Shenhar's model [8]. To make it more pragmatic, we will simplify the model, while maintaining its comprehensive nature. Three steps will be used to describe the model:

▲ Define project types.

▲ Describe how the two dimensions impact the SPM process of the project types.

▲ Describe the project management toolbox for the four project types.

Each of the two dimensions includes two levels:

▲ Technical novelty (levels: low, high)

▲ Project complexity (levels: low, high). Note that here we use the system scope (very similar to project size) as the measure of project complexity.

This helps create a two-by-two matrix that features four major types of projects (see Figure 16.4):

▲ Routine

▲ Administrative

▲ Technical

▲ Unique

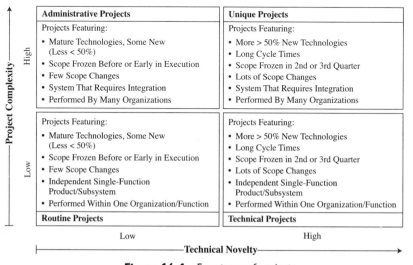

Figure 16.4 Four types of projects.

Reprinted by permission, Shenhar, J. Aaron, "One Size Does Not Fit All Projects: Exploring Classical Contingency Domains," *Management Sci.* 47(3) 394–414, 2001. Copyright (c) 2001, The Institute for Operations Research and the Management Sciences (INFORMS), 901 Elkridge Landing Road, Suite 400, Linthicum, Maryland 21090-2909, USA.

Routine projects have a low level of technical novelty and assembly type of system scope. At the time of project initiation, these projects mostly use the existing or mature technologies or adapt the familiar ones. Sometimes, some new technology or feature may be used, but not to exceed more than 50 percent of the total number of technologies used. Because of the known technologies, scope is frozen before or early in the execution phase, and few scope changes may occur. Essentially, these are low- to medium-tech projects. The system scope—the assembly type—means that the project can produce a product that is a subsystem of a larger system or a stand-alone product capable of performing a single function. Typically, the project is performed within a single organization or organizational function (e.g., engineering or marketing function) [8]. Because it takes the routine to deal with the existing technologies and modest scope, we dubbed these projects routine. Examples include the following:

- ▲ Continuous improvement project in a department
- ▲ Upgrading an existing software package or existing product
- ▲ Adding a swimming pool to an existing hotel
- ▲ Developing a new model of the traditional toaster
- ▲ Expanding an established manufacturing line

Administrative projects are similar to routine projects in terms of the technical novelty—they are low- to mid-tech. Therefore, these projects also use less than 50 percent of new technologies, enjoy an early scope freeze, and have the benefit of few scope changes. However, they differ in the system scope domain. Unlike the routine projects, they produce project products consisting of a collection of interactive subsystems (assemblies) that are capable of performing a wide range of functions [9]. As a result, many organizations or organizational functions are involved, generating a strong need for the integration of both the subsystems and organizations. Such integration will call for more administrative work, which is why we termed these projects administrative. Some examples are as follows:

- ▲ Corporate-wide organizational restructuring
- ▲ Deploying a standard information system for a geographically dispersed organization
- ▲ Building a traditional manufacturing plant
- ▲ Developing a new automobile model
- ▲ Upgrading a new computer or upgrading a software suite.

The major property of technical projects is their technical content, hence, their name. In particular, more than 50 percent of the technologies that these projects use are new or not developed at the project initiation time. This naturally makes them high-tech and tends to create a lot of uncertainty, requiring long project cycle times. Because of the challenging nature of new technology deployment or development, scope often changes and is typically frozen in the second or third quarter of the project implementation period. Like routine projects, the technical projects build single-function independent products or subsystems of larger systems. For this reason, they are of a low complexity level, calling for a single organization to execute the project. Here are some examples:

- Reengineering a new product development process in an organization
- Developing a new software program
- Adding a line with the latest manufacturing technology to a semiconductor fab
- Developing a new model of computer
- Developing a new model of a computer game.

Like technical projects, unique projects feature high-technology content. What makes them unique is that they push to the extreme on both system complexity and technological uncertainty. More than 50 percent of their technologies are new or nonexistent at the time of the project start. This level of uncertainty, combined with high system complexity, is destined to prolong project cycle times and cause a lot of scope changes. Adding to this challenge is the need for integration of the multiple uncertain technologies. In such an environment, normally the scope would freeze in the second or third quarter of the project cycle time. Further complications are instigated by the involvement of many performing organizations that also need to be managerially integrated. Among such projects:

- Building a city's transportation system
- Developing a new generation of microprocessors
- Building a new software suite
- Constructing the latest technology semiconductor fab
- Developing a platform product in an internationally dispersed corporation.

Now that we have defined the four project types, we can move on to the next step—describe how the two dimensions impact the SPM process of these project types. Taken overall, the growing technical novelty in projects generates more uncertainty, which consequently requires more flexibility in the SPM process. As a result, the SPM process [9]

- Takes more cycles and time to define and freeze the project scope
- Needs to make use of more technical skills
- Intensifies communication
- Requires more tolerance toward change

System scope as a measure of the project complexity also has a unique impact on the SPM process. In summary, an increasing system scope leads to an increased level of administration, requiring the SPM process to feature [9]

▲ More planning and tighter control

▲ More subcontracting

▲ More bureaucracy

▲ More documentation

To support these features of the SPM process, you need to adapt the project management toolbox accordingly. In Figure 16.5 we show examples of several project management tools that have to be adapted to account for different SPM processes in these four types of projects. Detailed explanations about how these tools differ are provided in Appendix D, along with more examples of toolboxes for the four project types.

A summary comparison of the tools for the four project types reveals that they use very similar types of the tools. For example, all use WBS. Still, when the same type of tool is used, there are differences in their structure and the manner how they are used. Consider, for example, Gantt and Milestone Charts. Both are used in the routine and unique projects, but terms of use are significantly different. This is the situational approach—as the nature of the SPM process changes, so does the project management toolbox. In other words, there may be as many toolboxes as SPM processes and the project types they support (we simplified this to four project types, SPM processes, and toolboxes).

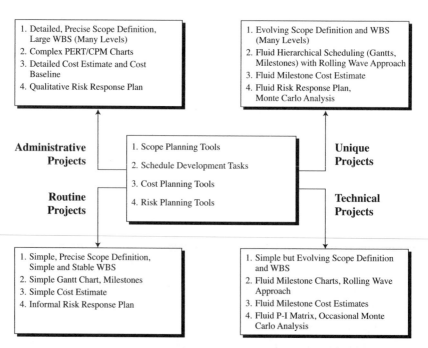

Figure 16.5 Customizing project management toolbox by the project type.

Customization of the toolbox by the project type has its advantages and risks. The advantages are that this customization approach is

- ▲ Sufficiently comprehensive. It includes two dimensions that account for major sources of contingencies.
- ▲ Unifying. This approach may be used for the large majority of projects, providing a unifying framework for all projects in an organization.

Risks from using the approach of customizing the toolbox by the project type need to be recognized as well:

- ▲ Not entirely comprehensive. To make this approach relatively easy to use, we built only two dimensions into it, leaving out others. For example, speed in implementing projects is another such dimension that could be a source of variation.
- ▲ May be difficult to implement. The demanding framework with two dimensions may be difficult to apply in certain corporate cultures. Examples are companies that are not mature in project management or have a long tradition with their industry's dominant project management models.

In summary, when customizing the project management toolbox by project type, follow these guidelines:

- ▲ Use the two dimensions we described or adapt them to your needs.
- ▲ Classify your projects and their SPM processes into four types.
- ▲ Match each SPM process with the proper toolbox, supporting each managerial deliverable with specific tool(s).

Which Customization Option to Choose?

We offered three options for the customization of the project management toolbox. Each has its advantages and risks and fits some situations better than the others. To decide each one to select, refer to Table 16.6. Customization by project size is the right option when an organization has projects of varying size and needs a simple start toward more mature forms of the toolbox customization. In addition, projects of varying size characterized by mature technologies lend themselves well to this customization. If an organization has a stream of projects that feature both mature and novel technologies but project size is not an issue, customization by project family may be the first option. This is also the option to go for when projects are dominated by a strong industry or professional culture.

Table 16.6: Project Situations That Favor Each of the Three Options for Customizing PM Toolbox

Situation	Favoring Customization by Project Size	Favoring Customization by Project Family	Favoring Customization by Project Type
Simplest start in customizing PM Toolbox	√		
Projects of varying size with mature technologies	√		
Projects with both mature and novel technologies; size is not an issue		√	
Projects with strong industry or professional culture		√	
Projects of varying size with both mature and novel technologies			√
Need a unifying framework for all organizational projects			√

Customization of the project management toolbox by project type is also favored in certain situations. One of those situations is when an organization has lots of projects that significantly vary in size but also in technical novelty, spanning from low-tech to high-tech. Organizations searching for a unifying framework that can provide the customization for all types of their projects—from facilities to product development to manufacturing to marketing to information systems—may find customization by the project type an appropriate choice for this purpose.

Once one of the customization options has been selected, there is a need to plan for its implementation. Here are steps for it:

- ▲ Set the desired level of the customization
- ▲ Establish the current level of the customization.
- ▲ Identify gaps between the current and desired customization.
- ▲ Act to provide the desired customization.

These steps are very similar to those for strategic alignment of the project management toolbox. They require a clear definition of where the company wants to go with the customization and a shared knowledge of where the current toolbox is.

Then establishing the gaps between the two is possible, prompting the development of a management action plan to guide the customization as described in each of the options. When the toolbox has been customized, it will be more effective if continuously improved.

Improve Project Management Toolbox Continuously

The purpose of this task is to put in place the rules for continuous maintenance and improvement of the customized toolbox. Without such improvement, the toolbox will gradually deteriorate, losing the ability to support the SPM process, project management strategy, and eventually, the competitive strategy of the organization [10]. Avoiding such a predicament and instead sustaining an effective toolbox can be achieved through the following steps:

- ▲ Form a project management toolbox improvement team.
- ▲ Identify mechanisms for improvement ideas.
- ▲ Follow improvement process.

Form an Improvement Team. The toolbox improvement team is usually part of the process team responsible for designing and managing the project management process. This team has the total responsibility for simplifying, improving, and managing the implementation of the project management toolbox. Each team member owns a piece of the toolbox, and overall, the responsibility should be distributed as evenly as possible across the team. When forming a team, it is important to understand that management enforces, while the team operates and owns the toolbox. Because it is mostly project managers that must use the toolbox, we recommend that the majority of the toolbox team members should come from the ranks of project managers.

Identify Mechanisms for Improvement Ideas. Ideally, there should be a continuous stream of suggestions and ideas to improve the customized toolbox. To secure such a stream, you can require that project teams at the end of the project perform postmortem reviews. If the reviews find a need to change the toolbox, the team should submit a change request. Besides, change requests may come at any time from anyone involved in projects. Note that requests are not the only way to collect the toolbox improvement ideas. A survey or management's periodic talk with members of the organization or small focus groups may serve this purpose as well.

Follow Improvement Process. This process defines steps in acting upon change and deviation requests, as well as an escalation procedure, in case the requests are turned down and those proposing them want them to be evaluated by management. Change requests are suggestions for making changes to the toolbox, most frequently coming from project teams. Quickly collecting and responding to them is of vital importance. Also significant are requests to deviate from the customized toolbox. If appropriate, the deviation requests should be allowed in order to make sure the toolbox is flexible. These are requests to deviate one time only from some part of the project management toolbox. Since they are submitted while the project is in progress, it is important to respond as soon as possible. Later, the requests can be evaluated to determine if the toolbox should be adjusted.

Final Thoughts

When it comes to managing projects, the toolbox has been one of the most critical things to learn. We would like to point out a few lessons we believe are most important from this book.

First, companies adopt project management to support the company's competitive strategy, which, in turn, has the task of creating competitive advantages that are so vital in outperforming their rivals in the market place. However, this support is often delivered through the SPM process, a disciplined and interrelated set of phases, deliverables, and milestones that each project goes through.

Second, companies can use project management tools to produce specific managerial deliverables in the SPM process. As a matter of fact, this is what the conventional purpose of each of the tools was. Examples, of course, can help get the point across. If we need to define the project scope, we can use the Scope Statement and WBS tools for this purpose. Similarly, faced with the problem of speeding up the project schedule, we can enlist the help of the Schedule Crashing tool or perhaps the Critical Chain Schedule. This, of course, calls for good knowledge of the individual tools, a reason for our decision to cover so many tools in this book.

Third, companies can use the tools to build the project management toolbox. In this manner, the tools become basic building blocks of the SPM process. Conceived as a set of predefined tools that are capable of completing the whole set of SPM process's managerial deliverables, the toolbox helps us go the full circle—the toolbox supports the SPM process that helps deliver project management strategy to support the competitive strategy that generates competitive advantages.

We have argued in this book that companies have a clear choice in using project management tools: They can go for the tools one at a time, or they can use the toolbox. Project managers can reach a certain managerial deliverable with one tool at a time, or they can attempt to leverage the toolbox through SPM process across all their projects. There are organizational costs to the project management toolbox as we have described it. This mode of operation calls for making a conscious choice of tools for the toolbox and aligning it with the competitive strategy of the company. It also requires making a smart decision on how to customize the toolbox— by project size, family, or type. In addition, companies need to be smart in how they continuously improve the toolbox.

But these are no more than details, as important as they may be. On a fundamental level, companies need to decide whether they want to think and manage in a project management toolbox mode. To us, the choice is clear. Individual tools used one at a time provide an outstanding systematic procedure that organizations have used to produce individual deliverables. But such a focus may no longer be suitable for organizations to win in the marketplace. The demands have changed as organizations manage more and more projects—faster, cheaper, better. Such changes necessitate new best practice; one that puts a strategically-aligned and customized project management toolbox in the center of the company's competitive stage.

References

1. Cooper, R. and E .J. Kleinschmidt. 1994. "Determinants of Timeliness in Product Development." *Journal of Product Innovation Management* 11(5): 381–396.

2. Shenhar, A. J. 2001. "Contingent Management in Temporary, Dynamic Organizations: The Comparative Analysis of Projects." *The Journal of High Technology Management Research* 12: 239–271.

3. Smith, P. and D. Reinertsen. 1990. Developing Products in Half the Time. New York: Van Nostrand Reinhold.

4. Pinto, J. K. and J. G. Covin. 1989. "Critical Factors in Project Implementation: A Comparison of Construction and R&D Projects." *Technovation* 9: 49–62.

5. Harrison, J. S. and C. H. S. John. 1998. *Strategic Management of Organizations and Stakeholders*. St. Paul, Minn.: South-Western College Publishing.

6. Brown, S. L. and K. M. Eisenhardt. 1997. "The Art of Continuous Change: Linking Complexity Theory and Time-Paced Evolution in Relentlessly Shifting Organization." *Administrative Science Quarterly* 42: 1–34

7. Tatikonda, M. V. and R. S. Rosenthal 2000. "Technology Novelty, Project Complexity, and Product Development Project Execution Success: A Deeper Look at Uncertainty in Product Innovation." *IEEE Transactions on Engineering Management* 47(1): 74–87.

8. Shenhar, A. J. 2001. "One Size Does Not Fit All Projects: Exploring Classical Contingency Domains." *Management Science* 47(3): 394–414.

9. Shenhar, A. J. 1998. "From Theory to Practice: Toward a Typology of Project-Management Styles." *IEEE Transactions on Engineering Management* 45(1): 33–48.

10. Juran, J. M. 1992. "Managing for World-Class Quality." *PM Network* 6(4): 5–8.

Linking Project Management Tools to PMBOK

When tasked with selecting individual project management (PM) tools or designing PM toolbox or PM process, many PM practitioners refer to *A Guide to the Project Management Body of Knowledge* (PMBOK) as their major source of PM knowledge. This is a logical choice, since PMBOK is a widely accepted PM standard developed by the Project Management Institute. To make the task of the PM practitioners easier, this appendix links PM tools covered in this book with PMBOK's PM processes and knowledge areas in tables that follow. More specifically, for each PM tool we will first indicate in which of the following PM processes it can be used:

- ▲ Initiating Processes (Init.)
- ▲ Planning Processes (Plan.)
- ▲ Executing Processes (Exec.)
- ▲ Controlling Processes (Cont.)
- ▲ Closing Processes (Clos.)

Then, we will show in which of the following knowledge areas the tool can be used:

- ▲ Project Integration Management (Int.)
- ▲ Project Scope Management (Scope)
- ▲ Project Time Management (Time)
- ▲ Project Cost Management (Cost)
- ▲ Project Quality Management (Qual.)
- ▲ Project Human Resource Management (HR)
- ▲ Project Communications Management (Com.)
- ▲ Project Risk Management (Risk)
- ▲ Project Procurement Management (Proc.)

Symbols in parentheses are used as a key for the following tables.

Project Management (PM) Tool	Can Be Used in PMBOK PM Processes					Can Be Used in PMBOK PM Knowledge Areas							
	Init.	Plan.	Exec.	Cont.	Clos.	Int.	Scope	Time	Cost	Qual.	HR	Com.	Risk
Ch. 2 Scoring Models	✓						✓						
Analytic Hierarchy Process	✓						✓						✓
NPV (Payback Period, IRR)	✓						✓						
Portfolio Selection	✓						✓						
Real Options Approach	✓						✓						
Ch. 3 Bar Graphs	✓						✓		✓				✓
Bubble Charts	✓						✓		✓				✓
Ch. 4 Customer Roadmap		✓				✓	✓			✓			
Focus Statement		✓				✓	✓			✓			
Sample Selection		✓				✓	✓			✓			
Discussion Guide		✓				✓	✓			✓			
Quality Function Deployment		✓				✓	✓			✓			
Ch. 5 Project Charter		✓				✓	✓				✓		
Project SWOT Analysis		✓				✓	✓						✓
Scope Statement		✓				✓	✓						
Work Breakdown Structure		✓				✓	✓						

continued

Project Management (PM) Tool	Can Be Used in PMBOK PM Processes					Can Be Used in PMBOK PM Knowledge Areas							
	Init.	Plan.	Exec.	Cont.	Clos.	Int.	Scope	Time	Cost	Qual.	HR	Com.	Risk
Ch. 6 Gantt Chart		✓				✓		✓					
Milestone Chart		✓				✓		✓					
Time-scaled Arrow Diagram		✓				✓		✓					
CPM Diagram		✓				✓		✓					
Critical Chain Schedule		✓				✓		✓					
Hierarchical Schedule		✓				✓		✓					
Line of Balance		✓				✓		✓					
Ch. 7 Cost Planning Map		✓				✓			✓				
Analogous Estimate		✓				✓			✓				
Parametric Estimate		✓				✓			✓				
Bottom-up Estimate		✓				✓			✓				
Cost Baseline		✓				✓			✓				
Ch. 8 Project Quality Program		✓				✓				✓			
Flowchart		✓				✓				✓			
Affinity Diagram		✓				✓				✓			
Ch. 9 Risk Response Plan		✓				✓							✓
Monte Carlo Analysis		✓				✓							✓
Decision Tree		✓				✓							✓

	Init.	Plan.	Exec.	Cont.	Clos.	Init.	Scope	Time	Cost	Qual.	HR	Com.	Risk
Ch. 10 Four-Stage Model		✓	✓	✓	✓						✓		
Stakeholder Matrix		✓									✓	✓	
Skill Inventory		✓									✓		
Commitment		✓	✓								✓		
Ch. 11 Change Coordination Matrix				✓		✓	✓						
Project Change Request				✓		✓	✓						
Project Change Log				✓		✓	✓						
Ch. 12 Jogging Line Method				✓				✓				✓	
B-C-F Analysis				✓				✓				✓	
Milestone Prediction Chart				✓				✓				✓	
Slip Chart				✓				✓				✓	
Buffer Chart				✓				✓				✓	
Schedule Crashing				✓				✓				✓	
Ch. 13 Earned Value Analysis								✓	✓			✓	
Milestone Analysis								✓	✓			✓	

continued

Project Management (PM) Tool	Can Be Used in PMBOK PM Processes					Can Be Used in PMBOK PM Knowledge Areas							
	Init.	Plan.	Exec.	Cont.	Clos.	Init.	Scope	Time	Cost	Qual.	HR	Com.	Risk
Ch 14 Quality Improvement Map			√	√						√			
Pareto Chart			√	√						√			
Cause and Effect Diagram			√	√						√			
Control Charts			√	√						√			
Ch. 15 Risk Log				√						√			√
Progress Report				√			√	√	√	√		√	
Postmortem					√							√	

B

Project Management Toolboxes per Project Size

As discussed in Chapter 16, the customization of the project management (PM) toolbox by project size is the right option when an organization has projects of varying size with mature technologies. Here, we provide examples of toolboxes for such option with the following goals:

- ▲ Goal 1: Tier PM toolbox per project size
- ▲ Goal 2: Tier each size's toolbox into more and less important tools

Goal 1: Tier the Toolbox per Project Size

A simple typology of projects by their project size—small, medium-sized, and large projects—is a convenient and well-accepted framework by organizations to reflect on different management needs of their project managers. Following this logic, in the tables that follow, we built examples of

- ▲ PM toolbox for small projects
- ▲ PM toolbox for medium-sized projects
- ▲ PM toolbox for large projects

Since the size is the only criterion in building the toolboxes, note that our examples are *non-project family/industry/type specific*. Also, because size means different things to different organizations, we chose the generic labels of small, medium, and large without specific criteria. For this reasons, you may choose to adapt our examples per your own definition of project size.

Goal 2: Tier Each Size's Toolbox into More and Less Important Tools

To project practitioners all PM tools are not created equal. Some are more, others less important in accomplishing their project goals. Acknowledging this reality and using Kano methodology, the PM toolbox for each size is divided into the following:

- ▲ Must tools
- ▲ Optional tools
- ▲ Delighter tools

Must tools, the first tier, are crucial for a project process to be successful, and if they are not present in the project process, there is a significant likelihood for the project to go wrong. Unlike them, tools in the second tier—optional tools—are not required, although their use may offer project managers more choices and satisfaction. The delighter tools, the third tier, are not even expected to be seen in the project process. Their application, however, may create a special value, which may delight project managers and lead to higher quality of the standardized project management (SPM) process. The hope is that this approach will help project managers be flexible in using the tools. Since we advocate the use of the toolbox as a support for SPM process, note that for the sake of simplicity, only the PM tools are shown, leaving out technical deliverables.

continued

Project Size	Type of Tools	Project Phases			
		Concept	Definition	Execution	Finish
Small	Must	Scope Statement WBS Responsibility Matrix Gantt Chart	Progress Report Gantt Update	Progress Report	
	Optional	Scoring Model Project Charter	Cost Estimate Risk Resp. Plan	Jogging Line	
	Delighter	Bar Graphs/ Bubble Charts Project Quality Program	QFD	Milestone Analysis	Postmortem
Medium	Must	Project Charter	Scope Statement WBS Responsibility Matrix Cost Estimate Gantt/Milestone Chart Project Qual. Program Risk Resp. Plan Commitment	Progress Report Gantt Update Cost Estimate Update Quality Program Update Change Request Commitment	Progress Report Gantt Update Cost Estimate Update Quality Program Update Change Request
	Optional	Scoring Model Bar Graphs/ Bubble Charts Four-stage Model	Discussion Guide QFD Risk Resp. Plan CPM/TAD Cost Baseline	B-C-F Method Milestone Analysis/ Earned Value Change Request Log Risk Log	Postmortem

Project Size	Type of Tools	Project Phases			
		Concept	Definition	Execution	Finish

	Delighter	NPV
		Stakeholder Matrix
		Skill Inventory
			Customer Roadmap
			Process Flowchart
			Monte Carlo
				Milestone Pred. Chart
				Slip Chart
				Change Coordination Matrix
				Quality Improvement Map
					Progress Report
					CPM/TAD Update
					Milestone Analysis/
					  Earned Value
					Slip Chart
					Cost Estimate Update
					Quality Program Update
					Change Coordination Matrix,
					  Change Request and Log
					Risk Log
					Postmortem

Large	Must	Scoring Model
		NPV
		Project Charter
		Four-stage Model
		Stakeholder Matrix
		Skill Inventory
			Scope Statement
			WBS
			Responsibility Matrix
			Cost Estimate
			Cost Baseline
			CPM/TAD
			Project Qual. Program
			Risk Response Plan
			Customer Roadmap
			QFD
			Commitment
				Progress Report
				CPM/TAD Update
				Milestone Analysis/
				  Earned Value
				Slip Chart
				Cost Estimate Update
				Quality Program Update
				Change Coordination Matrix,
				  Change Request and Log
				Risk Log
				Commitment	

	Optional	AHP Portfolio Selection	Monte Carlo Process Flowchart	Schedule Crashing Milestone Pred. Chart Quality Improvement Map Cause and Effect Diagram and Pareto	
	Delighter		Critical Chain	Buffer Management Control Chart	

Project Management Toolboxes per Project Family

As discussed in Chapter 16, some organizations favor customization of the PM toolbox by project family. This is the case when the organizations have a stream of projects featuring both mature and novel technologies, and project size is not considered an issue. Also, when projects are dominated by a strong industry or professional culture, this customization is a valid option. Here, we offer examples of PM toolboxes for the following families:

- ▲ Software development project family
- ▲ New product development project family
- ▲ Manufacturing project family

Because these families have a strong technological emphasis, we added technical deliverables to the PM tools (managerial deliverables) to create the complete PM toolbox. In this way, you can readily use our examples of the PM toolbox as an adaptable blueprint to create the PM toolbox for the specific family in an organization as a proxy for standardized project management process.

In presenting the toolboxes, our goals are as follows:

- ▲ Goal 1: Tier each family toolbox into classes of technical novelty (simple, medium, complex).
- ▲ Goal 2: Tier each class's toolbox into more and less important tools.

Goal 1: Tier the Toolbox into Project Classes

The degree of technical novelty in projects is often used as a criterion for project classification. Applying this criterion, you can group the projects into classes of simple, medium, and complex. We used this simple framework to distinguish between managerial needs of these classes' project managers. Note, however, that since the technical novelty is the only criterion in building the toolboxes, our examples are not in any way related to the project size. Note also that we chose the generic labels of simple, medium, and complex to describe degrees of technical novelty. This is a sufficient reason for organizations that have their own definition of technical novelty to adapt our examples per such definition.

Goal 2: Tier Each Class's Toolbox into More and Less Important Tools

Here we again use the Kano methodology to divide each class's toolbox into must, optional, and delighter tools. These types of tools are defined in Appendix B.

PM Toolbox for the Software Development Project Family

Project Class	Type of Tools	Initiation & Definition	Planning	Execution & Control	Closure
				Project Phases	
Simple Software Development	Must	Requirements Document Scope Statement	Progress Report	Functional Specs Update System Documentation Progress Report	Progress Report
	Optional	Gantt/Milestone Chart	Gantt/Milestone Update		
	Delighter	Responsibility Matrix Risk Resp. Plan			Postmortem
Medium/ Complex Software Development	Must	General Assessment Document Requirements Document Project Charter Four-Stage Model Skill Inventory Scope Statement WBS Responsibility Matrix Gantt/TAD Chart Cost Estimate Risk Resp. Plan	Progress Report Gantt Update Cost Estimate Update Commitment	Functional Specs Constructed System Test Strategy Training Strategy System Documentation Implementation Plan Progress Report Gantt Update Cost Estimate Update Risk Log Change Request Commitment	Progress Report Gantt Update Cost Estimate Update Risk Log Change Request

continued

PM Toolbox for the Software Development Project Family (Continued)

Project Class	Type of Tools	Project Phases			
		Initiation & Definition	Planning	Execution & Control	Closure
	Optional	Scoring Model Bar Graphs/ Bubble Charts NPV Stakeholder Matrix Discussion Guide	Milestone Prediction Chart Change Request Log	Milestone Prediction Chart Quality Program Update Change Request Log	Quality Program Update Postmortem
	Delighter	Monte Carlo Customer Roadmap	Quality Improvement Map Process Flowchart	Milestone Pred. Chart Quality Improvement Map Change Coordination Matrix	Milestone Pred. Chart Quality Improvement Map Change Coordination Matrix

Note: Shaded areas indicate technical deliverables

PM Toolbox for the New Product Development Project Family, High Technology

Project Class	Type of Tools	Planning	Development	Verification	Pilot	Mfg. Release
				Project Phases		
Simple	Must	Prod. Summary Project Charter Gantt/Milestone Chart Scoring Model Bar Graphs/Bubble Charts	Specs Update Progress Report Gantt/Milestone Update	Progress Report Gantt/Milestone Update	Mfg. Release Progress Report Gantt/Milestone Update	
	Optional	Responsibility Matrix Cost Estimate	Prototype Jogging Line	Alpha Test Jogging Line	Beta Test Jogging Line	
	Delighter	Risk Resp. Plan	Risk/Issue Log	Risk/Issue Log	Risk/Issue Log Postmortem	
Medium	Must	Prod. Summary Prod. Definition Scoring Model NPV Bar Graphs/ Bubble Charts Project Charter Four-Stage Model Skill Inventory Scope Statement WBS Responsibility Matrix Cost Estimate Gantt Chart Risk Resp. Plan Discussion Guide	Eng. Documents Prototype Test Pass BOM Mfg./Service Review Progress Report Jogging Line Cost Estimate Update Change Request Risk/Issue Log Commitment	Alpha Test Mfg. Documents Progress Report Jogging Line Cost Estimate Update Change Request Risk/Issue Log Commitment	Beta Tests Pilot Run Complete Mfg. Signoff Mfg. Release 4 Progress Report Jogging Line Cost Estimate Update Change Request Risk/Issue Log Postmortem	

continued

PM Toolbox for the New Product Development Project Family, High Technology (Continued)

Project Class	Type of Tools	Project Phases			
		Planning	**Development**	**Verification**	**Pilot**
	Optional	Prod. Dev. Plan	B-C-F Method	B-C-F Method	B-C-F Method
		Stakeholder Matrix	Milestone Analysis	Milestone Analysis	Milestone Analysis
		Project SWOT Analysis	Change Coordination	Change Coordination	Change Coordination
		CPM/TAD	Matrix and Log	Matrix and Log	Matrix and Log
	Delighter	QFD	Milestone Analysis	Schedule Crashing	Schedule Crashing
		Monte Carlo	Schedule Crashing	Milestone Pred. Chart	Milestone Pred. Chart
		Customer Roadmap	Milestone Pred. Chart	Quality Improvement	Quality Improvement
			Quality Improvement Map	Map	Map

Note: Shaded areas indicate technical deliverables

PM Toolbox for the New Product Development Project Family, High Technology

Project Class	Type of Tools	Project Phases			
		Planning	**Development**	**Verification**	**Pilot**
Complex	Must	Prod. Proposal	Eng. Documents	Verification Material	Mfg. Checklist
		Prod. Definition	Software API	Tests (Alpha, Device, SW,	Beta Tests
		Prod. Dev. Plan	Prototype	Regression, Configuration,	Pilot Run Complete
		Scoring Model	Test Pass	Mfg., Mechanical, etc.)	Pilot Run Margins
		Bar Graphs/	Mfg. Tools and Documents	Prod. Release Cycle	Mfg. Signoff
		Bubble Charts	BOM	Mfg. Documents	Reliability Analysis
		NPV	Long Lead Items Ordered	Service Documents	Mfg. Release
		Project Charter	Mfg./Service Review	Technical Pubs	Progress Report

PM Toolbox for the New Product Development Project Family, High Technology

Project Class	Type of Tools	Project Phases			
		Planning	Development	Verification	Pilot
		Four-Stage Model Skill inventory Stakeholder Matrix Scope Statement WBS Project SWOT Analysis Responsibility Matrix Cost Estimate Cost Baseline CPM/TAD Hierarchical Schedule Risk Response Plan Discussion Guide	Marketing Plan Progress Report CPM/TAD Update Slip Chart Milestone Analysis Cost Estimate Update Change Coordination Matrix, Change Request and Log Risk/Issue Log Commitment	Marketing Documents Progress Report CPM/TAD Update Slip Chart Milestone Analysis Cost Estimate Update Change Coordination Matrix, Change Request and Log Risk/Issue Log Commitment	CPM/TAD Update Slip Chart Milestone Analysis Cost Estimate Update Change Coordination Matrix, Change Request and Log Risk/Issue Log Postmortem
	Optional	Customer Roadmap QFD Project Quality Program Monte Carlo	Process Flowchart Quality Program Update Schedule Crashing Milestone Pred. Chart Quality Program Update Quality Improvement Map Cause and Effect Diagram and Pareto	Quality Program Update	
	Delighter	AHP Real Options Critical Chain	Buffer Management	Buffer Management	Buffer Management

Note: shaded areas indicate technical deliverables

PM Toolbox for the Manufacturing Project Family

Project Class	Type of Tools	Project Phases			
		Plan	**Procure**	**Install**	**Ramp Up**
Simple	Must	Implementation Summary Cost Estimate Gantt/Milestone Chart NPV	Purchase Order Progress Report Gantt Update	Training Erection Pilot Runs Progress Report Gantt Update	Mfg. Release Progress Report Gantt Update
	Optional	Scoring Model Project Charter Scope Statement	Cost Estimate Update Risk/Issue Log	Cost Estimate Update Risk/Issue Log	Risk/Issue Log
	Delighter	WBS Responsibility Matrix	Jogging Line	Jogging Line	Postmortem
Medium	Must	Mfg. Capacity Analysis Implementation Plan Project Charter Four-stage Model Scope Statement Responsibility Matrix Cost Estimate Gantt Chart NPV	Eng. Specs Purchase Order Progress Report Jogging Line Cost Estimate Update Change Coordination Matrix, Change Request and Log Commitment	Pilot Runs Op. and Maint. Manuals Progress Report Jogging Line Cost Estimate Update Change Coordination Matrix, Change Request and Log Commitment	Mfg. Checklist Mfg. Release Progress Report Jogging Line Cost Estimate Update Change Coordination Matrix, Change Request and Log

PM Toolbox for the Manufacturing Project Family

Project Class	Type of Tools	Project Phases			
		Plan	Procure	Install	Ramp Up
	Optional	Scoring Model Skill Inventory WBS Cost Baseline Project Quality Program	Earned Value Quality Program Update Risk/Issue Log	Change Coordination Matrix, Change Request and Log Earned Value Quality program Update Risk/Issue Log	Change Coordination Matrix, Change Request and Log Earned Value Risk/Issue Log Quality Program Update Postmortem
	Delighter	Stakeholder Matrix Discussion Guide QFD	Quality Improvement Map	Quality Improvement Map	Quality Improvement Map

Note: Shaded areas indicate technical deliverables.

A Manufacturing Project Family

Project Class	Type of Tools	Project Phases			
		Plan	**Procure**	**Install**	**Ramp Up**
Large	Must	Soft Proposal	Eng. Specs	Training	Mfg. Checklist
		Mfg. Capacity Analysis	Request For Proposal	Erection	Mfg. Signoff
		Feasibility Study	Vendor Evaluations	Pilot Runs	Mfg. Release
		Implementation Plan	Purchase Order	Capability Tests	Progress Report
		Project Charter	Progress Report	Op. and Maint. Manuals	TAD Update
		Scope Statement	TAD Update	Progress Report	Earned Value
		Four-stage Model	Earned Value	TAD Update	Change Coordination
		Skill Inventory	Change Coordination	Earned Value	Matrix, Change
		Stakeholder Matrix	Matrix, Change Request	Change Coordination	Request and Log
		Responsibility Matrix	and Log	Matrix, Change Request	Postmortem
		WBS	Commitment	and Log	
		Cost Estimate		Commitment	
		Cost Baseline			
		TAD			
		Project Quality Program			
		Scoring Model			
		NPV			

A Manufacturing Project Family

Project Class	Type of Tools	Project Phases			
		Plan	Procure	Install	Ramp Up
	Optional	Discussion Guide	Process Flowchart Risk/Issue Log	Quality Improvement Map Cause and Effect Diagram and Pareto Control Chart Risk/Issue Log	Risk/Issue Log
	Delighter	Customer Roadmap QFD	Schedule Crashing	Schedule Crashing	Schedule Crashing

Note: Shaded areas indicate technical deliverables

Project Management Toolboxes per Project Type

The discussion in Chapter 16 indicated that some organizations might favor customization of the PM toolbox by the project type. Typically, such organizations have lots of projects that significantly vary in size but also in technical novelty, spanning from low tech to high tech. Also, they may see this customization type as a unifying framework for all types of their projects—from facilities to product development to manufacturing to marketing to information systems. For these organizations, this appendix offers examples of the toolboxes for the following project types:

- ▲ PM toolbox for routine projects
- ▲ PM toolbox for new administrative projects
- ▲ PM toolbox for technical projects
- ▲ PM toolbox for unique projects

While definitions of these project types are provided in Chapter 16, our goals in presenting the toolboxes are as follows:

- ▲ Goal 1: Show only PM tools
- ▲ Goal 2: Tier each project type's toolbox into more and less important tools
- ▲ Goal 3: Describe how some PM tools have to be adapted

Goal 1: Show PM Tools Only

Because these project types are of an across-industry, across-project-family nature, we show only PM tools in the toolbox, leaving out technical deliverables, as well as managerial deliverables that the tools support. You can readily use the examples as an adaptable blueprint and add your technical and managerial deliverables to form the PM toolbox for your organization. As you do that, you should first read our criteria for defining the four project types. If your criteria are different, you should adapt the offered examples of the toolboxes accordingly.

Goal 2: Tier Each Project Type's Toolbox into More and Less Important Tools

Here we again use the Kano methodology to divide each project type's toolbox into must, optional, and delighter tools. These types of tools are defined in Appendix B.

Goal 3: Describe How Some PM Tools Have to Be Adapted

In Chapter 16, Figure 16.5 we showed examples of several PM tools— scope, schedule, cost, and risk planning tools—that have to be adapted to account for different SPM (standardized project management) processes in the four types of projects. These tools are presented in each of the following toolbox examples. Later in this appendix we provide more details on the adaptation.

The Toolbox for Routine Projects

Type of Tools	Project Phases			
	Concept	Definition	Execution	Finish
Must	Scoring Model	Scope Statement	Progress Report	Progress Report
	NPV	WBS	Jogging Line	Jogging Line
	Project Charter	Responsibility Matrix	Cost Update	Cost Update
		Cost Estimate	Change Request	Change Request
		Gantt Chart	Commitment	
		Commitment		
Optional	Four-Stage Model	Risk Resp. Plan	B-C-F Analysis	B-C-F Analysis
	Stakeholder Matrix	B-C-F Analysis	Milestone Analysis	Milestone Analysis
	Skill Inventory	Milestone Analysis	Change Coordination	Change Coordination
	Bar Graphs/ Bubble Charts	Change Coordination Matrix	Matrix and Log	Matrix and Log
		and Log	Risk/Issue Log	Risk/Issue Log
				Postmortem
Delighter		QFD	Schedule Crashing	Schedule Crashing
		Project Qual. Program	Milestone Pred. Chart	Milestone Pred. Chart
		Milestone Pred. Chart		

The Toolbox for Administrative Projects

Type of Tools	Project Phases			
	Concept	Definition	Execution	Finish
Must	Scoring Model NPV Project Charter Four-Stage Model Skill Inventory Bar Graphs/ Bubble Charts	Scope Statement WBS Responsibility Matrix Cost Planning Map Cost Estimate (Analogous, Parametric, Bottom-up) Cost Baseline CPM/TAD Project Qual. Program Risk Response Plan Customer Roadmap Discussion Guide Commitment	Progress Report CPM/TAD Update Milestone Analysis/ Earned Value Quality Update Change Coordination Matrix, Change Request and Log Risk Log Commitment	Progress Report CPM/TAD Update Milestone Analysis/ Earned Value Quality Update Change Coordination Matrix with Change Request and Log Risk Log Postmortem
Optional	Stakeholder Matrix AHP Portfolio Selection	QFD Monte Carlo Process Flowchart	Schedule Crashing Milestone Pred. Chart Quality Improvement Map Cause and Effect Diagram and Pareto	Schedule Crashing Milestone Pred. Chart Quality Improvement Map Cause and Effect Diagram and Pareto
Delighter		Critical Chain	Buffer Management Control Chart	

The Toolbox for Technical Projects

Type of Tools	Concept	Project Phases		
		Definition	*Execution*	*Finish*
Must	Scoring Model NPV Project Charter Four-Stage Model Skill Inventory Bar Graphs/ Bubble Charts Discussion Guide	Scope Statement WBS Responsibility Matrix Cost Estimate CPM/TAD/Milestone Chart (rolling wave) Affinity Diagram Risk Response Plan Change Coordination Matrix, Change Request and Log Commitment	Progress Report CPM/TAD/Milestone Update Slip Chart B-C-F Analysis Change Coordination Matrix, Change Request and Log Risk/Issue Log Commitment	Progress Report CPM/TAD/Milestone Update Slip Chart B-C-F Analysis Change Coordination Matrix, Change Request and Log Risk/Issue Log Postmortem
Optional	Customer Roadmap Stakeholder Matrix	Project SWOT Analysis QFD Monte Carlo	Schedule Crashing Milestone Pred. Chart	
Delighter	AHP Real Options	Critical Chain	Buffer Management	Buffer Management

The Toolbox for Unique Projects

Type of Tools	Concept	Definition	Execution	Finish
		Project Phases		
Must	Scoring Model	Project SWOT Analysis	Progress Report	Progress Report
	NPV	Scope Statement	CPM/Tad Update	CPM/TAD Update
	Project Charter	WBS	Slip Chart	Slip Chart
	Bar Graphs/ Bubble Charts	Responsibility Matrix	Jogging Line/Milestone Update	Jogging Line/ Milestone Update
	Four-Stage Model	Cost Estimate	Milestone Analysis	Milestone Update
	Skill Inventory	Cost Baseline	Change Coordination Matrix, Change Request and Log	Milestone Analysis
	Stakeholder Matrix	CPM/Tad Chart	Risk/Issue Log	Change Coordination Matrix, Change Request and Log
		Gantt/Milestone Chart (rolling wave)	Commitment	Risk/Issue Log
		Hierarchical Schedule		Postmortem
		Affinity Diagram		
		Risk Response Plan		
		Discussion Guide		
		Commitment		
Optional	AHP	Customer Roadmap	Schedule Crashing	Schedule Crashing
	Portfolio Selection	Focus Statement	Milestone Pred. Chart	Milestone Pred. Chart
	Real Options	Sample Selection	Buffer Management	Buffer Management
		QFD		
		Critical Chain		
		Monte Carlo		
Delighter		Decision Tree	Decision Tree	

Adapting Project Management Tools for Four Types of Projects

Scope, schedule, cost, and risk planning tools shown in Figure 16.5 in Chapter 16 and also included into the all examples of toolboxes in this appendix have to be adapted per requirements of SPM processes in the four types of projects. In the following, we describe the adaptation. Similar adaptation may be necessary for other tools in these toolboxes, as well as for toolboxes developed with other types of customization—customization by project size and customization by project family.

Routine Projects. With their low complexity and technical novelty, routine projects need a precise Scope Statement that will be based on a well-defined scope that is frozen before or early in the execution phase of the project. That enables us to use a simple WBS with few levels. Because of the precise scope, the WBS is also stable over time. In such a situation, using a template WBS is a viable choice. Good knowledge of mature technologies to be deployed in simple projects also makes it possible for the project team to rely on simple Gantt Charts with milestones, informally recognizing the dependencies between activities without showing them. With precise scope and simple schedule, the team has no challenge in developing a simple and precise Bottom-up Cost Estimate. An apparent lack of surprises grounded in low complexity and technical novelty poses no significant risks either. As a result, the team can do well with a simple but informal Risk Response Plan that may address a risk event or two.

Administrative Projects. These projects' high complexity dictates a different approach to project management tools than for routine projects. Although still precise because of the low technical novelty, the Scope Statement is very detailed because of high complexity. As a consequence, WBS tends to be large, with many levels of hierarchy. Added complexity requires precise scheduling of many organizational participants and their dependencies, perfectly served by PERT/CPM Diagrams. The same added complexity generally means a larger project size, whereas the cost becomes a crucial managerial concern, justifying the need for Bottom-up Cost Estimates and Cost Baseline (Time-phased Budget). Risks are primarily related to the increased number of interactions between the project participants and activities, which, in turn, is a ramification of the increased complexity. This generates a need for a formal Risk Response Plan with numerous risk events of an essentially qualitative nature.

Technical Projects. The high-tech nature brings special requirements to the technical projects. With the scope planning tools, the emphasis is on evolving Scope Statement and WBS. Because there are many new technologies, it will take time to define scope and WBS, typically after several iterations. Still, the statement is simple, while WBS features few levels. For the same reasons, the schedule becomes fluid, readily reshaping to account for scope changes. The rolling wave approach, built on the Milestone Chart for the whole project and Gantt Charts for the present and near-present activities, is a good fit for such a scheduling situation. Similarly, Cost Estimates are for milestones, fluid as the schedule is. Those estimates for near-term milestones are usually less fluid than those further down the project schedule. The increased technical novelty is to credit for multiple risks, often little understood in the project beginning. For the fluidity of the whole project, frequently changing P-I Matrices are a good choice. To better understand the interactive nature of risks, the project team may occasionally use Monte Carlo Analysis.

Unique Projects. For their high technical novelty and complexity, unique projects require highly challenging project management tools. Similar to the technical projects, Scope Statement and WBS are evolving over project time, a result of the presence of many new technologies and related uncertainty. As a result of large complexity, though, the Scope Statement is detailed and WBS is large, just as with administrative projects. The evolving nature of scope leads to the need for fluid schedules. Again, the rolling wave concept is applied, using the Milestone Chart to provide a big-picture schedule of the project. Supporting the Milestone Chart are Gantt Charts for the present and near-present activities. Because of the complexity of the project and many participants, the principles of hierarchical scheduling are integrating all schedules. Similarly, Cost Estimates for milestones are more detailed for the near-term ones than for later ones. Combined novel technologies and complexity push risks to the extreme, making it the single most challenging element to manage. In response, a combination of tools that constantly reshape to follow the scope evolvement may be used in a comprehensive Risk Response Plan—P-I Matrix, semiquantitative approach, Monte Carlo Analysis. Still, the plan may include both informal and formal components.

Index